OIL ECONOMISTS' HANDBOOK

OIL ECONOMISTS' HANDBOOK

GILBERT JENKINS

British Petroleum Company, London, UK

APPLIED SCIENCE PUBLISHERS LTD
LONDON

APPLIED SCIENCE PUBLISHERS LTD
RIPPLE ROAD, BARKING, ESSEX, ENGLAND

ISBN: 0 85334 728 X

Printed in Great Britain by Galliard (Printers) Ltd, Great Yarmouth

Preface

The purpose of this book is to provide a ready reference for those working both inside and outside the oil and energy industries, especially those concerned with that part of it known as oil economics, which covers much of the commercial side of the oil industry.

The first part is concerned with the jargon, abbreviations and terms used in the oil industry. Important topics, such as the pricing of oil products, are considered in detail. There is extensive coverage of the energy field, and some technical material of interest to those in the commercial field is included.

In addition to the numerous entries in the first part, the second part is composed of energy and oil statistics and data useful to oil economists, some of which are not easy to find elsewhere. Much of the energy data was back-up material for the *BP Statistical Review* and was collated prior to publishing the more detailed form of the Review.

There is international coverage of energy resources, energy production, transportation, petroleum refining, petroleum products (including specialised products), storage, energy economics, pricing and energy companies.

Contents

DICTIONARY OF TERMS

A

AAODC American Association of Oil well Drilling Contractors.

ABADAN Abadan refinery was commenced in 1909 and is still one of the largest refineries in the world. It was owned by BP (then the Anglo–Iranian Oil Company) until 1951 when it was nationalised. It was unused for a short period and from 1954 has been run by a consortium of oil companies (*q.v.*) in which BP has a major stake.

ABCO A small company but one of the oldest UK oil companies; originally established as Arthur Brown & Co. Ltd in 1860. Bought by Sinclair Oil (*q.v.*) and then became part of ARCO (*q.v.*) when it took over Sinclair.

ABS Acrylonitrile–butadiene–styrene resins which are important petrochemical resins.

ABU DHABI For crude oil production *see* Table 1.

ACCOUNTING METHODS *See* Exploration Expenses, FIFO, LIFO.

ACCURACY For accuracy of energy and oil statistics, *see* Precision.

ACHNACARRY AGREEMENT In 1928 more than 50% of oil sales in the world outside North America were controlled by Jersey Standard (Exxon), Shell and BP. Secret conferences took place in 1928 to obtain closer co-operation between the three majors and, when agreement was close, Deterding of Shell invited Teagle of Jersey and Cadman of Anglo–Persian to Achnacarry House in the Scottish Highlands. The result of the meeting was an agreement for an international oil cartel to operate in all countries except the USA and USSR. Within a few months most of the other American companies had become party to the agreement. There were principles in the agreement covering market shares, use of facilities, expansion of facilities, production, movements and limiting costs.

ACIDISATION This is a process where acid is pumped into a well to enlarge the pore spaces and to allow the oil to flow more easily. It is effective in carboniferous structures.

ACTIVATED CARBON This may be manufactured from certain qualities of petroleum coke. It is also manufactured from a variety of materials including coal, lignite, peat, wood, bones, coconut shells and pulp mill black ash.

ADLA CRUDE *See* Murban.

ADMA Abu Dhabi Marine Areas; the company was formed in 1954 and owned two-thirds by BP and one-third by CFP. With participation the State took a 60% shareholding (ADNOC). Before participation BP sold a 30% interest in ADMA to JODCO (*q.v.*) Japanese Overseas Petroleum Corp. (*q.v.*). Production from the Zakum field commenced in October 1967 some five years after the Umm Shaif field which came on stream in June 1962. ADMA production reached 507 000 b/d in 1973 declining to 462 000 b/d in 1974.

ADMA CRUDE *See also* Umm Shaif, Zakum. El Bunduq and Mubarras are also fields contributing to the ADMA blend.

ADNOC The Abu Dhabi National Oil Company which participated in ADMA and ADPC (*q.v.*).

ADNTC The Abu Dhabi National Tanker Company which is a wholly owned subsidiary of the Abu Dhabi National Oil Company (ADNOC) and which was established in 1975.

ADPC The Abu Dhabi Petroleum Company, first registered in 1936 as Petroleum Development (Trucial Coast) Ltd. Original shareholdings were $23\frac{3}{4}$% BP, $23\frac{3}{4}$% Shell, $23\frac{3}{4}$% CFP, $23\frac{3}{4}$% Near East Development Corporation and 5% Participations and Exploration Corporation. With participation ADNOC took a 60% holding. Production commenced in 1962 with just two tankers being loaded and reached 920 000 b/d in 1974.

ADRIATIC OIL PIPELINE This pipeline, to be completed in 1979, was estimated in 1975 to cost $450–$550 million. The capacity of the line was to be 34 million tons p.a. and it was to serve Yugoslavia, Hungary and Czechoslovakia.

ADVANCED RECOVERY METHODS (OF CRUDE OIL) A term used to embrace secondary and tertiary recovery methods of crude oil. The amount of oil which may be recovered by advanced methods is an even more subjective matter than assessing the proved reserves of oil. An example was a report, in November 1974, that advanced recovery methods could increase US domestic oil supplies by 80 billion barrels thus increasing proved reserves from a 7-year supply to an 18-year supply (Congressional Research Service Report prepared for Chairman of House Science and Astronautics Energy Sub-Committee).

AEA The UK Atomic Energy Authority which built and operated one of the first nuclear power stations—Calder Hall, commissioned in 1956 and had a maximum output of 150 MW.

AEC The Atomic Energy Commission of Japan.

AFESD The Arab Fund for Economic and Social Development which began operations in April 1973. AFESD is a 21-member Arab regional financial organisation with headquarters in Kuwait. It was originally set up by the Arab Economic Council in May 1968. The members are Jordan, Tunisia, Algeria, Sudan, Iraq, Saudi Arabia, Syria, Libya, Egypt, North Yemen, Kuwait, Lebanon, Morocco, South Yemen, the UAE, Bahrain, Qatar, Somalia, Mauritania, Oman and Palestine. Its purpose is to participate in financing economic and social development in the Arab states by:

1. financing economic projects with loans granted on easy terms to governments and to public or private organisations and institutions, giving preference to those vital to overall Arab development and to joint Arab projects;
2. encouraging investment, both public and private, to ensure the development and growth of the Arab economies;
3. providing technical expertise and assistance in the various fields of economic development.

AFFREIGHTMENT In an affreightment contract the contract specifies the carriage of a quantity of oil or any other cargo between named ports. The owner tenders a rate for the contract and may use any suitable vessels.

AFGHANISTAN For crude oil production *see* Table 1.

AFM PRICES *See* Prices—AFM.

AFRA Average Freight Rate Assessment. This is an assessment of the level of freight rate actually paid for commercially chartered vessels engaged in the carriage of oil during the period over which the assessment is made. The assessment is made independently by a panel of London tanker brokers. Each assessment covers a period from the 16th of one month (e.g. January) to the 15th of the following month. The rate, expressed in Worldscale, is published at the beginning of the next succeeding month (March) for application during that month. The Worldscale rate is readily convertible into $/ton for any particular route.
AFRA was designed to be a reliable commercial and arms length indicator of tanker freight rates which could be used for the freight element of inter and intra company business. The terms of reference under which AFRA is assessed were compiled in consultation with the London Tanker Brokers Panel and are reviewed from time to time whenever the World tanker freight market indicates that circumstances have changed sufficiently to warrant a change in the form of assessment.
AFRA was commenced in early 1954 and originally was a weighted average of all sterling tanker freight paid for vessels of 10 000 tons dwt or more. It covered all types of charter and one index only was used until 1959. As vessel sizes and market conditions changed, AFRA was revised and expanded and AFRA indices were issued for different categories of vessel, from tankers of 20 000 tons dwt class to VLCCs. Two categories were introduced in 1959, three in 1964, four in 1968 and five in 1974.
Under the terms of reference which came into effect on 1st November 1968, AFRA was divided into four size categories, specified by their summer dead weight (sdw) ratings.

GP (General Purpose Vessels)	16 500–24 999 t dwt
MR (Medium Range Vessels)	25 000–44 999 t dwt
LR1 (Large Range 1 Vessels)	45 000–79 999 t dwt
LR2 (Large Range 2 Vessels)	80 000–159 000 t dwt

On 1st January 1974, a fifth category was added:

VLCC (Very Large Crude Carriers) 160 000–320 000 t dwt

Within each of these categories, vessels are classed into:

(i) Long term charter—tonnage on charter originally fixed for a period of 12 months or more, or for a number of consecutive voyages extending over 12 months or more;
(ii) Short term charter—tonnage on charters originally fixed for a period of less than 12 months or for a number of consecutive voyages extending over less than 12 months;
(iii) Single voyage—tonnage on single voyage (spot) charter.

Within each charter class and for each of these size categories, weighted average freight rates are calculated, using for the purpose of standardisation a notional voyage approximating to the average length of voyage performed by the World tanker fleet—taking into account where necessary such factors as size, speed and type of vessel. The resulting rates are then combined for each of the categories into a weighted average freight rate for each category. Only vessels operating under normal commercial charter are included in the calculations and care is taken to eliminate certain types of hire which might unduly inflate the assessment, e.g. vessels engaged in specialised trades and those engaged in protected coastal trade. It should be noted that the vessels owned by the oil companies that are not on arms length charters have no effect on the assessment.
AFRA includes the price of time chartered and spot chartered vessels in the proportions in which they are represented in the freight market. Thus time charter

rates continue to influence AFRA over the whole period of charter, whilst the influence of spot rates extends over a much shorter period of time. In this way AFRA tends to smooth out violent fluctuations in market rates. AFRA should therefore be regarded as representing the price of freight to an average buyer.

AGHA JARI *See* Iran Light Crude Oil.

AGIP Agip is the main operating company of ENI (*q.v.*).

AGRs *See* Nuclear Reactors.

ALASKA OIL COSTS *See* Tanzer Report.

ALBANIA For crude oil production *see* Table 1.

ALCIP This company is 51 % Sonatrach and 49 % Saipem (ENI) and was formed in 1974. There are many other companies in which the Algerian State has a majority interest named AL . . .

ALCOHOL Unless specified could be methanol, ethanol, isopropanol (IPA) or a number of higher alcohols. Alcohols can be used at up to 1 % concentration as mogas anti-icing additives.

ALGERIA First crude oil production was in 1944. Development of Algerian oil was an important part of French oil policy. On independence in 1962, Sonatrach (*q.v.*) became responsible for much of the Algerian oil industry. Algeria's reserves of oil are not high and the country can absorb all the revenue from petroleum that it can obtain. A member of OPEC and OAPEC. For crude oil production *see* Table 1.

ALLIED OIL CO. *See* Ashland.

ALUMINIUM Aluminium, of the common minerals and metals, is the one which requires most energy in its manufacture; *see* Energy Requirements. Aluminium does not occur in crude oils.

ALWYN A commercial UK North Sea oil field operated by Total. Peak production rate estimated to be 100 000 b/d with recoverable reserves estimated to be 500 million barrels.
The original shareholding was as follows:

3/14	Total Marine Ltd	33·33 %
3/15	Aquitaine Oil (UK) Ltd	22·22 %
	Elf Oil Exploration and Production (UK) Ltd	44·45 %
9/28	Hamilton Bros. Oil Co. (GB) Ltd	48·00 %
	Hamilton Bros. Petroleum (UK) Ltd	12·00 %
	Rio Tinto Zinc Corporation Ltd	25·00 %
	Blackfriars Oil Co. Ltd	12·50 %
	Trans European Co. Ltd	2·50 %

AMOCO *See* Standard Oil of Indiana.

AMORPHOUS COKE This description is sometimes applied to certain grades of petroleum coke.

AMPOL The Ampol Petroleum Company of Australia which was formed in 1936 as the Australian Motorists Petrol Co. Ltd, the name being changed to Ampol in 1949.

AMPTC The Arab Maritime Petroleum Transport Company.

AMYL NITRATE Amyl nitrate is an additive used in diesel fuels to improve the cetane number just as TEL is used in motor gasolines to improve the octane rating. Amyl nitrate, however, has been relatively little used and does not have the economic significance as a diesel fuel additive that lead alkyls have as motor spirit additives.

ANCAP The state oil company of Uruguay.

ANDREW This UK North Sea field was discovered by BP in July 1974. The structure straddles the line between block 16/28 and Phillips' block 16/27. Early estimates considered the field to have a production potential of 100 000 b/d. In 1975 the BP interest was 56 %.

ANGLO–IRANIAN OIL CO. LTD Formed from the Anglo–Persian Oil Co. Ltd (*q.v.*) in 1935 and the name was changed to BP (*q.v.*) in 1954.

ANGLO–PERSIAN OIL CO. LTD Formed on 14th April 1909, the name was changed to Anglo–Iranian Oil Co. Ltd in June 1935 and this name changed to The British Petroleum Company Ltd in December 1954. *See* BP.

ANGLO–SAXON PETROLEUM CO. *See* Shell.

ANGOLA For crude oil production *see* Table 1.

ANIC An Italian petroleum refining and chemical company formed in 1936, which is 73 % owned by ENI.

ANRO A BP oil trading company formed in 1975 to trade in the European barge and cargo markets at Rotterdam.

ANTAR Antar-Pétroles de l'Atlantique which was incorporated in 1933. A French refining and marketing company with three refineries in France. The French Government holding in the company amounted to about 90 % in 1976.

ANTICLINAL TRAPS These are formed as a result of foldings of strata enabling any gas, water or oil contained in a reservoir rock to collect at the crest of the fold. The material is prevented from escaping by a 'cap' rock which is impervious.

ANTIFREEZE This is likely to be ethylene glycol but could be glycerol or a number of other materials.

ANTIKNOCK COMPOUNDS The history of gasoline antiknock additives began in 1916. Early additives included iodine, selenium oxychloride and many mixtures of organic compounds. Eventually TEL (tetraethyl lead) was manufactured by a satisfactory high pressure process in 1924. *See also* Lead Alkyls.

ANTIKNOCK RATING *See* Octane Rating.

ANTITRUST *See* Exxon, Energy Antitrust Bill.

API The American Petroleum Institute, incorporated in 1919, under the laws of the District of Columbia, serves as a forum, information bureau, technical clearing house and national trade association.
The API publishes weekly statistics of the US petroleum industry and since 1933 a most informative (bi) year book. The 1971 edition of 'Petroleum Facts and Figures' included nearly all the information presented in previous editions including some data on an annual basis back to 1859.
A Statistical Bulletin of the US industry is published in April each year.

API Anonima Petroli Italiana, an Italian company which operates a refinery of about 80 000 b/d near Alcona.

API GRAVITY The API Gravity is calculated from the specific gravity at 60 °F using the formula:

$$\text{API Gravity} = \frac{141{\cdot}5}{\text{Specific Gravity at } 60\,^\circ\text{F}} - 131{\cdot}5$$

This parameter has considerable economic significance since it is the main quality criterion for crude oil pricing; the higher the gravity the higher the price of crude oil of a particular type.

API REGIONS Regions in the US designated by the API and the Bureau of Mines:

Region	1	Eastern Seaboard
	2	Midwest
	3	Southern
	4	Rocky Mountains
	5	Far West

See also PAD

APIC The Arab Petroleum Investments Company was set up in November 1975 after approval by the Organisation of Arab Petroleum Exporting Countries (OAPEC). The decision was announced in Riyadh on 24 November by OAPEC's Ministerial Council.
APIC was formed to provide finance by loans or equity participation in the development of all aspects of the oil and gas industry, including production.

AQUILA S.p.A. An Italian refining company formed in 1934 with a 3 million tons p.a. refinery near Trieste. Owned by CFP.

AQUITAINE The French exploration and production company which was formed in 1941. Also referred to as SNAP and SNPA (*q.v.*).
It is a private company and ELF has from time to time been interested in requiring the company to increase the French State's share of the French oil industry. This finally came about in 1976.

ARAB LEAGUE *See* League of Arab States.

ARABIAN GULF *See* Middle East.

ARABIAN LIGHT CRUDE OIL The World's most prolific crude oil and from 1970 probably qualified for the title as the World's marginal crude oil. It was marginal in the sense that Saudi Arabia was looked on in the early 1970s to provide the balance of production to satisfy demand and there was talk of expanding production to 20 million b/d and more. However, it was certainly not a marginal crude or energy source in the cost sense—with development costs of only about 10 cents/barrel it was the World's cheapest energy source.
Arabian Light crude oil is frequently used as marker crude and the prices of other crude oils are fixed by comparing their quality with that of Arabian Light. Posted Price Arabian Light 34–34·9 API Crude ex Ras Tanura ($/bbl):

1959	1960	1961–1971
$2·08	1·90	1·80
Jan. 1972	**Jan. 1973**	**Jan. 1974**
$2·29	2·59	11·65

See also Saudi Arabia.

ARAL Aral AG is the German marketing and distribution company first formed in 1898. It was known as BV-Aral from 1952 to 1962. Aral is jointly owned by Mobil, Veba, Gelsenberg and Wintershall.

ARAMCO The Aramco partners from 1948 to Government participation in 1974 were Socal 30%, Texaco 30%, Exxon 30% and Mobil 10%.

The original Aramco concession, taken by Socal alone, was signed in Jeddah on May 29th 1933 and became effective on July 14th 1933. In November 1933 the concession was assigned to Casoc, a Socal company. In 1936 Socal and Texaco merged their interests between Egypt and the Hawaiian Islands and so Texaco acquired a 50% share in Casoc. In 1946 when there was some idea of the extent of Saudi Arabian reserves, arrangements were made to bring Exxon and Mobil into the agreement. These were completed in 1948. For Aramco partners liftings and dividends 1963–1973 *see* PIW April 8th 1974.
See also Crude Oil Sales Profits.

ARCO The Atlantic Richfield Oil Company. A US multinational oil company which includes elements of the former companies: Atlantic Refining Company, which dated from 1870, Richfield Oil and Sinclair Oil.
Arco has oil production facilities in the US including Alaska, Canada, UK North Sea (Hewett field), Indonesia and Iran. The company has substantial refining, petrochemical, coal and marketing interests in the US but, compared with other oil companies of its size, is not strongly represented outside the US.

ARGENTINA The Argentinian state oil company is YPF. For crude oil production *see* Table 1.

ARGYLL A commercial UK North Sea oil field 220 miles east of Dundee. Peak production rate estimated to be 200 000 b/d with estimated recoverable reserves at 75 million barrels. Original shareholdings were:

Hamilton Bros Oil Co. (GB) Ltd	48·0%
Hamilton Bros. Petroleum (UK) Ltd	12·0%
Rio Tinto Zinc Corporation Ltd	25·0%
Blackfriars Oil Co. Ltd	12·5%
Trans European Co. Ltd	2·5%

The first oil from the UK North Sea was landed from the Argyll field by tanker at BP's Kent refinery on June 18th 1975.

ARJUNA (Indonesia) This is an export crude oil of 37·7°API gravity, 0·1% sulphur and a high wax content. Export is by SBM.

ARMS LENGTH An arms length deal is one which would take place between a willing buyer and a willing seller. The price involved should therefore be a true reflection of the market. Intra company deals are not always at spot market prices, e.g. at AFRA freight rates rather than spot market freight rates.

AROMATICS Aromatics occur in petroleum fractions boiling above 80°C. Refining processes are employed to increase the quantity of aromatics in some products, e.g. mogas, whilst reducing the aromatics content of other products, e.g. premium grade kerosenes. The most important individual aromatics are benzene, toluene, orthoxylene and paraxylene. These are important as chemical feedstocks and as motor gasoline components. In recent years the benzene content of mogas has been reduced, though it was once considered a premium component.
See also Benzol, National Benzole Company.

ARPEL The Asistencia Reciproca Petrolera Estatal Latino-Americana. The following are members: Argentina, Bolivia, Brazil, Chile, Colombia, Ecuador, Peru, Uruguay and Venezuela.

ARZEW (Algeria) This is an export crude oil of 44° API gravity and 0·1% sulphur. The crude blend has the name of the export terminal.

AS IS AGREEMENT *See* Achnacarry Agreement.

ASCOP AGREEMENT An agreement between the Algerian and French Governments signed on July 29th 1965. It covered all aspects of the oil industry giving extra revenue to Algeria whilst allowing a continuation of the French presence in Algeria. Exploitation of hydrocarbons was to be carried out by a co-operative association, ASCOP, which grouped together French companies and Sonatrach (the Algerian National Company).

ASHLAND A medium size US petroleum company which was formed from the Ashland Refining Co. and the Swiss Oil Corporation in 1936. Subsidiaries' names include Allied Oil Co. and Valvoline.

ASIATIC PETROLEUM COMPANY *See* Shell.

ASPHALT Asphalt is an alternative name for bitumen (*q.v*) which is favoured in the US.

ASPHALTENES Asphaltenes are the coal-like materials which occur in the residues obtained by distillation of crude oil. The asphaltenes are rich in the undesirable elements (sulphur, nitrogen, nickel and vanadium) that occur in fuel oils.

ASPHALTIC CRUDES *See* Crude Oils, Heavy.

ASSOCIATED GAS *See* Natural Gas.

ASTM The American Society for Testing Materials. The D-2 Committee deals with petroleum products and lubricants through many sub-committees, on which major oil companies and customers are represented. ASTM specifications are used worldwide.

ATIC The Association Technique d'Importation Charbonniere. The French state company controlling coal imports which was founded in 1944.

ATMOSPHERIC POLLUTION The main cause of atmospheric pollution from the burning of fossil fuels is by sulphur dioxide (*q.v.*). Pollution by oxides of nitrogen from car exhaust gases have been of considerably less economic consequence.

ATRS American Tanker Rate Schedule; *see* Spot Market Freight Rates.

ATTAKA (Indonesia) This is an export crude of 43° API gravity and 0·1% sulphur. It is shipped from the Scanlan terminal, Kalimantan.

AUK A commercial UK North Sea oil field 200 miles east of Dundee. Maximum production rates estimated to be 40 000 b/d with estimated recoverable reserves at 75 million barrels. It is owned 50/50 by Shell and Exxon. Production commenced in 1976.

AUSTRALIA For crude oil production *see* Table 1.

AUSTRIA The Austrian state oil company is OMV. For crude oil production *see* Table 1.

AVCAT Avcat is a special jet kerosine fuel for military use. It has a flash point of 140°F compared with 100°F for ordinary Jet fuel (A-1). The manufacturing cost is similar to that of Jet A-1.

AVGAS Avgas (or aviation gasoline) is manufactured in seven grades. Five of the grades have similar manufacturing costs to those of motor gasolines, the main difference being lower and strictly controlled volatility. The two high octane or performance grades cost a little more to manufacture than best quality gasoline.
The grades are Avgas 73, 80 unleaded, 80, 91, 100, 108 and 115. The three latter grades are frequently termed 100/130, 108/135 and 115/145. The first of the two numbers refers to the lean mixture rating and the second to the rich mixture rating.

AVIATION TURBINE GASOLINE (ATG) *See* Jet B.

AVIATION TURBINE KEROSINE (ATK) *See* Jet A.

AVTAG Avtag is an aviation turbine fuel also known as Jet B (*q.v.*).

AVTUR Avtur is a jet kerosine fuel more commonly known as Jet A (*q.v.*).

AZUR A marketing network in France which in 1967 was integrated into the Total network of CFP.

B

BAADE–FREIDENSBURG REPORT A report published in 1962 of West Germany's energy consumption 1960–1975. It was prepared by the Association of German Economics Institutes and leading the research team were Professors F. Baade and F. Freidensburg.

BACHAQUERO (Venezuela) This is a naphthenic crude oil of 16° API gravity and 2·4% sulphur exported from La Salino and also Puerto Miranda.

BACK TO BACK DEAL This is a deal where a purchase is made to satisfy a sale which has already been fixed. The object is to minimise risk and stock carried by effectively matching buying and selling orders.

BACKHAUL In general, tankers spend 50% of the time travelling in ballast. Sometimes it is possible to arrange for a cargo on part of the return voyage. For example, a tanker taking low sulphur North African or Nigerian crude to Japan may return with Indonesian crude to Rotterdam. This is not a crosshaul (*q.v.*) because some Indonesian crude is required as a petrochemical feedstock at Rotterdam.

BAGASSE Bagasse is the refuse from cane sugar making which is used as a fuel in sugar extraction plants. It has a heating value similar to that of wood (*see* Calorific Value) and is termed a non-commercial fuel (*q.v.*). Consumption of bagasse has been of the order of 10 million tonnes oil equivalent p.a.

BAGHDAD PACT *See* CENTO.

BAHRAIN For crude oil production *see* Table 1.

BAI HASSAN (Iraq) This is an export crude of 34° API gravity and 2·4% sulphur. Crudes from Iraq may be exported from a Mediterranean terminal (Tripoli) or from the PG.

BALANCING MARGIN A term used to describe the payments made by the Consortium to Iran in 1973/74. The retroactive payments made were $100 million for Sept./Dec. 1973, $700 million for Jan./Sept. 1974 and $50 million interest. The payments were agreed to in July 1973 to equate payments in Iran with those accruing from participation agreements made elsewhere in the Middle East.

BARBADOS For crude oil production *see* Table 1.

BARGE PRICES Barge prices are quoted at Rotterdam which is one of the most important marginal oil markets. They refer to quantities of up to 5000 tonnes.
The barge trade on the Rhine from Rotterdam is substantial; there were about 1750 barges in operation in 1974 of total capacity about 1 500 000 tonnes.

BARTER DEALS These have been by no means uncommon in the petroleum industry. In the early days of ENI, Mattei bartered Italian goods for Russian crude oil increasing Italy's import of Russian crude from 4% to about 16%.
More recently (April 1975) it was reported that Spain had acquired 150 000 tonnes of Algerian crude in exchange for LPG cylinders.

BASE PRODUCTION CONTROL LEVEL In the US this level was used in connection with price control on domestic crude oil following the large increase in worldwide crude oil prices in 1973/74. It was the quantity produced in the corresponding month in 1972 in each property then in production.

BASEL PRICES Basel is supplied with products from Rotterdam and Strasbourg. Product prices are sometimes quoted in Platt's Oilgram Price Service.

BASRA (Iraq) This is a crude oil of 34° API gravity, 2% sulphur exported by SBM at Khor al Amaya.

BCD OR B/CD Barrels per calendar day; a term used to denote the rate of throughput of refinery plants. BSD, barrels per stream day, is also used.

BD OR B/D Barrels per day; frequently used to express capacity of a refinery (in which it refers to the crude oil distillation unit). The measure is also used for other refinery processing plants. The term is ambiguous for refinery capacity where rates should be expressed as barrels per calendar or stream day.
Also used for crude oil production, liftings, consumption and demand forecasts when it is of course referring to the calendar day.

BEA The British Electricity Authority which with the CEB (Central Electricity Board) were forerunners of the CEGB, Central Electricity Generating Board of the UK.

BEB Gewerkschaften, Brigitta und Elwerath. Betriebsfuhrungsgellschaft mbH. Formed in 1969; 50% Deutsch Shell AG, 50% Esso AG.

BECKETT REPORT J. A. Beckett was the Chairman of the OECD Oil Committee which was responsible for the report 'Oil. The Present Situation and Future Prospects'. The report was completed in June 1972 and published by OECD, Paris in 1973. Forecasts were provided to 1980.

BEKAPAI (Indonesia) This is a 41° API gravity, 0·1% sulphur offshore crude oil exported by SBM.

BENZENE Benzene has been an important motor gasoline blending component but since the 1950s it has been more important as a petrochemicals feedstock.
It is a minor component of most crude oils (ca 0·1%), some is formed in the cat. reforming process and larger quantities are obtained by the various cracking processes.
See also Benzol.

BENZOL Also known as benzole and, when intended for blending into petroleum spirit, as motor benzole. The chemical composition is 90 to 100% benzene. Because of its high octane number (110 research method) it was favoured as a blending component for motor spirits until the 1960s.
Its decline as a component of motor gasoline is well illustrated by the quantity blended into US motor gasolines:

	1000 *US barrels*
1940	3167
1960	275
1965	13

During the period it was used as a motor gasoline blending component, its price was always much in line with best quality motor spirit.

BENZOL: YIELD FROM COAL Disregarding losses, which are a few per cent, during coal gas manufacture one tonne of coal (300 therms) yields about 80 therms of gas, 20 therms of liquids and 200 therms of coke. The yield of liquids is about 11 UK gallons of which 2 to 3 gallons is benzol.
High yields of liquid fuels from coal are obtainable only by indirect methods such as the Fischer–Tropsch synthesis from water gas.

BERGIUS PROCESS A catalytic high pressure hydrogenation process to produce liquid fuels from coal. It was operated in Germany in the 1930s and in World War II.

BERYL A commercial UK North Sea oil field 210 miles north-east of Aberdeen. Peak production is estimated to be 100 000 to 200 000 b/d with estimated recoverable reserves at 600 million barrels. Original shareholding:

Mobil Producing North Sea Ltd	50%
Amerada Exploration Ltd	20%
Texas Eastern (UK) Ltd	20%
Gas Council (Exploration) Ltd	10%

The crude is 39·5° API gravity with a sulphur content of 0·4%.

BGC The British Gas Corporation. A UK state-owned company which is just about a monopoly buyer of gas from the companies producing in the UK areas of the North Sea.

BHC *See* BP Chemicals International.

BILLION This term is a source of ambiguity in economic literature since it is not always clear if it is the US or UK billion. The US term occurs more frequently and is one thousand million (10^9). The UK billion is a million million (10^{12}).

BITUMEN Bitumen is the term used to describe the heavy petroleum residue which is used for a variety of road-making and roofing applications. Bitumen is classified by the penetration test; hard brittle bitumens have a penetration of 10 or 20 whilst very soft bitumens have a penetration of up to 300 units. The price of bitumens is not correlated with their penetration grade. Because of the relatively small quantities produced, prices are influenced by brand names and gimmicks having a pseudo technological basis.
With handling and marketing elements removed from the price, bitumens should be priced less than heavy fuel oils.

BITUMINOUS COAL ANNUAL This publication is one of the best sources of US energy statistics and was first published in 1948.
Included in early editions were energy statistics dating from 1800. It is published by the Bituminous Coal Institute of the NCA.

BKB Brown coal briquettes which were of importance for industrial and domestic use in Germany.

BLM The US Bureau of Land Management; the agency which amongst its other activities controls offshore oil leases.

BLS PRICES Prices published by the US Bureau of Labor Statistics. Included in the statistics are monthly national averages of wholesale prices for light and middle distillates.

BLUE BOOK The UK Government annual publication of National Income and Expenditure.

BLUE WATER GAS *See* Water Gas.

BNOC The British National Oil Corporation, established on 1st January 1976. On that date it acquired the shareholding and personnel of NCB (Exploration) Ltd.

BOLIVIA For crude oil production *see* Table 1.

BOM The US Bureau of Mines. BOM publications are a most reputable source of US oil statistics. The monthly petroleum statement has a lag of about three months, the annual statement about fifteen months.

BONNY *See* Nigerian Crudes.

BONUS PAYMENTS Bonus payments were sometimes part of concession agreements in which the operating company made specified payments to the state if and when a certain level of production had been reached.

BOPEC The Bonaire Petroleum Corp. NV which owns the Bonaire trans-shipment terminal in the Netherlands Antilles. BOPEC is owned by Northville Industries Corp. of New York and Paktank BV of Rotterdam. The Bonaire trans-shipment terminal can receive tankers of up to 500 000 tons dwt.

BOSCAN (Venezuela) A crude oil of 10° API gravity and 5·4% sulphur, being one of the heaviest crude oils in world trade.

BP One of the World's seven major oil companies. It was formed in 1909 as the Anglo–Persian Oil Company (*q.v.*) and became the British Petroleum Company in 1954.
It operates some companies on a 50/50 basis with Shell. Some of these companies use the BP name first, e.g. the BP–Shell Petroleum Development Co. of Kenya, whilst others use the Shell name first, e.g. Shell and BP (Sudan) Ltd. The most important joint company was the UK company Shell–Mex and BP Ltd. which operated joint marketing in the UK until 1975.
BP grew out of the activities of D'Arcy who was granted a concession in Iran in 1901, found oil at Majid-i-Suliman in 1908 and the company was formed to operate the concession in 1909. In 1909 the construction of Abadan refinery (*q.v.*) was also commenced.
In 1914 the British Government took a stake in BP and since that time the Government has had a major stake in the company without interfering with the commercial management of the company. The company has always operated in a manner akin to

the US majors rather than as a European State Oil Company. The following statistics described BP operations in 1974:

650 subsidiary and associate companies;
operates in 70 countries;
operates 25 000 service stations;
operates at 500 international airports;
operates at 300 seaports;
170 000 employees;
113 ships of total 8·1 million tons dwt;
200 million tons crude oil and product sales;
£9300 million gross proceeds.

BP CHEMICALS INTERNATIONAL Wholly owned by BP, it is one of three big chemical companies in the UK. The company combined in the UK the activities of Distillers Chemicals, Mobil Chemicals (UK) and BHC (British Hydrocarbon Chemicals) which BP owned jointly with ICI. It does not publish separate accounts.

BPC The Basrah Petroleum Company of Iraq.

BRAND SUPPORT FACTOR A price which includes a premium to identify the lot as a private or major brand. This premium should guarantee that the product conforms to the brand specification, i.e. quality control may be more stringent especially on items such as gum in gasolines and cloud point in gas oils.

BRANDED DISTRIBUTOR In the US a Branded Distributor is a gasoline wholesaler who provides his own finance, transport and storage and perhaps also retails. He uses his supplier's brand name.

BRANDED TANK TRUCK DISTRIBUTOR In the US the function is very similar to that of a Branded Distributor (*q.v.*) except that a Truck Distributor does not have permanent storage.

BRASS RIVER (Nigeria) A crude oil of 43° API gravity and 0·1% sulphur exported by SBM from the Brass River.

BRAZIL The Brazilian state oil company is Petrobras. For crude oil production *see* Table 1.

BREEDER REACTOR *See* Nuclear Fission.

BREGA (Libya) A 40° API gravity, 0·2% sulphur crude exported from Marsa el Brega.

BRENT A commercial UK North Sea oil field 300 miles NNE of Aberdeen. Peak production is estimated to be 450 000 b/d with estimated recoverable reserves at 1750 million barrels. Brent is owned 50/50 by Shell and Exxon. Production from Brent is due to be landed at the Sullom Voe terminal.

BRIGHT STOCK One of the heavier grades of lub oil base stocks which may be used as a motor oil blending component at about 10% concentration.

BRITISH PETROLEUM CO. LTD *See* BP.

BROWN BOOK The annual 'Digest of UK Energy Statistics' (*q.v.*) published by the UK Government; early editions by the Ministry of Power, later editions by the Ministry of Technology, and the Department of Trade and Industry. The first edition published by the Department of Energy (1974) appeared in a mainly white cover thus breaking with tradition.

BROWN COAL This is a low grade coal which has a similar calorific value (*q.v.*) to that of lignite. *See also* BKB and Lignite.

BROWN PAPERS Papers published by the UK Department of Energy (formerly Ministry of Power, Ministry of Technology, Department of Trade and Industry).

BRP The French company Bureau de Recherches de Pétrole which was formed in 1945 with the object of carrying out petroleum research for the French Government.
In 1966 ERAP (*q.v.*) was formed by merging BRP and RAP.

BRUNEI For crude oil production *see* Table 1.

BSD OR B/SD Barrels per stream day; a measurement expressing the rate of throughput on a refinery plant of any type. Unless the downtime of the plant is specified there is ambiguity. Probably the most common number of stream days in a year is 330 but 300 and 340 are also used.

BSW Bunkers, stores and water which take up about 10% of the carrying capacity of an oil tanker.

BTX Benzene, toluene and the xylenes; the main aromatic compounds used as petrochemicals feedstocks.

BU ATTIFEL (Libya) A waxy crude oil of 41° API gravity and 0·1% sulphur exported from Zueitina.

BULGARIA For crude oil production *see* Table 1.

BULK CONSIGNEE In the US a bulk consignee is a gasoline wholesaler who works on commission from the supplier.

BULK SALES *See* Inorganic Trade.

BULK SUPPLY TARIFFS Tariffs for large scale users of energy. In the case of gas supplies a special tariff is one in which the supplier has the right to cut off supplies in periods of peak demand. In this instance the consumer will use standby facilities such as oil or LPG.

BULLETIN MENSUEL The Bulletin Mensuel of the Comité Professionel du Pétrole, 51, Bd. de Courcelles, 75008 Paris. A monthly French language journal published around the 20th of the month giving detailed coverage of the previous month's oil activity in France and a good source of other European oil and gas statistics.

BUNJU (Indonesia) A waxy, 32° API gravity, 0·1% sulphur crude exported from Balikpapan.

BUNKERS A general term used for supplies of fuels and lubricants to ships and aircraft. Bunkers do not include military use.

BUREAU OF MINES, US *See* BOM.

BURGAN (Neutral Zone) A 23° API gravity, 3·4% sulphur crude oil exported from Mina Saud.

BURMA For crude oil production *see* Table 1.

BURMAH OIL A major UK oil company founded in 1902 and an early backer of BP. Until 1974, Burmah retained a major shareholding in BP when it sold the BP shares to the Bank of England. Burmah also owns the UK lubricants company Castrol, bought in 1966, Rawlplug, bought in 1967, Halfords, bought in 1970, and Quinton Hazel, bought in 1972. In 1971 Burmah proposed a merger with Conoco but the plan fell through on account of Burmah's inability to disengage itself from the BP holdings. In 1974 Burmah bought Signal Oil and Gas Co. (*q.v.*) for $480 million but was unable to hold on to all this investment because of financial difficulties largely associated with its poor handling of the oil tanker market. The name of the company originated from activities in Burma which date back to exploration activity as early as 1886.

BUSHCO An Iranian exploration company whose shareholders are NIOC and Amerada Hess.

BUTANE *See* LPG.

BUTTES GAS AND OIL CO. An American exploration and production company which was formed in 1952.

BUYBACK CRUDE OIL *See* Crude Oil Prices, Buy-back Oil.

BV-ARAL *See* Aral.

C

CABINDA (Angola) A 33° API gravity, 0·15% sulphur crude oil from the Molongo offshore field; exported by SBM.

CALCINED COKE *See* Calcining.

CALCINING Petroleum coke is calcined by heating to about 2500 °F in order to reduce the volatile content (down to about 1%), increase density and improve crystalline structure.

CALIFORNIA Companies with California in the name could well be subsidiaries of Standard Oil of California (*q.v.*).

CALIFORNIA REPORT An early report discussing the cost of nuclear power concluding that electricity from nuclear power would cost $1\frac{1}{4}$ times thermal costs. It was one of the lowest estimates ever made. ('Atomic energy—its future in power production', *Chemical Engineering*, October 1946). *See also* Nuclear Power Costs.

CALORIFIC VALUE The calorific value of a fuel is the quantity of heat released on combustion. It is usually expressed in Btu/lb or kcal/tonne for solid fuels or Btu/ft^3 or kcal/m^3 for gases. Values may be quoted on a gross or net basis, the technical difference being that gross values assume that the water formed during combustion is liquefied whilst the net values assume that the water stays in the vapour phase. The calorific value is the basis upon which the relative monetary values of fuels are based.

	Btu/lb gross	
Natural gas	26 000	1·00
Coal gas	11 000	0·42
Water gas	6 750	0·26
Propane	21 500	0·83
Butane	21 200	0·82
LPG	21 350	0·82
LDF	20 500	0·79
Kerosine	20 000	0·77
Gas oil	19 500	0·75
Light fuel oil	18 750	0·72
Heavy fuel oil	18 500	0·71
Anthracite	15 000	0·58
Bituminous coal	12 000	0·46
Coke	12 000	0·46
Lignites*	6 000	0·23
Peat*	5 000	0·19
Bagasse*	5 000	0·19
Wood*	5 000	0·19

* Water content about 50%.

CALPAM The company is owned by Socal.

CALTEX COMPANIES These companies are owned 50% by Standard Oil of California and 50% by Texaco.

CAMBRIDGE TANKERS LTD UK tanker brokers; the company was formed in 1949. Cambridge Tankers frequently publishes reports on the oil tanker market.

CAMPSA Cia. Arrendataria del Monopolio del Petroleo SA. The Spanish state monopoly marketing company which also has other petroleum interests. Formed in 1927.

CANADA The Canadian state oil company is Petrocan. For crude oil production *see* Table 1.

CANAL TOLLS Canal tolls are designed to give a potential user some cost advantage over alternative routes and are not based on construction or maintenance costs. Before the Suez Canal closed in 1967, Suez Canal and Panama Canal rates were about 90 cents per ton whilst the St Lawrence Seaway rate was about half that rate. *See also* SDRs.

CANDU *See* Nuclear Reactors.

CANNEL COALS Cannel coals are better known as kerogen (*q.v.*). They are similar to soft bituminous coals but are of little economic importance in the world energy scene.

CAP ROCK This is an impervious layer of rock or sediment above a reservoir which prevents leakage of oil or gas.

CARACAS PETROLEUM CO. This company, formed in 1938, is a subsidiary of Ultramar Petroleum (*q.v.*).

CARBON BLACK This is a product which is largely carbon and is consumed mainly as an elastomer (for tyres, etc.) with minor uses for paints, paper, printing ink, etc. It is a general name used for acetylene black, attrited black, channel black, flame black, furnace black, lamp black and thermal black. The main country of manufacture has been the US where, prior to 1950, the main production process was the Channel process employing natural gas. From 1950 liquid hydrocarbon feedstocks predominated using the Furnace process. In 1972 US production of carbon black by the Channel process was less than 1% of the total. The liquid feedstocks are highly aromatic materials from other refinery operations such as decant oil (*q.v.*) and the fuel oil residue from the production of ethylene.

CARBON PRODUCTS This is a group of materials which are composed largely of elemental carbon or of very high carbon content, which includes petroleum cokes, carbon blacks and graphites.

CARBONISATION Essentially a distillation process whereby the residue acquires a higher carbon content than the feed. Carbonisation of coal distils off materials which are relatively rich in hydrogen whilst the coke residue has a higher carbon content than the coal.
Carbonisation of wood provides charcoal.

CARBURETTED WATER GAS Also sometimes known as carburetted blue gas. Made by enriching water gas (*q.v.*) with a hydrocarbon oil. It has a calorific value of about 50% that of natural gas.

CAREP Cie. Algèrienne de Recherches et d'Exploitations Pétrolieres, formed in June 1955. SN.REPAL 33%, Algerian Government 18%, BRP and SNIP 26%.

CARGO PRICES Cargo prices are usually quoted for lots of greater than 5000 tonnes and are therefore usually a little below Barge Prices (*q.v.*) *See* Platt's European Prices.

CARTEL A cartel is a group which has been formed to influence prices, usually to put prices up and keep them up by eliminating competition between the members of the cartel.
The best known example in the oil industry is OPEC (*q.v.*) which has had the advantage of being led by Saudi Arabia. Saudi Arabia has the advantage of controlling the world's largest oil fields and not being under any pressure to produce since its revenue is always in excess of its requirements.
See also Achnacarry Agreement, Club Agreements and Swedish Cartel.

CASING A casing is steel lining of a well, the main purpose of which is to support the sides of the well.

CASOC The Californian Arabian Standard Oil Company.
See Aramco.

CASTROL The company was registered in the UK in 1918 as C. C. Wakefield and Co. and changed the company name in 1960 to its lub oil brand name, Castrol. It was bought by Burmah in 1966, being the initial acquisition by Burmah at the start of its expansion period.

CATUG A vessel which is propelled by a detachable propulsion unit. Such vessels may be up to about 30 000 tons dwt. They are a very small proportion of

the world's tanker fleet and are employed largely in US coastal movements.
Their economic advantage over traditional vessels will depend on the economies of storage of oil.

CDF Charbonnages de France; the French state coal company formed after World War II.

CEB The obsolete Central Electricity Board of the UK which with the British Electricity Authority was a forerunner of the CEGB (*q.v.*).

CECA The French language designation of ECSC (*q.v.*).

CEGB The UK Central Electricity Generating Board.

CEILING PRICE A ceiling price is a maximum price level. A particular application in the US in the immediate post-1973 period was when crude oil ceiling prices were defined as the posted price operative at 15th May 1973 plus $1·35 per barrel. Maximum prices for oil products have been stipulated in many European countries.

CELTIC SEA The offshore area between SW Wales and SE Ireland which became of interest to the oil industry in the early 1970s following the intensive activity in the North Sea.

CENTO The Central Treaty Organisation, formerly the Baghdad Pact, began in 1955 when Turkey and Iraq signed a mutual defence treaty. The UK, Pakistan and Iran acceded to the pact the same year, but Iraq withdrew in 1959.
Some of the work undertaken by CENTO has been devoted to economic development, including improving transport and communications.

CEP The Compagnie d'Exploration Pétroliere was formed in March 1944 with BRP and SNIP 58% initial holding. The French exploration company went into liquidation in June 1972 and the assets transferred to ERAP (*q.v.*).

CEPCEO The Association of the Coal Producers of the European Community. A report by CEPCEO, 'Energy in Europe, The Importance of Coal', published in 1974 by the NCB, Hobart House, Grosvenor Place, London SW1X 7AE, gives the views of the Association and lists previous reports by the coal industry.

CEPE Cia. Estatal Petrolera Ecuatoriana; the Ecuador state oil company. Its main interest is in the Texaco–Gulf field of which it had acquired a 25% interest by 1974.

CEPSA Cia. Espanola de Petroleos SA. The Spanish exploration, production and refining company formed in 1929.

CETANE INDEX (CI) *See* Cetane Number.

CETANE NUMBER An arbitrary scale from 1 to 100 is used to specify the ignition quality of diesel fuels just as octane numbers are used to specify the antiknock qualities of motor spirits. Most diesel fuels have cetane numbers in the range 35 to 60. The more paraffinic the fuel, the higher the cetane number but there is no purpose in blending fuels with cetane numbers higher than 60 since the correlation between ignition and cetane number is not close. Certainly fuels prepared from paraffinic crudes (all Middle East crudes for example) give better performance than fuels prepared from US Gulf crudes (naphthenic) and materials obtained by cracking which are highly aromatic in nature. Because the measurement of the cetane number in a CFR engine is expensive and really not required for quality control, two indices are employed which correlate quite well; the diesel index (DI) which is calculated from the gravity and aniline point and the cetane index (CI) which is calculated from the gravity and the 50% distillation point of the fuel.
Four grades of diesel fuels are quoted. A grade 52 CI minimum; 53 to 57 DI (50 CI min.); 48 to 52 DI (45 CI min.) and 43 to 47 DI (40 CI min.). There is usually only about 0·1 cent/gallon difference between the grades. Large buyers are sometimes prepared to pay for a minimum specific gravity specification in a diesel fuel but again the sum involved is small.

CETANE NUMBER IMPROVER *See* Amyl Nitrate.

CFP Compagnie Française des Pétroles; formed in 1924. The French Government holds 35% of the equity and has 40% of the voting rights. The company is an integrated national oil company and is often considered along with the seven majors. The name Total is used for many of its operating companies. In the UK it has a 50% share in Lindsey Oil Refinery (*q.v.*) and in Italy controls the Aquila Refinery near Trieste.

CFP(A) Cie. Française des Pétroles (Algérie) formed in January 1953. It was owned 85% by CFP.

CFR Compagnie Française de Raffinage which was formed in 1929 and is an affiliate of CFP. The two major refineries of CFR are near Le Havre and at l'Etang de Berre, near Marseille.
The CFP holding in 1974 was 54·7%.

CHANNEL BLACK *See* Carbon Black.

CHASE MANHATTAN BANK A US bank which is much involved in the international petroleum industry. It produces a monthly statement of the activity of the US petroleum industry, forecasts of economic activity and oil supply demand and since 1945 an annual analysis of the financial performance of a large number of petroleum companies.

CHEVRON *See* Standard Oil Company of California.

CHILE For crude oil production *see* Table 1.

CHINA For crude oil production *see* Table 1.

CHINESE PETROLEUM CORPORATION This is the state oil company of Taiwan.

CHURCH COMMITTEE This was the US Senate Multinational Corporations Subcommittee of 1974 chaired by Senator Church. During the investigations, information on crude oil sales profits was released; this information was previously regarded by the petroleum industry as confidential. The release of Aramco financial data was taken as a guide to crude oil sales profits but both Exxon and Texaco (two of the Aramco partners) released qualifying statements. Exxon stated: 'It is not correct to consider that Aramco reported profits in any way represent the real profits of Aramco shareholders or are indicative of other operations in the Middle East. Exxon's net profits on crude oil offtake from Aramco based on sales by Exxon at world market prices were approximately 34¢/bbl in 1973 (in contrast to $1·21/bbl reported for Aramco). Profits comparable to the 1973 levels were realized by Exxon in 1971 and 1972, the years immediately following the Tehran Agreement. In the decade prior to Tehran, Exxon profits on Aramco crude ranged up to a level above 60¢/bbl.'
Texaco stated: 'The Aramco statements are based on posted prices which the Saudi Arabian Government requires be used for the determination of Aramco's income upon which its income tax and royalty payments are made. Such posted prices are much higher than the actual market prices at which such oil is sold in international trade. The posted prices therefore give an inflated impression of the ultimate dollar values of Saudi Arabian oil after it has gone through the entire process required to reach the ultimate consumer. Contrary to the implications contained in the (subcommittee's) press account, Texaco does not receive foreign tax credits as a result of any foreign taxes paid by Aramco. Texaco accounts to US taxing authorities for taxes on dividends received from Aramco on the same basis as any other US corporate shareholder receiving dividends from a domestic corporation with foreign source income.'

CIA The UK Chemical Industry Association which is the trade association of the UK chemical industry.

CIEPSA Cia. de Investigacion y Exploitationes Petroliferas SA. A Spanish exploration company formed by CEPSA and Socony–Vacuum with Socony as the operator. Socony sold its 50 % share to CEPSA in 1948.

CIF Cost including freight and insurance but excluding customs duty. *See also* Platt's European Prices.

CINTA (Indonesia) A 32° API gravity 0·1 % sulphur offshore crude oil exported by SBM.

CITIES SERVICE COMPANY An integrated US oil company registered in 1910.

CITY GATE MEASUREMENT The measurement taken when a gas utility company takes gas from a pipeline or grid system.

CLAPPIER REPORT This was a plan for restructuring the French chemical industry which much influenced the state holdings. Three state groups emerged in 1965 from the original five; of particular interest to the petroleum sector was the amalgamation of the chemical sectors of SNPA and CFP to form ATO.

CLARKSON H. AND CO. London tanker brokers, the company being formed in 1918. A source of frequent reports on tanker market activity.

CLAYMORE A commercial UK North Sea oil field with an estimated peak production rate of 100 000 b/d, and reserves of 450 million barrels. Operated by Occidental. Production planned for 1977.

CLEAN PRODUCTS Clean products include all the petroleum distillates. Products containing residual materials, i.e. most fuel oils, are termed 'dirty' or 'black'.

CLOUD POINT The cloud point is measured on distillate fuels and is an indication of the temperature at which filters may become blocked and cause malfunctioning or stalling of a diesel engine. The material which first separates out from a straight run fuel as the temperature is lowered is paraffin wax material, i.e. almost completely comprised of *n*-paraffins. To manufacture fuels with acceptable cloud points it is necessary to limit the end point and hence limit the availability. Cloud point restriction therefore limits the amount of fuel that can be manufactured and, depending on the product market price structure, the cost could be negligible or considerable (a few dollars per ton).

CLOW REPORT A White Paper on energy—'Report of the Committee to Study the Electricity Peak Load Problem in Relation to Non-Industrial Consumers', Cmd 7464, HMSO, London, 1948.

CLUB AGREEMENTS In effect these are limited cartels. For example, in Italy there was a club agreement on fuel oil which was more concerned with market shares than prices. The implications of a club agreeement could go beyond the direct object of the agreement. In Italy the natural gas price (gas being a monopoly of ENI) was linked to the market fuel oil price.

CMB The Chase Manhattan Bank (*q.v.*).

CMH The Compagnie Marocaine des Hydrocarbures which was formed in May 1976 by the merger of BP Maroc and Agip Maroc.

CNAN Compagnie Nationale Algérienne de Navigation. In 1974 its fleet consisted of five vessels of total 200 000 tons dwt.

CNP Cooper Nutrition Products which was acquired by BP in November 1974 to extend its activities in the animal feed markets.

COAL *See* Hard Coal, Lignite, Brown Coal, BKB, Calorific Value.

COAL AND ENERGY QUARTERLY A publication of the UK NCB; first issued in Summer 1974. Scope of the publication was intended to be the worldwide energy scene.

COAL GAS A gas manufactured by heating a bituminous coal in a retort. The liquid distillate products include benzene and other aromatic hydrocarbons and phenols and the residue in the retort is coke. Coal gas is about 25% methane and 50% hydrogen and has a calorific value about 50% that of natural gas.

CODU A crude oil distillation unit which may be more precisely specified as the atmospheric distillation stage. The letters CODU are not used to describe vacuum distillation units.

COFIRIP The French state sponsored finance company Compagnie Financiere des Recherches Pétrolieres.

COKE *See* Petroleum Coke.

COKE BINDERS These are pitches manufactured from coal or petroleum which are employed to bind the coke when electrodes are manufactured.

COKE OVEN GAS *See* Coal Gas.

COKING Coke is made from petroleum feedstocks by heating to about 950 °F and then holding in the coking drum for an extended (delayed) period. An indication of coke yield is provided by the Conradson carbon residue content of the feedstock.

COLLECTIVE HEATING Now usually referred to as District Heating.

COLOMBIA For crude oil production *see* Table 1.

COLOMBIAN PETROLEUM CO. A Colombian production company first formed in 1917 and almost 50% owned by Texaco.

COMBINED CARRIER A combined carrier is a vessel which may trade either in the bulk trades or in oil. Such vessels cost more to manufacture but the additional expense is justified by the owner on the grounds of their versatility. When spot rates are high, most combined carriers will be trading in oil, otherwise they are likely to be in the bulk trade. In the 1973 peak period for freight rates just over 80% of combined carrier tonnage was trading in oil. The average size of the combined carrier at that time was 100 000 tons dwt (307 vessels of total tons dwt 29 900 000).

COMMON CARRIER PIPELINES In the US in the early 1900s, the Standard Oil Trust placed many obstacles in the way of competitors building or acquiring use of pipelines. To limit the abuse of power of those controlling pipelines, successful efforts were made for many lines to acquire common carrier status. This was effected by a variety of methods such as granting permission to build lines only if they were to function as common carriers. However, operation of pipelines as common carriers has never been quite equitable and cases have been made for federal ownership of crude oil and product pipelines.

COMRATE The US Committee on Mineral Resources and Environment.

CON. VOYAGE A charter of freight running for two or three consecutive voyages as opposed to the usual single voyage charter in the spot market. Such charters are frequently seen when the spot market for freight is tight. *See also* Time Charter.

CONCESSIONS Concessions awarded to individuals and companies to explore for and produce

oil have taken many forms. A cash payment on being granted the concession; an agreement to drill a certain number of wells by a specified date; to return areas to the state by specified dates; to pay royalties to the state on production and clauses covering the ultimate ownership of any oil fields discovered.
For the concession granted to D'Arcy in Persia in 1901 the Government received £20 000 in cash, the promise of £20 000 in shares and 16% of any future profits. The concession granted was exclusive rights over 480 000 square miles.

CONDENSATES These are liquid hydrocarbon mixtures recovered from non-associated gas reservoirs. Until the 1920s they had a value equivalent to that of the straight run gasoline distilled from crude oil, both being used as motor fuels. The development of pressure cracking in 1915 and thermal cracking in 1921 made condensates less important as motor fuels. Subsequently, condensates have been used as refinery feedstocks and since the 1950s as petrochemical feedstocks. For these applications their price is in line with naphtha prices. *See also* NGL and LPG.

CONGO For crude oil production *see* Table 1.

CONNALLY HOT OIL ACT This was 1935 US Federal Legislation prohibiting interstate trading in oil produced in violation of State prorationing laws. The Act was a device which was intended to stabilise wellhead prices of crude oils.

CONOCO *See* Continental Oil.

CONSERVATION—OIL Following the large increase in oil prices in 1974, considerable conservation measures were encouraged by governments to save oil. Conservation was practised in all sectors of the economy regardless of the type of energy consumed. Thus oil was saved directly and other energy savings also substituted for oil. High oil prices caused energy forecasts to be revised downwards and oil took the full burden. As an example, consider freeworld oil forecasts. Before the large oil price increase was announced, the oil demand forecast for 1980 was 70 million b/d and this was revised to 51·7 million b/d. The saving of 18·3 million b/d was assigned 10·7 million b/d to lower economic growth, 2·8 million b/d to net substitution by other fuels and 4·8 million b/d to energy conservation. Source: J. Birks, IISI meeting, Munich, October 1974.

CONSERVATION—Summer Time Hours Introduction of summer time, a change in the clock of one hour, has been estimated to save about 0·3% of total energy consumption (France 1976). Assuming that all the saving would ultimately be reflected on oil the saving would be about 0·5% or 1 million tonnes of oil. The saving on the oil import bill would therefore be about $90 million in 1974 values.

CONSOLIDATED PETROLEUM CO. A company 50% owned by Shell and 50% by BP marketing petroleum products in the Middle East and Africa.

CONSORTIUM Following the nationalisation of BP's interests in Iran in 1951, a consortium of companies was formed to run Abadan refinery.
On 29th October 1954 an agreement came into force between BP, Gulf, Mobil, Esso, Socal, Texaco, Shell and CFP and the National Iranian Oil Company. These eight companies, increased to seventeen in April 1955 by the addition of nine smaller American companies, became known as the Consortium in which BP had a 40% interest.

CONTINENTAL OIL A major US integrated oil company formed in 1920 as the Marland Oil Co., changing its name to Continental Oil in 1929. It has worldwide interests in all aspects of the oil industry. Its major brand name is Conoco; in the UK it owns the Jet filling stations chain.

CONTROLLED CRUDE OIL Controlled crude oil can have a variety of meanings. It could be company or government controlled oil and had a special meaning in the US in the post-1973 period. Controlled oil was the volume of oil equal to the base production control level (*q.v.*) limits on amount of released oil equal to the new oil production (*q.v.*) on that property. Stripper wells were exempt. Controlled oil was subject to the defined ceiling price (*q.v.*).

COPAREX Cie. de Participations des Recherches et d'Exploitations Pétrolieres formed in September 1957. No direct French state participation.

COPEFA Cie des Pétroles France-Afrique, formed in March 1957. BRP held 67% on formation.

COPESEP Formed in November 1956. Royal Dutch Shell 85%; SAFREP 15%.

COPETAO Cie. des Pétroles Total d'Afrique Occidentale, formed in April 1959 and controlled by CFP.

COPEX The Polish state exploration company.

CORCO The Commonwealth Oil Refining Company Inc. which was registered in 1953 in Puerto Rico. A refining and marketing company with a refinery near Puerto Rico.

CORING In drilling operations a process which is used to obtain rock samples. The drilling bit is replaced by a core barrel. This cuts a cylinder of rock and the rock is held within the cylinder whilst it is pulled up out of the well.

CORMORANT A commercial UK North Sea oil field with estimated peak production of 100 000 b/d and estimated recoverable reserves of 400 million barrels. The field is operated by Shell, owned by Shell and Esso with production planned for 1978.

CORRADINI REPORT A report entitled 'The Outlook for Oil in the World and in the Community', EEC, Luxemburg, 1962, which gave forecasts to 1975. A subsequent report, with forecasts to 1980, was published by the EEC in mid-1968. Lucio Corradini was then Director of the EEC Energy Commission.

CORTEMAGGIORE The Cortemaggiore oil field in the Po Valley gas field was insignificant in size but the name was exploited by ENI. Agip service stations carried the sign 'Supercortemaggiore La Potente Benzina Italiana' accompanied by the six-legged dog.

COST OF CAPITAL As in other industries there is not a single figure that can be quoted with confidence. A survey carried out in 1974 reported the before tax cost of capital as being between 11 and 15%—'Economic Decision Making Practices in the US Petroleum Industry', S. B. Mather, O. L. Carey, *49th Meeting of Soc. Pet. Eng. of AIME*, Houston, October 1974.

COST OIL A term used for Equity Cost Oil (*q.v.*).

COSTS OF CRUDE OIL For details of Alaskan Crude, *see* Tanzer Report.

COSTS OF ENERGY PRODUCTION Energy production costs vary widely and comparisons are usually not meaningful. They are produced and published, however, and a reasonably comparable set of average energy costs for the EEC of Nine was published by the Commission for the year 1974:

	Production, tonnes petroleum equivalent	*Cost \$1974/tonne oil equivalent*
Coal	157	65
Lignite	26	26
Oil	11	11
Natural gas	130	8
Nuclear power	16	25

COUNCIL OF EUROPE The Statute of the Council of Europe was signed in London on 5th May 1949. The members were Austria, Belgium, Denmark, France, Federal Republic of Germany, Greece, Iceland, Italy, Luxemburg, Netherlands, Norway, Sweden, Turkey and United Kingdom. Relations were established with Switzerland, Portugal and Spain, certain British Commonwealth countries and the USA.
The Seat of the Council is in Strasbourg, France. The Council was partly responsible for setting up the ECSC and Euratom.

CPA *See* DTW.

CPA Cie. des Pétroles d'Algerie, formed in 1953. Royal Dutch Shell Group 65%; SOGERAP 24%; BRP 4%.

CREOLE PETROLEUM This company, formed in 1920, is engaged in all aspects of the petroleum business in Venezuela and is 95% owned by Exxon (*q.v.*).

CREPS Cie. de Recherches et d'Exploitation de Pétrole au Sahara, formed in April 1953. Royal Dutch Shell Group 35%, the rest was mainly controlled by RAP.

CRISTAL The Contract Regarding an Interim Supplement to Tanker Liability for Oil Pollution. CRISTAL was intended not as a substitute but as a supplement to TOVALOP liability. It was introduced in April 1971 and increased the liability for a single incident to \$30 million.

CROSSHAUL A crosshaul occurs when vessels carrying similar crudes or products make the same voyage but travel in opposite directions. International exchanges (*q.v.*) between companies eliminate some of the need but crosshauling must occur because it is not possible to utilise the world tanker fleet with 100% efficiency.
It is not unknown for LP solutions (*q.v.*) of a single company's transportation programme to propose crosshauling.

CRUDE OIL ASSAY This is a procedure for determining the distillation curve and quality characteristics of a crude oil. Using modern analytical techniques it may be done using samples as small as a single drop of oil.

CRUDE OIL COSTS *See* Development Costs, Production Costs.

CRUDE OIL COSTS, ESCALATION Estimates prepared by Salomon Brothers, New York Investment Bankers, suggested that costs rose between 40% and 150% in 1976 from historic levels.

Average Investment Cost (\$) *Per Daily Barrel* *Country*	1960–72	1976
United States	3 700	5 200
Canada	3 100	5 175
Venezuela	1 000	1 265
UK North Sea	3 500	7 200
Nigeria	800	1 270
Libya	200	500
Saudi Arabia	200	500
Indonesia	1 400	2 000

CRUDE OIL PRICES, BUY-BACK OIL Following participation, a proportion of all of the crude oil production was owned by the government. This was termed participation crude oil. Such oil could be sold by the government on the open market or sold under special arrangements to the producing company. When it was sold to the company it was termed buy-back crude oil. Buy-back crude, of course, would be at a discount on the posted price and usually on the open market price.

CRUDE OIL PRICES, ECA SYSTEM Crude oil prices outside the USA were influenced by the Gulf-plus system until 1944. From 1944 to 1949 ECA prices operated, ECA being the Economic Co-operation Administration. In 1944 a price of \$1·05/bbl was fixed for Arabian Light at Ras Tanura which was an equivalent price to Texas 36° API gravity crude in the US Gulf. The ECA system was replaced in 1950 with the system of posted prices for PG crudes.

CRUDE OIL PRICES, EQUITY CRUDE The price of equity crude would be no different from unspecified crude oil since, following participation, companies selling crude oil continued to sell at prices dictated by market conditions.

CRUDE OIL PRICES, GULF-PLUS SYSTEM A non-exclusive cartel arrangement whereby the posted prices of crude oils throughout the world would be in line with crude oil prices as quoted in the US Gulf. The main point of the agreement was that customers would pay for crude and for delivery as if the oil had been shipped from the US Gulf regardless of where the oil had actually come from. Thus, India buying oil shipped from the PG would pay US Gulf/India freight not the much smaller PG/India freight. The system lasted until 1944 when it was replaced by ECA prices.

CRUDE OIL PRICES, HISTORICAL The following prices, in current dollars, show the significant changes since commercial crude oil production commenced.

	\$/bbl
1860	9·59
1861	0·49
1864	8·06
1874	1·83
1893	0·64
1920	3·07
1933	0·67
1947	2·20
1973	5·12
1974	11·25

CRUDE OIL PRICES, PARTICIPATION CRUDE Participation crude oil was sometimes sold at special rates to the producing companies at a discount on market prices. This crude oil was then termed buy-back oil (*q.v*).

Participation crude sold by the producing country itself to a third party would be at the normal arms length price which would not differ significantly from an equity crude sale price.

CRUDE OIL PRICES, UNITARY CRUDE Following participation by the producing country governments, there existed a variety of prices for crude oil; posted price, equity price, buy-back price. The average price paid by a producing company depended on the degree of participation and on buy-back arrangements.

In 1975 OPEC adopted a new single price (unitary price) system intended to give the producing government proceeds of about \$10/barrel.

CRUDE OIL, PRODUCER COUNTRY REVENUE Payments to the producing countries consist of royalty and income tax payments. Before participation, the royalty payment was usually $12\frac{1}{2}\%$ and income tax was charged on the selling price or posted price, after deducting an allowance for production costs and the royalty. In the period 1955 to 1970 payments to most Middle East producers, Abu Dhabi excepted, was in the range of 75 to 95 cents/barrel. In Abu Dhabi payments were lower in the years 1962 to 1965 due to relatively high costs. In 1971 average payments to Middle East countries rose to \$1·25 and in 1972 to \$1·40 rising rapidly in 1973 to of the order of \$10/barrel for owned oil in 1974.

CRUDE OIL PRODUCTION, SURPLUS CAPACITY Whilst crude oil production in the freeworld is reported monthly on a regular basis, installed capacity values are not. The amount of surplus production capacity available is an important item in determining sentiment in crude oil markets and price changes of crude oils.

Substantial surplus capacity has been the norm throughout most of the history of commercial crude oil production. When there has existed only a relatively small surplus of crude oil production capacity there have been sharp price increases (*see* Crude Oil Prices, Historical).

In the late 1960s and early 1970s spare production capacity amounted to a few million b/d. However, in 1975, with the sharp fall in freeworld crude oil production from the peak of 50 million b/d reached in 1973, surplus production capacity was much higher. US Treasury estimates put the value at about 12 million b/d in early 1975 and 17 million b/d in mid-1975.

CRUDE OIL PROTECTION DUTY When, under the Treaty of Rome in 1957, crude oil imports to the EEC of Six were declared to be duty free, under a special protocol Germany was allowed to retain crude oil protection duty until 1st January 1964.

CRUDE OIL QUALITY There are two main aspects of crude oil quality which influence the price; the distillate content and the sulphur content. Additionally the price of a crude oil varies with its location, the price differential naturally reflecting transportation costs.

There have been various attempts to relate crude oil quality and hence price to the average market demand, e.g. European oil products demand, but the work has had limited success. This is because oil product prices in the market are set not by average demand, but by marginal demand and supply. Thus, although light distillates usually command a price premium over fuel oils, there are times in Europe when naphtha has been used as refinery fuel and as much as possible 'degenerated' to gas oil and fuel oil. Nevertheless, light crude oils do command higher prices than the heavier crudes and this may be expressed as a gravity differential. Gravity correlates with the quantity of heavy fuel oil. This differential can be as high as 5 cents per °API gravity.

The difference between a 1% sulphur fuel oil and a 3·5% sulphur fuel oil may be about $1/barrel or more, which would suggest 30 cents/barrel sulphur premium on crudes. Desulphurisation of a crude oil would certainly cost more but sulphur premia are not as important as gravity premia.

CRUDE OIL, RESERVES Estimates of crude oil reserves for fields and countries and areas are revised (and published) at frequent intervals. The following shows the growth over the last 40 years of the world figures:

World 'Published Proved' Oil Reserves

Year	*Reserves at 31st December (thousand million barrels)*	*Production in year (thousand million barrels)*
1935	22·0	1·7
1955	160·7	5·2
1974	720·4	21·2

Sources: *Science of Petroleum*, A. E. Dunstan *et al.*, OUP, 1938, Vol. I, p. 532, and *BP Statistical Review*.

CRUDE OIL RESERVES RATINGS To avoid quoting exact numbers which might give a false impression of precision, numerical ratings are sometimes employed. A system employed by the US Geological Survey specifies seven categories:

Category	*Quantity of oil or gas*
I	1 000–10 000
II	100–1 000
III	10–100
IV	1–10
V	0·1–1
VI	0·01–0·1
0	less than 0·01

Oil is in billions of barrels of 42 US gallons or 150 million metric tons. Gas is in trillions of cubic feet or 30 billions of cubic metres (*See* US Geological Survey Professional Paper 817, Published by US Government Printing Office, Washington, 1973.)

The categories take into account the estimates of the total reserves in place and the estimate of the recovery factor.

CRUDE OIL SALE CREDITS *See* Time Credits.

CRUDE OIL SALES PROFITS In the US there have been many independent operators who have bought or leased land, to sell at a dollar a barrel profit even after taxes. On the other hand, some operators have made only cents a barrel whilst others have incurred large losses.

Two of the major companies, BP and Gulf, have traditionally been sellers of crude oil on a large scale and since the 1960s large quantities of Arabian Light crude oil have also been sold by Aramco partners. In 1974 a full disclosure was made of dividends to the Aramco partners 1963 to 1973 (*PIW*, April 8th 1974). The profit to producer country governments has traditionally been the total tax element which was only a few cents a barrel in the US in depressed times and up to 1970 it would have been unusual for a government to be able to extract more than one dollar a barrel in total proceeds. However, in the 1970s the position changed and many governments saw a total take of more than $10/barrel. *See* Crude Oil, Producer Country Revenue.

CRUDE OILS, HEAVY Crude oils of API gravity in the range 7° to 22·0°. Only of the order of 1% of freeworld crude oil production is in this category. Heavy crude oils are favoured for bitumen making although most bitumen is probably manufactured from residues of medium crude oils. Venezuela is the main source of heavy crude oils in international trade.

CRUDE OILS, LIGHT A crude oil of API gravity more than 30°. Oils at the light end of this range will have similar properties to condensates (*q.v.*). Light crude oils command a higher price than other grades because of their high distillate contents and also because they invariably have low sulphur contents. Light crude oils may also be classified as paraffinic crude oils, this distinction being important for the quality of the distillates. Important light crude oils in international trade include Arabian Light, Iranian Light, Iraq crudes, Qatar, Abu Dhabi and Libyan and Algerian crudes. The definition of a light crude oil does present problems because Kuwait and Iranian Heavy, although they have API gravities of more than 30°, are not usually regarded as light crudes.

CRUDE OILS, MEDIUM Crude oils of API gravity in the range 22·1° to 30·0°. Two well-known crudes in this class are Nigerian Medium and Tia Juana Medium. The differential between Nigerian Light and Nigerian Medium from 1965 to 1971 was 14 cents/barrel on the posted price equivalent to about 2 cents/°API gravity. On Venezuelan crudes the differential was about 5 cents/°API gravity. On US indigenous crudes, 5 cents/°API gravity was a representative value.

CRUDE OILS, NAPHTHENIC Naphthenic crude oils do not usually command premiums over equivalent paraffinic grades since there are few petroleum products that can be obtained only from naphthenic crudes. *See also* Naphthenic Acids.

CRUDE OILS, PARAFFINIC Most of the world's crude oil production is composed of paraffinic grades. However, within paraffinic grades the relative degrees of naphthenicity and paraffinicity are of importance. For naphthas of the same boiling range, the more paraffinic would be preferred as a feedstock for an ethylene cracker whilst the more naphthenic would be preferred as feedstock for a cat. reformer. This is because the paraffinic material breaks down more readily under steam-cracking conditions to give a better yield of ethylene. The object of cat. reforming is to produce aromatics, and in this process a better yield would be obtained from the naphthenic naphtha. In practice the maximum differential in price for specifying naphtha quality is likely to be about $5/ton (post-1973 values).
In the kerosene boiling range, paraffinic fractions yield jet kerosenes that meet the specification limits more readily than fractions from naphthenic crude oils. However, there is no premium for jet kerosenes of particular qualities within the specification limits. In the gas oil range, paraffinic crudes yield diesel fuels of high cetane number which is a desirable characteristic. However, the differential between a poor cetane index diesel fuel and a good one is only about 0·5 cent/gallon.
Most lubricating oils are manufactured from paraffinic crudes; only relatively small quantities of naphthenic lub oils are required.

CUBA For crude oil production *see* Table 1.

CURFEW PETROLEUM A small UK gasoline chain, first registered in 1965, and soon bought by Burmah.

CVP Corporacion Venezolana del Petroleo. *See* Venezuelan Oil Corporation.

CYLINDER STOCK A very heavy lub oil base stock obtained by deasphalting petroleum residues. A very specialised product relatively little of which is manufactured.

CYRUS (Iran) An offshore crude of 19° API gravity, 3·5% sulphur exported by SBM.

CZECHOSLOVAKIA For crude oil production *see* Table 1.

D

DAILY PEAK A term mostly used with natural gas being the maximum volume of gas sent out during one day in a calendar year.

DARIUS (Iran) An offshore crude of 34° API gravity, 2·5% sulphur exported from Kharg Island.

DAVIES AND NEWMAN London Tanker Brokers; the company being registered in 1922. Also well known for its DAN air services charter operations.

DECANT OIL A heavy aromatic oil (SG about 1·00 to 1·05) obtained from cracking operations. This oil comes from the bottom of the fractionating column used to separate the cracked products. The word 'decant' refers to the fact that the oil is decanted from the catalyst used in the cracking operation. The oil when still contaminated with catalyst is known as slurry oil. Decant oil is a good feedstock for coke production.

DEEP-SEA VESSEL CAPACITY The carrying capacity of a vessel including fuel, water, provisions as well as the cargo.

DEFLATORS The rate at which inflation is exported to oil producing countries is a key issue in negotiating oil prices. Two deflators commonly used are amongst those published by the IMF: the index of export prices of the industrialised countries and the index of import prices to Iran. Both these indicators are available with a lag of only a few months in 'International Financial Statistics', IMF, Washington, US.
The deflator of the GNP of total OECD countries is also employed.

DEGREE DAY A measure of the extent the mean daily temperature falls below an assumed base temperature. Different base temperatures are used in different locations; 65°F, 62·5°F and 60°F are all commonly employed.
The degree day measurement is used to assess the quantity of fuel required for space heating. It is used by all the sectors of the energy industry both to monitor consumption and to provide forecasts of fuel demand in connection with weather forecasts. Detailed degree day records for the US and for major cities in Europe and Japan are published in *Platt's Oilgram*.

DELAYED COKING *See* Coking.

DEMINEX Deutsche Erdolversorgungs GmbH. Deminex was formed on 27th October 1966 with the following participants: 21% Wintershall (private), 21% Gelsenberg (private), 16% Scholven-Chemie (part of VEBA, 36% Government owned), 12% Preussag (26% owned by VEBA), 10% Schachtbau (subsidiary of state owned mining group, Salzgitter AG), 10% Wesseling (private), 5% Deilmann (private), 5% Saarbergwerke (owned by Government in partnership with Saar authorities).
The participants retained their corporate identities but co-ordinated through Deminex exploration and exploitation of overseas oil concessions for which government financial assistance was given (including 30% of the costs of purchasing oil concessions). Loans were to be repayable only if exploration was successful. The Government undertook to support Deminex with 'initial aid' of DM 575 m. (£65·5 m.) during the period 1969–1974.

DENMARK For crude oil production *see* Table 1.

DEPLETION ALLOWANCE An allowance made to take into account that production of a crude oil reservoir depletes an asset. The depletion allowance has been an integral part of the US taxation system on oil. In 1974 the US House Ways and Means Committee recommended that the allowance be reduced from 22 to 15% and to zero in 1975. This was the thirteenth occasion that such a recommendation had been made and the bill was not passed by Congress in 1974. The usual comment was made in the press that the oil lobby once again had proved too powerful. The first failure had been in 1951.
In March 1975 a compromise agreement was reached in which the depletion allowance was to be eliminated for the larger companies and phased down for the smaller producers. The new allowance was agreed at 22% on the first 2000 b/d in 1975 with a reduction of the Unit by 200 b/d each year to 1980. In 1981 the allowance on the first 1000 b/d was to be lowered to 15% except on secondary and tertiary recovery production, which would be reduced to 15% in 1984. The final bill repealed depletion allowances as from January 1st 1975 and it was estimated that the additional tax would cost the industry $2000 million a year.

DESULPHURISATION The removal of sulphur from petroleum products which may be achieved by a variety of processes during petroleum refining. *See also* Fuel Oil *and* Crude Oil Quality.

DEVELOPMENT BANK OF SAUDI ARABIA A bank set up in 1974 to help recycle the surplus oil revenues. The bank is managed by the US Chase Manhattan Bank (*q.v.*).

DEVELOPMENT COSTS, OECD ESTIMATES There is never close agreement in the energy industry on development costs and a 1975 OECD report 'Energy R & D', which received some publicity, quoted costs which were lower than many in the industry were prepared to accept.

1980 *Development costs in* 1973 $

	($/*barrel*)
North Sea and Alaska	1·25–1·50
Offshore oil, 300–400 m deep	4
Offshore oil, 400–2000 m deep	5–8
Shale oil, existing technology	3–7
Shale oil, *in situ* extraction	3–5
Canadian Tar Sands	4
Solvent refined coal	5–7
Low Btu gas	3·20–4·60
High Btu gas (Lurgi Technology)	6–9
Methanol (from natural gas)	10–12

DEVELOPMENT COSTS, OIL FIELDS The amount of capital required to develop an oil field depends upon its location and on its productivity. The following costs were quoted in the *Petroleum Economist*, December 1974, p. 447 (estimates based on 1973 data):

Oil fields' capital requirements	(£ *sterling per barrel per day*)
North Sea Field A*	800–1 000
North Sea Field B†	1 000–1 200
North Sea Field C‡	1 500–2 000
Middle East—Land	100–150
Middle East—Offshore	150–200
West Africa—Land	300–400

* Field A, 50 million barrels reserves and 15 000 b/d plateau output
† Field B, 100 million barrels reserves and 30 000 b/d plateau output
‡ Field C, 500 million barrels reserves and 150 000 b/d plateau output

For data applicable to the 1960s see PPS, May 1966, pp. 177–9. Sometimes costs are published for an area which has been extensively exploited in a relatively short time. For example, in April 1975 the Norwegian Information Service stated that in the 10 years the Norwegian North Sea had been opened 104 wells had been drilled at a cost of $360 million and 7500 million barrels of recoverable crude discovered.

DEVELOPMENT COSTS, SHALE These costs, like those quoted for tar sands, change frequently and are invariably revised upwards.
Estimates in early 1975 were as follows:

	Capital costs (*$ per barrel per day*)	*Production cost per barrel* ($)
Oil shale	8 000	10
Tar sands	10 000–15 000	10
North Sea oil	4 000–7 000	4–5
Alaska oil	4 000	6
Middle East oil	250	0·12

One of the reasons for the revision of shale and tar sand estimates is that a large amount of energy is required to extract the oil. Some estimates have been as high as 0·7 barrel oil equivalent being required to provide one barrel of oil from shale.
See also Production Costs.

DEVELOPMENT COSTS, USA 1975 The FPC provided detailed figures for finding and producing a barrel of 'new' crude oil:

	Cost/barrel 1975 $
Oil well drilling cost	0·53
Lease acquisition cost	0·46
Production facilities	0·33
Sub-total	1·32
Dry hole drilling cost	0·26
Other exploration costs	0·24
Exploration overheads	0·07
Sub-total	0·57
Operating expense	0·88
Casinghead gas credit	0·61
Return on working capital	0·14
Return on investment	2·84
Royalty at 16%	0·98
Total, excluding production tax	6·12

If Federal income tax (corporation tax at 48%) and 1975 depletion allowances are taken into account, the production cost increases to between $6·76 and $9·17/bbl.

DEVELOPMENT WELL *See* Production Well.

DEVIATION WELL This is a well drilled at an angle to the vertical. Deviation wells are drilled so that as large an area as possible of an offshore field can be produced from only one production platform.

DFN The Danish sales organisation of Nafta, the Soviet Union's oil marketing company.

DFSC The US Defense Fuel Supply Center. The regular reports of prices paid by the DFSC are good indications of the prices of grades of fuels and lubricants that do not appear in regular market reports.

DHT The distillate hydrotreating process. *See* Sour Oils.

DICA The French Government authority 'Direction des Carburants' which has wide reaching powers of intervention in the day-to-day operation of commercial oil companies in France. For example, companies must submit investment and refining programmes for DICA approval. DICA can refuse approval for oil companies' plans on the grounds that they are not in the French national interest.

DIESEL FUELS Diesel engines may be classified into three types and each uses a different fuel:

Type	*Application*	*Fuel*
Low speed (less 300 rpm)	Marine engines Electricity generation	Heavy Fuel Oil (HFO) (Residual)
Medium speed (300–1 000 rpm)	Marine auxiliary Pumps	Heavy distillate or mixture of residual and distillate
High speed	Road and rail transport	Distillate gas oil (GO)

DIESEL INDEX (DI) *See* Cetane Number.

DIGEST OF UK ENERGY STATISTICS Since 1948 an annual UK Government publication which is a source book on UK energy facts.
The early numbers were particularly strong on the coal industry and numbers such as 1944 (Cmd. 6639), 1945 (Cmd. 6920) and 1946, 1947 (Cmd. 7548) included data back to 1938.
The Digest is published with a lag of about 12 months. From January 1946 monthly statistics were published by the UK CSO (Central Statistics Office) which was supplemented in 1974 by *Energy Trends* (*q.v.*).

DIN Deutsche Industrie Norm; the German Industry Standard. The equivalent of the US ASTM and the UK BSI.

DIRECT MARKETING Sales by an integrated oil company or refining company directly to the public, i.e. no jobber or independent dealer is involved. Sales through company owned filling stations are of course in this category.

DISCOVERY WELL A wildcat well that was successful in striking oil.

DISTRESS CARGO A cargo of a product which is in long supply and for which buying offers may be made below the spot market quoted rates.

DIVERSITY FACTOR The diversity factor is a concept which may be used for any power or energy supply system, but it is commonly used with electrical power. It is the ratio of the sum of the maximum demands of the separate consumers divided by the maximum demand made by an area (or station). The ratio is usually much greater than unity because the maximum demands of individual consumers are not coincident.

DOWNSTREAM Activities stated to be downstream are with reference to the refinery, i.e. product distribution and marketing operations.

DREWRY (H. P.) Shipping and oil brokers in Paris and London. The original company was registered in Paris in 1932. The company frequently publishes forecasts of tanker market activities.

DRILLING COSTS—OFFSHORE Offshore costs are much higher than land costs and the main reason is the expense involved in the drilling platforms:

	Cost per day ($)	
	Jack-up	*Semi-submersible*
1965	10 000	15 000
1975	30 000	40 000

DRILLING RIGS The charter rates for offshore drilling platforms varies, of course, with supply and demand. In the North Sea rates of about $40 000/day were paid for semi-submersible platforms in 1974 but rates of $30 000/day were quoted in the Spring of 1975.

DRY GAS FIELD A field which produces methane gas with possibly some ethane but excluding substantial amounts of propane, butanes and higher boiling hydrocarbons (condensates). Not to be confused with Dry Hole (*q.v.*).

DRY HOLE A well which has been drilled without showing signs of oil or gas. It is 'dry' in respect of hydrocarbons only.

DSM The Dutch State Mines which was formed in 1902 when the Dutch nationalised their mines to prevent them coming under foreign control. Since World War II, DSM has been a prominent European petrochemical company.

DTW (DEALER TANK WAGON) The US DTW price is that quoted for delivery to a (retail) dealer in lots of approximately 5000 (US) gallons. The dealer adds taxes and his margin to arrive at the retail price. For example:

DTW	33
State, federal and local taxes	12
Dealers margin	10
	—
Service station retail price	55 cents per gallon

A temporary DTW is one offered for a short time period to stimulate sales. Other allowances to dealers are: cda, competitive dealer allowance; tva, temporary voluntary allowance.
Each month Platt's Prices contains a survey of DTW prices in 55 key US cities. In the UK the standard quotation for motor spirits is in 1000–1499 (Imperial) gallon lots. A lot of less than 1000 gal will incur an extra charge of about 0·5 gal; a lot of more than 2000 gals a discount of 0·5 gal. The standard UK quotation for gas oils and fuel oils is in 500–999 gallon lots.
Price quotations appear in a number of sources. The main US and international daily publication is Platt's Price Service (*q.v.*), for the UK the weekly *Petroleum Times* (*q.v.*), for France the monthly *Bulletin Mensuel* of the CPDP.

DUBAI An offshore crude of 32·5° API gravity and 1·7% sulphur exported by SBM. For crude oil production *see* Table 1.

DUBAI MARINE AREAS A company registered in 1954 producing from the Fateh field with a capacity of 300 000 b/d. Owned 50% by CFP and 50% by Hispanoil.

DUCKHAM Alexander Duckham & Co. Ltd was registered in London in 1920 and has operated as a lubricants company with a major share of the UK market. The company was bought by BP in 1972.

DUNLIN A commercial UK North Sea oil field 310 miles NNE of Aberdeen. Peak production rate estimated to be 125 000 b/d with estimated recoverable reserves at 500 million barrels. Owned 50/50 by Shell and Esso. Production planned for 1978.

E

EAC The estimated annual consumption (of gasoline by a filling station or of oil product purchases by a consumer). A concept which is used to evaluate a site or business.

EARNEST MONEY A deposit which bidders for a lot may be asked to make; commonly 5% of the price offered.

ECA The Economic Co-operation Administration which was set up in 1948 to see that funds provided by the Marshall Plan were well spent. Since much of the aid was spent on oil, the ECA was much concerned in probing oil prices. The ECA was succeeded by the MSA; the Mutual Security Agency. The MSA was active in oil policy until 1952.

ECA SYSTEM *See* Crude Oil Prices, ECA System.

ECE French language abbreviation for EEC (*q.v.*).

ECE The Economic Commission for Europe. This organisation which is part of the United Nations, is based in Geneva. Economic and Energy reports by the Secretariat of the ECE are published by the UN in New York. The reports cover both Western Europe (OECD) and Eastern Europe. The economic reports have appeared annually since the late 1940s.

ECLA The Economic Commission for Latin America.

ECO The European Coal Organisation which in 1946 grew out of the Solid Fuels Division of the Supreme Headquarters Allied Expeditionary Force (SHAEF). ECO later became the Coal Committee of the UN's ECE.

ECONOMIC RENT *See* Rent.

ECSC The European Coal and Steel Community, *see* Schuman Plan.

ECUADOR Although oil production dates back to 1917, production has been small for much of the period. The rapid development of the Gulf–Texaco field (and the trans-Andean pipeline) in the early 1970s made Ecuador an exporter of some prominence. *See also* Table 1. The state company is CEPE. The Oriente export crude is of 30° API gravity, 0·9% sulphur content.

EDF Electricité de France; the state power company.

EEC—ENERGY POLICY The first draft agreement by the EEC of Six was approved in April 1964; this was however merely a treaty of good intentions. The Treaty of Rome had specified a common external tariff to be operated from 1st November 1964. Import duties are levied on product imports of non-EEC origin but not on crude oils or feedstocks. However no common energy policy emerged during the lifetime of the EEC of Six and it has always appeared that national interests, especially French national interests, would outweigh EEC common energy policy interests.

EEC OF NINE The EEC of Six was joined in January 1973 by the UK, Eire and Denmark. By a narrow margin Norway decided not to join, a decision of significance to the EEC in view of Norway's crude oil export potential from the North Sea fields.

EEC OF SIX The Messina Conference of June 1955 decided the need for the EEC and on 25th March, 1957, the Treaty of Rome was signed. In the late 1950s and early 1960s the EEC (of Six) was functionally divided into the ECSC (responsible for steel and coal), Euratom (responsible for nuclear power) and the European Economic Community, or Common Market, which covered all other economic spheres forthwith. Thus in 1957 when a co-ordinated view of energy in Europe was desired, an inter-executive working group was set up chaired by M P. O. Lapie, with members drawn from the ECSC, Euratom and the EEC executive commissions. This 1957 report is sometimes referred to as the Lapie Report.

EECE The Emergency Economic Committee for Europe which was created in May 1945. Much of the work of the EECE was later taken over by the Industry and Materials Committee of the UN's ECE.

EENI Energy Equivalent of Network Inputs. The energy required to produce one unit of a commodity can be calculated by considering all the inputs. A form of energy accounting.

EFFICIENCY IN THERMAL POWER STATIONS The efficiency of power stations varies from one station to another and generally the newest stations are more efficient, i.e. use less oil or coal to generate a unit of electricity. A rule in general use is that 3 tons oil equivalent of fossil fuels are required to generate 1 ton o.e. of electricity. Before approximately 1960, it is necessary to change the factor every few years. For example the coal consumption per kWh generated has changed substantially; between 1925 and 1939 the quantity of coal required per unit of electricity production fell by 40%.

EGERTON REPORT A report on the Heating and Ventilation of Buildings which assessed the contributors to, and the cost of, atmospheric pollution. It was published in the UK in 1945 by HMSO. Electricity generation in the UK in 1938 consumed 14·9 million tons of coal and produced 0·38 million tons of sulphur dioxide. Gas production used 19·1 million tons of coal and produced 0·34 million tons of sulphur dioxide.

EGOCO An Iranian exploration company whose shareholders are NIOC, Elf–Erap, ENI, Hispaniol, Petrofina and OMV. Elf–Erap as operator announced a major gas discovery at Kangon in 1974. Subsequent reports have suggested that it may be the biggest gas field in the world.

EGPC The Egyptian General Petroleum Company; the Egyptian state oil company which was formed in 1957.

EGYPT For crude oil production *see* Table 1. The state oil company is EGPC. The export crude oil is a Gulf of Suez blend of 31·5° API gravity and 1·4% sulphur shipped from Ras Shukheir.

EIFE The Economica Internationale delle Fonti di Energia of Milan.

EKOFISK This Norwegian North Sea crude oil is landed in the UK and shipped from Teesside. It is of 36° API gravity, 0·2% sulphur content.

EL BUNDUQ *See* ADMA and ADMA Crude.

ELAN A marketing company of OMV (*q.v.*).

ELASTICITY OF DEMAND The elasticity of demand (E) is a proportional change in the quantity sold (Q) which corresponds to a proportional change in price (P):

$$E = \frac{dQ}{Q} \div \frac{dP}{P}$$

Thus if the price of gasoline increases from 50 to 60 units and the consumption falls by 5%,

$$E = \frac{5}{100} \div \frac{10}{50} \qquad \text{i.e. } E = -0{\cdot}25$$

In certain areas and at certain price levels, motor gasoline demand has been quite inelastic. The elasticity of demand has proved to be a function of the price level and the availability of alternative transport.

ELASTICITY OF SUPPLY The elasticity of supply may be defined as for demand. The large rise in oil and energy prices in 1974 gave a new lease of life to coal and fostered the development of high cost oil, e.g. the offshore development. The lead times required for development of new energy supplies are usually quite long; of the order of ten years. Whereas the elasticity of demand may be observed to act in a relatively short time, supply is relatively inelastic in the short-term. Thus in the US proved reserves of both oil and natural gas were lower at 31st December 1974 than they were at 31st December 1973 in spite of the huge price rises.

ELF *See* ERAP.

ELF UNION The company was formed in Paris in 1960 as the Union Generale des Petroles, the name being changed in 1967. The company is 80% owned by ERAP and 20% by Aquitaine. The company is the main operating arm of ERAP; transporting, refining and marketing petroleum products.

ELFO Extra light fuel oil; the fuel used in domestic heating systems in Germany. It is essentially gas oil. The Germany gas oil market is the most important in Europe for influencing prices. German ELFO prices and Rotterdam GO prices are closely related.

ENAP Empresa Nacional del Petroleo (Chile). The Chilean state oil company which was formed in 1950.

ENEA The European Nuclear Energy Authority which was formed in 1957 with 18 European country members. The US, Canada and Japan were Associate members.

ENERGETIC ANALYSIS *See* Energy Accounting.

ENERGETIC COSTS The total cost in energy terms (solar energy, geological pressures, etc.) that have gone into the production of a fossil fuel. Human exploitation of that fossil fuel resource usually renders the fossil energy unrecoverable.

ENERGY ACCOUNTING There are various forms of energy accounting but all are techniques which provide balance sheets in energy units just as cash accounting provides balances in money units. To avoid ambiguity it is necessary to be explicit on the convention used. Although energy accounting methods have long been employed, the technique became popular following the 1974 increase in energy prices. Energy accounting is also referred to as energy analysis, energy balancing, fossil fuel accounting, energy thrift analysis, energy input/output analysis and energy flows analysis.

ENERGY ANTITRUST BILL A bill introduced to the US Senate in March 1975 with the intention of prohibiting multiple ownership by one company of oil, natural gas, coal and uranium reserves, requiring enactment by 1980. The bill was backed by the AFL–CIO and the United Mine Workers.

ENERGY POLICY All countries and governments have energy policies but they are not always explicitly stated and sometimes when they are, they are unrealistic.

In the post-World War II period to 1973, most countries were concerned with protecting their own energy industries whilst importing cheap oil. Economic activity in this era benefited greatly from cheap energy and cheap energy policies.

Post-1973 policy was concerned not so much with protecting indigenous energy supplies but promoting them to minimise the cost of oil imports.

Security of supply, many aspects of the shipping industry, OPEC and international politics are all factors which influence energy policy. Conservation, or the sensible use of energy, will become increasingly important as the cost of energy increases in real terms.

ENERGY REQUIREMENTS The following values are an approximate guide to the quantity of energy required to obtain 1 tonne of the following materials:

	Energy requirements (10^6 *Btu*)
Sand or gravel extraction	0·05
Bricks	3·5
Timber (ex-sawmill)	4
Cement	8
Steel	24
Aluminium	250

A table such as the above is unambiguous for ranking in general terms but special grades of bricks (ceramics) can require more energy than cement. *See also* Energy Accounting.

ENERGY STATISTICS The following publications are amongst the most useful for oil and energy statistics:

API bulletins and Handbook (US and World)
Bituminous Coal Annual (US)
BP Statistical Review (World)
Brown Book (UK)
Bulletin Mensuel (France and Europe)
Bureau of Mines (US)
Chase Manhattan Bank (US)
Energy Trends (UK)
Erdol Dienst (Stahmers) (Germany)
Petroleum Economist (World)
PIW (World)
Platt's Oilgram News and Prices
UN Series J (World)

ENERGY TRENDS A monthly publication of the UK Department of Energy started in 1974. It covers all aspects of energy consumption and indigenous production on a monthly basis.

ENHANCED RECOVERY OF CRUDE OIL *See* Tertiary Recovery.

ENI Ente Nazionale Idrocarburi; the Italian state oil company built up to a major World oil company by Mattei. The company was formed in 1953 but was based largely on Agip which itself dated from 1924. Mattei's efforts were directed at making ENI a major integrated company in the same league as the 'Seven Sisters' (as he termed the largest seven majors) but his efforts were largely frustrated by the failure to find major oil reserves. The important shareholdings of ENI include Agip, Snam, Anic, Snam Progetti and Saipem.

ENJAY *See* Exxon.

ENP Empressa Nacional de Petroleos SA was formed in 1974 and is owned 72% by the Spanish Government cartel INI. It was formed from the three refineries of Encaso, Repesa and Entesa. Minority shareholders are Chevron, Texaco, Union Exploaivos, Rio Tinto and a group of Spanish banks.

ENVIRONMENTAL PROTECTION COSTS The API provides estimates of what the US oil industry spends on environmental protection. In 1966 the amount spent was $271 million which rose to $1200 million in 1973.

EOCENE (Neutral Zone) An export crude of 18·5° API gravity, 4·5% sulphur shipped from Mina Saud.

EQUALISATION PROGRAM The US oil industry in 1974 was forced to accommodate international oil prices much higher than domestic prices. The FEA's Price Equalisation Program, which was introduced on 1st December 1974, dealt with crude oil and certain products. The main items of this program were reported in Platt's Oilgram Price Service on 11th November 1974.

EQUITY COST OIL Following participation, producing countries sold oil to the former producing companies at discounts as the posted price. The cheapest oil was equity cost oil and dearer oil, but still cheaper than the posted price, was buyback cost oil. The prices were not fixed for long periods of time and the producers wished to move to a single price structure. The following prices ($/bbl) were quoted for Arabian Light in 1974:

	October	*November*
Posted price	11·65	11·25
Equity cost price	8·37	9·93
Buyback cost price	10·84	10·67

EQUITY CRUDE Following participation in a producing company by the host government, equity crude is that proportion of the crude oil which is company crude oil. Royalty and tax payments are of course made on equity crude. The term was not used before participation when all crude oil production strictly, was equity crude.
See also Equity Cost Oil.

EQUITY OIL A US exploration and production company registered in 1923. In 1968 it sold some of the shale oil interest to ARCO.

ERAP L'Enterprise de Recherches et d'Activités Pétrolieres. A wholly state owned company which was a main instrument for French oil policy. It was formed by a Government decree dated 17th December 1965, which merged BRP and RAP (*q.v.*) to form ERAP as from 1st January 1966. ERAP's main objectives were to be to provide France with a cheap and independent supply of oil and to compete with the majors. ERAP has a majority holding in ELF which operates in several European countries.

ERAP AGREEMENT This agreement between NIOC and ERAP in 1966 evoked considerable comment in the industry. It was very favourable to NIOC; ERAP taking all the risk and NIOC possibly taking 90 % of the profits. The Iranians regarded it as a triumph for their oil policy and it was a landmark in the transfer of power from the exploration companies to the producing countries. For the text of the agreement, see Platt's Oilgram 30th December 1966. For comment 'The Erap Agreement, A Study in Marginal Taxation Pricing', T. R. Stauffer, Paper 72 (A-1) given at Sixth Arab Petroleum Congress, March 1967.

ERG The Raffineria Edoardo Garrone. Garrone is a prominent independent Italian refinery operating near Genoa, at 7·5 million tonnes p.a.

ES SIDER (Libya) An export crude of 37° API gravity, 0·45 % sulphur shipped from Sidra.

ESAB *See* OPB.

ESCRAVOS (Nigeria) An offshore export crude of 36° API gravity, 0·2 % sulphur shipped by SBM from the Escravos River.

ESSO On 1st November 1972 the name of Standard Oil Company (New Jersey) was changed to Exxon Corporation. However, many of the divisions of Exxon outside the US retained the name of Esso.

ESSO REP The Sté. Esso de Recherches et d'Exploitation Pétrolieres formed in May 1955. Esso Standard SAF 89 %, BRP and SNIP 10 %, FINAREP 1 %.

ETHANE Occurs in some sources of natural gas and is a prime feedstock for ethylene production.

EUR The European Unit of Account is sometimes so designated and has been used for recording oil product prices in Western Europe by the SOEC. In the 1960–1972 period one unit of account was equivalent to one US dollar.

EURAFREP The Sté. de Recherches et d'Exploitation de Pétrole which was formed in September 1957.

EURATOM Euratom, la Communauté Européenne de L'Energie Atomique, was formally set up by a treaty signed in Rome on 25th March 1957. Euratom was not a success; it lacked strong government backing and it was created just as the cheap energy era began.

EUROP-OIL PRICES, (LONDON) A weekly publication covering mainly European oil prices.

EUROPA POINT *See* Quoin Island.

EUROPEAN ENERGY PRICES One of the best sources of European energy prices is that published by the Statistical Office of the European Community as 'Energy Statistics'. A special number published in 1974 compared all energy prices in the EEC of Six over the period 1955–1970.

EXPLORATION EXPENSES There is no standard treatment for funds spent on exploration. Exploration expenditure may be fully written off in the year incurred, fully capitalised, or some in-between course adopted. Additionally the treatment adopted by one company may change from one year to the next. Thus company accounts are not always comparable.

EXPLORATION LICENCE Similar to an exploration permit but includes the right to develop the area to production if conditions are favourable. The production area will then become a concession (*q.v.*).

EXPLORATION PERMIT A non-exclusive permit for seismic work, covering a defined area for a limited period.
See also Exploration Licence.

EXPLORATORY WELL *See* Wild Catting.

EXPORT REFINERY Many refineries, including of course all seaboard refineries have product export potential. Some export refining locations are Singapore, the Persian Gulf (Bandar Mah-Sahr, Ras Tanura and Bahrain), Rotterdam, Italy (particularly Genoa and Sicily) and the Caribbean (Amuay, Aruba, Cardon, Curacao and Penuelas).
See also Marginal Markets.

EXPORTS Exports of crude oil are well documented. For a record of when production of oil commenced, which is usually an indicator of when exports commenced, *see* Table 1; World Crude Oil Production. *See also* Imports, for UK product export values.

EXXON The following is Exxon's own account of the adoption of the name. 'On November 1, 1972 the name of Standard Oil Company (New Jersey) was changed to Exxon Corporation. On January 1, 1973 two affiliates, Humble Oil & Refining Company and Esso Chemical Company Inc. merged into the parent corporation and became operating divisions. At the same time they adopted the Exxon name. Humble became Exxon Company, USA and Esso Chemical Company Inc. became Exxon Chemical Company. In addition, Enjay Chemical Company, became Exxon Chemical Company USA, a division of Exxon Chemical Company'.
Exxon's interests are worldwide, with production and/or exploration in 34 countries, refining in 38, and marketing in more than 100. Within the freeworld, Exxon has 14 % of the total production of crude oil and natural gas liquids, including other crude oil take-offs under special arrangements, 12 % of refinery runs and supplies 13 % of product demand. Exxon is owned by almost 755 000 shareholders. Shares held for employees in the Trustee-Thrift Fund of Exxon Corporation and its associated affiliates, represent the largest shareholder account with approximately 2·2 % of the total outstanding shares.

F

FABRICATION TAX A tax levied in Italy at the refining stage. It is levied on weight; typical values in 1975 were 23 070 lire/100 kg on motor spirits, 5976 lire/100 kg on automotive gas oil, 50 lire/100 kg on heating gas oil and 80 lire/100 kg on fuel oil for heating purposes.
It will be noted that the tax on motor spirit is on a weight basis but since motor spirit is sold by volume, there is an incentive to manufacture low specific gravity fuels. Thus there is an incentive to use low specific gravity blending components whilst aromatics are penalised. The isomerisation process, which produces a high octane/low specific gravity motor spirit blending component, is therefore more attractive in Italy than in other European locations.

FAS Free Alongside Ship at export terminal.

FBRS *See* Nuclear Reactors.

FEREIDOON (Iran) An offshore crude of 31 ° API gravity, 2·6 % sulphur exported from Kharg Island.

FIFO The method of stock valuation first in–first out. For details *see* LIFO.

FINA Fina is a brand name of the Belgian integrated oil company Petrofina (*q.v.*).

FINANCIAL STATISTICS An annual survey of the major petroleum companies is produced by the Chase Manhattan Bank (*q.v.*).

FINLAND The Finnish state oil company is Neste Oy.

FISCHER–TROPSCH PROCESS This is a process for producing liquid fuels from coal developed by F. Fischer and H. Tropsch in Germany in 1925. The coal is first gasified, then the gas is converted to liquids using high pressures and temperatures with a suitable catalyst. The general name for the technology is hydroliquefaction.
Plants were operated in Germany during World War II and are now by SASOL in South Africa. The efficiency of these plants is about 50 % and they have been justified only when coal has been a secure source of energy and there were real or potential problems with oil supplies.
In the future, hydroliquefaction technology could become important worldwide if crude oil supplies fall well short of potential demand.

FISSION POWER Nuclear power generated using fast breeder reactors is derived by fission as opposed to fusion and the process may increase natural uranium utilisation by about 50 times compared with thermal nuclear reactors. However sensible economic evaluation of the process is not possible since its commercial significance will not be evident before the 21st century.

FLAG OF CONVENIENCE Tankers registered in countries such as Liberia, to enable the owners to pay less tax than if the vessel was registered in one of the main industrial countries, are said to be carrying a flag of convenience. Lower direct operating costs, besides taxation, may also be an attraction. At the end of 1957 Liberia and the UK both had 8·4 million tons dwt tankers registered out of a total of 46·9 million tons dwt. In 1974 UK registration had increased to 32·2 million tons dwt whilst Liberia's tonnage had increased to 74 million tons dwt.
Panama is another country which has a large amount of flag of convenience tonnage.

FLARING During the production of crude oil, associated gas which is largely methane is separated from the oil and, when there are no adjacent markets, the methane is flared.
In 1973 Saudi Arabia was flaring about 2 million Mcf natural gas a day as the course of oil production. This quantity is equivalent to about 1/3 million bbl oil per day.

FLASH POINT The flash point of an oil is some indication of the risk of fire or explosion associated with its use or storage. Thus the vapour above a gasoline may flash at 0 °F, a kerosine at 120 °F, a gas oil at 150 °F and a lub oil at 400 °F. Flash point limits are included in the flash point specifications of most products above the gasoline boiling range but the test does not have any economic significance so long as the value recorded is inside the specification limit.

FLUSH PRODUCTION The yield of an oil well during the early period of its production.

FO Fuel Oil (*q.v.*).

FOD Fuel Oil Domestique—the distillate oil used for domestic central heating in France. The product is the same as gas oil. In Germany the product is termed ELFO (*q.v.*).

FORCADOS (Nigeria) An export crude of 30·5° API gravity, 0·2 % sulphur shipped from Forcados.

FORCE MAJEURE Many contracts for crude oil and petroleum products have 'force majeure' clauses which enable the obligations of the supplier or the purchaser to be suspended in the event of certain unexpected events. These could be political acts interfering with normal business, strikes which prevent supplies reaching the purchaser or the purchaser being unable to take delivery of the product.

FORTIES A commercial UK North Sea oil field 110 miles ENE of Aberdeen, Block 21/10. Maximum production rate estimated to be 500 000 b/d with estimated recoverable reserves at 1800 million barrels. The BP equity is 98 %. The crude is of 36·6° API gravity, sulphur content of 0·3 %. Production started in 1975.

FOSSIL FUEL ACCOUNTING *See* Energy Accounting.

FOSSIL FUELS The fossil fuels include natural gas, oil, coal, shales and tar sands. The production of particular resources depends on the cost of extracting the fuel and its cost of transportation to a market. There are many reserves of fuels that are never likely to be produced, e.g. deep coal in small seams and relatively remote small fields of low pressure oil and gas.

FPC The US Federal Power Commission. *See also* Natural Gas.

FRANC ZONE CRUDE OIL One aspect of French oil policy was to provide France with a supply of crude oil from French controlled territories. Until the 1960s Algerian oil was considered a more secure source of oil for France than most others. Oil from Algeria and France itself, was designated Franc Zone crude. Through the 1960s this oil was more expensive than that available from the Middle East and companies marketing in France, as part of the licencing system, were required to purchase a certain quantity of Franc Zone crude, although this oil did not necessarily have to be marketed in France.
In 1968 Franc Zone crude accounted for about one-third of the demand for oil in France so the foreign exchange or balance of payments aspect was also of importance.
Possibly the most important single act in ending the importance of Franc Zone crude was the cutting off by Algeria of ERAP exports for a short while in late 1968. Algeria joining OPEC ended the Franc Zone crude concept but Algeria continued to supply considerable quantities of oil to France.

FRANCAREP The Cie. Franco–Africaine de Recherches Pétrolieres, formed in July 1957. No direct French government participation.

FRANCE For crude oil production *see* Table 1. The main state company is ERAP.

FREIDENSBURG REPORT This report, named after one of its authors, was published by the Association of German Economic Institutes in 1962. The full title was 'Survey on the Development of Present and Future Structure of Supply and Demand of the Energy Economy in the Federal Republic with Special Reference to the Hard Coal Mining Industry'. Forecasts of US and Ruhr coal prices to 1975 and of energy consumption in 1975, were made.

FREIGHT RATES *See* Tanker Rate Schedules, *and* Spot Market Freight Rates.

FROPEX France-Outre-mer de Prospections et d'Exploitations Pétrolieres, formed in July 1954. Wholly owned by CEP.

FSH The 'Fonds de Sontien aux Hydrocarbures' set up by the French Treasury in 1950 to give financial assistance to French companies searching for oil in France. In 1951 this special fund was extended to the entire Franc Zone (*q.v.*).
The FSH derived funds from taxes on motor spirit, gas oil and light fuel oil. BRP and ERAP benefited from this fund.

FUEL OIL There are many grades of fuel oil which are classified according to viscosity and sulphur levels. Except for special cases (such as ELFO (*q.v*)), fuel oils are fuels containing petroleum residue materials. The lighter grade fuel oils are a mixture of about 50% gas oil and 50% heavy fuel oil. The heaviest (thickest) grade oils tend to be those used under ship's boilers. Bunker fuel oils will also have the highest sulphur levels. Since these bunker oils are treated as sulphur sinks, they are priced lower than the standard HFO 3·5%S, the price of which is quoted very widely.
Two other sulphur grades which are quoted: the HFO 1%S and the 0·5%S grade, used in steel manufacture. One of the reasons that desulphurisation capacity has not been more extensively installed in petroleum refineries is that the historic and anticipated price differences between 1%S and 3·5%S fuel oils have not justified the investment.
See also US Fuel Oil Grades.

FURFURAL An organic solvent which is used in the refining of lubricating oil basestocks to remove some of the polar materials present (heterocyclic compounds containing nitrogen and sulphur). The treatment is relatively cheap (about 2 cents/b in 1974, solvent cost) since the furfural is recycled.

FURNACE BLACK *See* Carbon Black.

FWONA The Free World Outside North America

G

GABON For crude oil production *see* Table 1.

GAETA PETROLI This is a company of the Monti Group which bought BP Italiana.

GAO The General American Oil Company of Texas; a US production company formed in 1936. GAO has exploration interests in Canada, the Red Sea and the North Sea.

GAS In the US, motor gasoline; full boiling range 15 to 190 °C. Not to be confused with LPG used for motor vehicles. *See also* PMS, RMS.

GAS ACT (UK) The UK Gas Act of 1948 established the structure of the UK Gas Industry into a Gas Council and 12 Area Boards.

GAS CYCLING OR RECYCLING *See* Gas Injection.

GAS/DIESEL OIL Additional terms include; for domestic heating use—heating oil, middle distillates, American No. 2 Fuel, automotive gas oil, compression ignition engine fuel, derv fuel (diesel engine road vehicle fuel) and high speed diesel oil.

GAS INJECTION Gas production from a crude oil reservoir is sometimes reinjected to help maintain pressure in the field; this may be done as an alternative to flaring. The procedure is used when the gas available in that location has little if any value. When a non-associated gas field is operated for recovery of the condensates only, the reinjection of the gas may be referred to as cycling or recycling.
See also Secondary Production.

GAS OIL Gas Oil is one of the 'main products' manufactured from crude oil and is the main component of the middle distillate range of products. Gas oils are manufactured by the distillation process (straight run gas oils) and also by various cracking and coking processes (cracked gas oils). The domestic central heating and the diesel fuels for road vehicles (DERVS) markets account for most of gas oil consumption.
Relatively small amounts of gas oils are used to manufacture ethylene by the steam cracking process since naphtha is a more attractive feedstock from the technical and usually from the economic point of view.
Most gas oils have a relatively low sulphur content, about 0·5% or less.
See also ELFO, Fuel Oil *and* US Fuel Oil Grades.

GASUNIE The Dutch company, formed in 1963 to manage the distribution of the gas produced by NAM (*q.v.*). The state controls 50% and Exxon and Shell the rest equally.

GATHERING LINES These are crude oil lines from individual wells to a central storage area.

GDP *See* Gross Domestic Product.

GEARING The ratio of debt to equity is the traditional capital gearing. Sometimes income gearing is of more importance when companies become over-extended.
See Lease Commitments.

GELSENBERG AG A German industrial company formed in 1953. Interests in the oil industry include an 18·5% stake in Deminex (*q.v.*).

GENERAL PURPOSE TANKER A tanker of 15 000 to 24 999 tons dwt as defined for the AFRA index in 1964. Also known as 'handy size'.
In 1968 the size was redefined as 16 500 to 24 999 tons dwt. GP tankers are mostly used as product carriers. For cost of operation *see* Tankers, Large Range.

GENEVA AGREEMENT The Geneva Agreement of 1972 was intended to compensate the oil producing countries for exchange rate adjustments.

GENOA PRICES *See* Platt's European Prices.

GEOPHYSICAL COSTS Each year the US Society of Exploration Geophysicists publishes estimates of the cost of survey work in the previous year. In 1973 actual petroleum geophysical spending in the non-communist world was near $1000 million. Seismic work was the most important. The following estimates are for 1973 in $ million; seismic-land 676, seismic-marine 278, surface gravity/magnetic 23, airborne magnetic 4, sonic/velocity logging 2. Total = $983 million.

GERMAN BUREAU OF STATISTICS *See* German Price Indices.

GERMAN OIL INFORMATION SERVICE (Erdoel-Informationsdienst) This service, commenced in 1947, is available in an English language edition. It is published twice a month and is a useful source of German oil industry information and statistics. The publisher is A. E. Stahmer, Hamburg.

GERMAN PRICE INDICES The German Bureau of Statistics publishes each month detailed statistics of German product prices. The indices include mogas, diesel fuel and ELFO (*q.v.*) for wholesalers and consumers and heavy fuel oil for delivered quantities of 15 to 200 tons/month, 210 to 2000 tons/month and more than 2001 tons/month. Also quoted are price indices for ex-refinery tank cars and ex-refinery barge lots of 650 tons or more. The standard quantity of ELFO for domestic deliveries is 5000 litres. The indices are usually available with a time lag of 50 days.

GERMANY For Crude oil production *see* Table 1. Germany has not relied on a state oil company for its oil policy to the extent that many other European nations have. Deminex was a vehicle for some exploration activity and an interest in the Veba-Gelsenberg company (as late as 1975) was the main state interest in marketing.

GETTY OIL COMPANY A large integrated US oil company with extensive worldwide interests.

GHAWAR *See* Saudi Arabia, Crude Oil Reserves.

GIGA The symbol is G and the multiple is 10^9.

GLYCOL *See* Antifreeze.

GO *See* Gas Oil, ELFO, Fuel Oil.

GOLDEN EAGLE *See* Ultramar.

GOMT A Libyan National Company, General Organisation for Maritime Transport. In 1974 its fleet consisted of five tankers totalling 272 000 tons dwt.

GOSP A Gas (from) Oil Separator Plant. These are installed in crude oil production areas to separate excess gas from crude oil before shipment (stabilisation of crude).

GRAPHITISATION A heat treatment process to improve the quality of petroleum coke.

GRAVITY DIFFERENTIAL *See* Crude Oil, Quality.

GREAT CANADIAN OIL SANDS LTD A company formed in 1953 to exploit Athabasca tar sands. Sun Oil has been much involved with the company.

GREEN COKE *See* Sponge Coke.

GREGORY COMMISSION This was the report of the Commission on Energy for France's Seventh Plan. It forecast that oil consumption in France in 1985 would amount to only 96 million tonnes compared with 111 million tonnes in 1973. The forecast for 1985 made in 1973 had been for 178 million tonnes.

GROSS DOMESTIC PRODUCT (GDP) The GDP of a country is the value of the goods and services produced by resident producers less the value of their intermediate consumption; i.e. the GDP is assessed on a value added basis. The real GDP (GDP in volume terms) is well related to primary energy consumption for long time periods within a country and also holds using data for several countries. Over five-year periods the correlation coefficient can be high, 0·99, but the statistical relationship is not useful when forecasting by quarter or for periods of less than five years ahead. Then forecasts are based on economic and energy sector detail.

GULF OIL CORPORATION One of the big seven of the world's major oil companies which grew from the Mellon family oil interests. The company has major interests in coal and atomic energy. It developed the Kuwait oil fields in 50/50 partnership with BP and like BP was a substantial seller of crude oil to the industry.

GULF OF SUEZ For quality of this export crude *see* Egypt. *See also* GUPCO.

GULF-PLUS SYSTEM *See* Crude Oil Prices, Gulf-Plus System.

GUPCO The Gulf of Suez Petroleum Company, formed in 1966. GUPCO is a joint venture between the General Egyptian Oil Company and Amoco. Operates the Morgan, July and Ramadan fields in the Gulf of Suez.

H

HALL–HEROULT PROCESS The process by which aluminium is manufactured by dissolving alumina in cryolite which is contained in a carbon lined cell. The carbon forms the cathode whilst the anode, also carbon, dips into the electrolyte. Molten aluminium is liberated at the cathode. The carbon anodes and cathodes are manufactured from petroleum coke.

HALLSTEIN REPORT An EEC report, presented on the 1st March 1959, on the subject of European Free Trade. At the time Hallstein was the Chairman of the European Economic Commission.

HAMILTON Hamilton Brothers Oil Co. (GB) Ltd and Hamilton Brothers (Petroleum) UK Ltd are parts of the Hamilton Bros. Group; a US oil company. The Group was prominent in UK North Sea exploration; *see* Argyle.

HANDIL (Indonesia) An offshore crude of 31 ° API gravity, 0·1 % sulphur content exported by SBM.

HANDY SIZE TANKER OR HANDY SIZED CARGO *See* General Purpose Tanker.

HANSARD Hansard is the journal which contains detailed reports of the actions of the UK Parliament.

HARD COAL The term hard coal always excludes brown coal and lignite but there is ambiguity associated with its usage. In Europe bituminous coals and anthracite are termed hard coals but in the US anthracite only is termed hard coal. The situation is further complicated in that in some localities the term hard coal is not used and would not be understood. For example, in the South Wales coalfield the coal on the west side of the field is termed anthracite, in the middle—steam coal and on the eastern side—bituminous.

HARTLEY REPORT A major OEEC report concerned with energy forecasts and policy: *Europe's growing needs of energy, how can they be met?* OEEC, Paris, 1956.

HASSI MESSAOUD (Algeria) A crude oil of 44 ° API gravity, 0·15 % sulphur content, exported from Bougie.

HCE Hard Coal Equivalent. One ton of hard coal is equivalent to about 0·7 ton of oil, the exact factor depending on the calorific values of the coal and oil specified.

HEAT PUMP A heat pump is a device which can be used to increase the efficiency of usage of fuels. The pump can be used to extract useful heat from low grade heat (e.g. water that has been used for cooling). Technically a heat pump is a device which drains heat from a relatively cold body and transmits it to a relatively hot one. A refrigerator is a heat pump but its sole purpose is of course to extract heat, i.e. keep the contents cool. If the purpose of a similar device was merely to transmit the heat taken from the cold body to another body for some useful purpose it would be termed a heat pump.

HEATHER A commercial UK North Sea oil field with estimated maximum production of 50 000 b/d and recoverable reserves of 150 million barrels. The field is operated by Unocal with production planned for 1978.

HERON The London property company Heron Corporation Ltd operates a small number of filling stations in the UK pursuing an aggressive marketing policy.

HFO Heavy fuel oil. HFO is the petroleum product which is in most direct competition with coal and consequently there is a close link between the prices of HFO and solid fuels. *See also* Fuel Oil.

HISPANOIL The Spanish State Oil Exploration Company which is active in oil exploration and production outside Spain.

HOECHST The German chemical company Farbwerke Hoechst AG which was first formed in 1863 when it was concerned with dyestuffs from coal. Now the company is prominent in petrochemicals and fibres as well as the traditional general chemistry field.

HOME OIL CO. Home Oil (of Canada) was incorporated in 1929 and is an integrated oil company operating in Canada and several other countries. Included in its production/exploration activities are interests in Alaska, the North Sea and Oman.

HOMS At Homs is the state owned refinery of Syria. The refinery has been supplied with crude oil from Iraq through the IPC pipeline and the throughput of this line is sometimes quoted with the refinery supply broken out.

HONEYCOMB COKE This is produced from relatively low asphaltenes, low sulphur, petroleum feedstocks and upon calcining (*q.v.*) and graphitisation (*q.v.*) can produce satisfactory electrodes. A major use of the coke is in aluminium refining: it takes about 0·4 lb coke to help produce 1 lb of aluminium.

HOPECO An Iranian exploration company whose shareholders are NIOC and Mobil.

HOT OIL *See* Connally Hot Oil Act.

HOUT (Neutral Zone) A crude oil of 34° API gravity, 1·7 % sulphur exported from Ras Khafji.

HOVIC The Hess Oil Virgin Islands Corp. which operates the Virgin Island Refinery of Amerada Hess. In many ways, oil refined in the Virgin Islands is treated as US domestically refined oil.

HTRS *See* Nuclear Reactors.

HUDSON'S BAY OIL AND GAS This company was incorporated in Canada in 1926. Its main interests are oil and gas production in Canada.

HUGHES RIG COUNT A report of the number of oil rigs operating in the US and Canada. Published weekly in *O. and G. J.*

HUMBLE OIL AND REFINING This company became an operating division of Exxon (*q.v.*) on 1st January 1973.

HUNGARY For crude oil production, *see* Table 1.

HUSKY OIL LIMITED The company was incorporated in 1953 as Canadian Husky Oil Ltd. It has production, refining and marketing interests in Canada and the US.

HUTTON A commercial UK North Sea oil field with estimated maximum production of 100 000 b/d and recoverable reserves of 300 million barrels. The operator is Conoco.

HYDROGEN Hydrogen is an important component of the gases resulting from cracking and reforming processes. It has important uses in the petroleum industry in the hydrogenation processes, e.g. hydrotreating and hydrocracking.
Of greater long-term significance is the fact that the world's fossil fuel reserves that are relatively rich in hydrogen (natural gas and crude oil) are being depleted at a relatively fast rate compared with the world's total hydrocarbon reserves (largely in coal). Thus in the future, fossil fuels rich in hydrogen may be expected to command more of a premium over fuels of low hydrogen content until cheap hydrogen becomes available from electrolysis of water using nuclear electricity.

I

ICI Imperial Chemical Industries of the UK which was formed by the amalgamation of four major chemical companies in 1926. One of the world's major chemical companies. Has extensive petrochemical interests and markets motor spirit in some parts of the UK.

ICOO The Iraq Company for Oil Operations; state owned. *See also* INOC. Reports of the Iraq oil industry are published in the Arabic language journal *al-Naft Wal-Aalam.* Translations of these reports appear in English language press, e.g. *MEES* 8 Nov. 1974 for report of 1972 and 1973 operations.

ICOR The Incremental Capital/Output Ratio which is used as a measure of the capital investment

required to achieve an anticipated volume of additional output.
ICORs, which decrease if new capital is employed more efficiently, are much used in the petrochemical industry. The usefulness of the concept in planning is a matter of opinion.

ICPA The International Co-operative Petroleum Association Inc. which was formed in 1947 in the US.

IDENTIFIED RESOURCES Specific resources whose location, quality and quantity have been specified. *See also* Resource.

IEA The International Energy Agency (IEA) which had first been proposed in February 1974 by the US Secretary of State Dr Henry Kissinger, was established by decision of the OECD Council on 15th November 1974, as an autonomous body within the framework of the Organisation. On 18th November 1974, the sixteen Members of the Organisation then participating in the Agency entered into an Agreement on an International Energy Program.
The International Energy Program is implemented through the IEA. The principal aims of the Program are:

- (i) development of a common level of emergency self-sufficiency in oil supplies;
- (ii) establishment of common demand restraint measures in an emergency;
- (iii) establishment and implementation of measures for the allocation of available oil in time of emergency;
- (iv) development of a system of information on the international oil market and a framework for consultation with international oil companies;
- (v) development and implementation of a long-term co-operation programme to reduce dependence on imported oil, including: conservation of energy, development of alternative sources of energy, energy research and development, and supply of natural and enriched uranium;
- (vi) promotion of co-operative relations with oil producing countries and with other oil consuming countries, particularly those of the developing world.

The countries which participate in the IEA are: Austria, Belgium, Canada, Denmark, Germany, Greece, Ireland, Italy, Japan, Luxembourg, the Netherlands, New Zealand, Norway, Spain, Sweden, Switzerland, Turkey, the United Kingdom and the United States. France does not participate but the Commission of the European Communities takes part in the activities of the Agency.

IEP International Energy Programme, *see* IEA.

IFP L'Institute Français du Pétrole was created following a Government decision of 1944 and commenced activity in 1945.
Some of the activities are similar to those of the IP in the UK and the ASTM D2 Committee and API in the US.

IMCO The Intergovernmental Maritime Consultative Organisation. The 1969 IMCO standard was an agreement stating that no discharge of oil would be permitted within 50 miles of a coast except for 'clean ballast'.
See also Load-on-Top, TOVALOP and CRISTAL.

IMINOCO The Iranian Marine International Oil Company. The shareholders are NIOC (50 %); ENI, Phillips and the Indian Government holding the other half equally.

IMPERIAL OIL The company is engaged in all aspects of the petroleum business in Canada and is 70 % owned by Exxon (*q.v.*). Imperial Oil is one of the largest Canadian oil companies and was incorporated in 1880.

IMPORT LICENCE FEES These are fees levied on crude oil and/or products to provide consumer government revenue, protect local industry or implement government energy policy.

IMPORTED COST OF OIL To a company the imported cost of oil is the price paid for the crude oil plus the freight. The crude oil cost could have been only the tax paid cost or a market price or a transfer price. Freight could be the spot market rate, AFRA or a company freight rate. To the importing country the cost of crude could include a profit element to the importing company. If market conditions and/or government controls in the consuming country forces the company to operate at a real loss, then crude oil imports can be effectively at below the tax paid cost of crude and marginal cost of freight.

IMPORTS Imports of petroleum by quantity and value are trade figures which are usually easy to obtain going back for considerable periods of time:

	UK Imports of Petroleum (*cif*) (£)	*Exports* (*fob*) (£)
1872	433 472	—
1879	1 382 534	—
1923	36 m	5·2 m
1933	31·3 m	3·1 m
1943	154·3 m	0·6 m
1953	209·7 m	75·1 m
1963	556·8 m	115·4 m
1973	1 688·3 m	347·5 m

INA The Yugoslav state oil exploration company.

INDIA For crude oil production *see* Table 1.

INDIAN OIL CORPORATION The Indian state oil company which was incorporated in 1959 as Indian Oil Co. the name being changed when it amalgamated with Indian Refiners Ltd in 1964.
Indo–Burma Petroleum Co. is a wholly owned subsidiary. *See also* ONGC.

INDONESIA Crude oil has been produced since 1893 and, as one of OPEC's poorest nations, there has always been a need to maximise production and revenue. With a population in 1974 of about 130 million and a per capita income of the order of $100, its main aim is in establishing a sound economic base for the nation.
The Indonesian state oil company is Pertamina. For crude oil production *see* Table 1.

INDONESIAN CRUDE OILS These include Atlaka, Bekapai, Bunju, Cinta, Duri, Handil, Jatibarong, Klamono, Melahin, Minas (Sumatran Light), Poleng, Tarakan and Walio.

INECO An Iranian exploration company owned by NIOC, Mobil and a Japanese consortium.

INFLATION The contribution to inflation caused by the large increase in oil prices in 1974 was relatively modest. The following estimates were provided by the EEC Economic and Social Committee.

Country	*Overall inflation rate in* 1974 (%)	*Inflationary effect of oil prices in* 1974 (%)
France	14·9	2·5
Germany	5·9	1·5
Italy	25·2	3·6
UK	18·3	2·7
Belgium/Luxemburg	15·7	4·2
Netherlands	11·0	1·5
Denmark	16·5	1·9
Ireland	20·0	1·7

See also Price Inflation.

INFLATION INDICES Following the 1972 inflationary period, many indices and combination of indices were discussed and proposed for crude oil price linking.
An example of a combination index was reported to be used by Shell in its sale of North Sea LPG to the US. The index was weighted 40% Arabian Light Crude Oil, 40% US Wholesale Price Index and 20% UN Commodity Index.
Price indices of goods in world trade (published monthly by the IMF) were also discussed for index linking of crude oil prices. *See also* Deflators.

INOC The Iraq National Oil Company which was formed on 8th March 1964. It operates from Baghdad and its aims are to operate both inside and outside Iraq in all aspects of the oil industry.
By 1974 the Iraq Government controlled 80% of the country's production through INOC and ICOO.

INORGANIC TRADE These are sales which are usually made on a spot basis with no guarantee of subsequent sales to the same source. They are frequently in large quantities (1000 to 1 000 000 tonnes) and may be to other oil companies, chemical companies or large consumers. Some large consumers of petroleum products have a policy to obtain a proportion of their requirements from oil companies under annual contracts and buy the rest at spot market prices.
See also Organic Trade.

INSTITUTE OF FUEL The UK Institute of Fuel which was founded in 1927 is much concerned with fuel technology and is not orientated towards the petroleum industry.

INSTITUTE OF PETROLEUM *See* IP.

INTERNAFT Tanker and oil brokers in London, Riyadh and Madrid. A frequent source of tanker forecasts.

INTERNATIONAL EXCHANGES An international exchange occurs when one company supplies another with a product in country A and takes in exchange the same product of equivalent quality in country B. As with local exchanges (*q.v*) agreements may be for several years.

INTERRUPTIBLE GAS Interruptible gas supplies are charged at a reduced rate, usually to large consumers on the understanding that in periods of peak demand supplies may be cut off. Stand-by energy supplies, frequently LPG or oil, are then used. The reduction in the tariff obviously needs to be sufficient to pay for the cost of the stand-by facilities.

INTERRUPTIBLE SUPPLY TARIFFS *See* Bulk Supply Tariffs.

INTERTANKO The International Association of Independent Tanker Owners established in 1971 to further the interests of independent owners.

IOCC The US Interstate Oil Compact Commission which represents 30 of the states.

IOP Iranian Oil Participants. *See* Consortium.

IOPDA The Iraqi Oil Products Distribution Organisation.

IOTC The Iraqi Oil Tankers Company.

IP The UK Institute of Petroleum was founded in 1913 and its aim is to promote the scientific study of petroleum. Although it has some commercial and economic interests in the industry, it is heavily biased towards the scientific and technical side.

IPAA The Independent Petroleum Association of America. Amongst its activities it produces regular quarterly reports of the medium term US supply and demand for petroleum.

IPAC An Iranian exploration company whose shareholders are NIOC and Pan American Oil.

IPC The Iraq Petroleum Co., which was formed in 1911, was first named African and Eastern Concessions Ltd, a year later the Turkish Petroleum Company Ltd, and in 1925 the name was changed to IPC. IPC was owned 5% by the Participation and Exploration Corporation (Gulbenkian) and the remainder in equal shares by BP, Shell, CFP and the Near East Development Corporation.
In 1972 Iraq nationalised IPC with a settlement being reached between IPC and Iraq in early 1973. Nationalisation of about 99% of IPC's concession area had been carried out in 1961.

IPIECA The International Petroleum Industry Environmental Conservation Association which was formed in 1974.

IRAN Iran's oil production dates from 1913 and there are many accounts of D'Arcy's exploration work, crude oil production by BP and refining at Abadan. BP (first as the Anglo–Persian and later as the Anglo–Iranian Oil Co.) ran the oil industry until 1951 when it was nationalised by Mossadeq. Production and Abadan refinery were shut down for 18 months and then resumed, being run by the Consortium (*q.v.*). The state oil company is NIOC. For crude oil production *see* Table 1.

IRANIAN HEAVY CRUDE OIL Sometimes referred to as Gach Saran (the name of one of the major fields), it is of 31° API gravity and 1·6% sulphur content. It is exported from Kharg Island.

IRANIAN LIGHT CRUDE OIL Iranian light crude oil is one of the world's major crude oils. It is also known as Agha Jari.
Iranian Light is of 33·5° API gravity and a sulphur content of 1·4% and is exported from Kharg Island.

IRAQ For crude oil production *see* Table 1. *See also* ICOO, INOC.

IRICON A US holding company set up to hold a 5% interest in the Iran Consortium through IOP (*q.v.*).
Those in the agency are American Independent Oil Co. (one sixth), Atlantic Richfield (one third), Continental Oil (one twelfth), Getty Oil (one sixth), Signal (one sixth) and Sohio (one twelfth).

IRISH STATE OIL COMPANY On 4th November 1974 it was announced that a state oil company was to be set up. On 29th April 1975 it was stated that the Government's licencing terms for offshore oil exploration included a state right to 50% participation and a sliding scale of royalties from eventual production.

IROM Industria Raffinazione Oil Minerali was formed by ENI (Agip) 51% and BP (49%) in 1947 to operate a refinery near Venice. BP disposed of its interest when it left the Italian market in 1973.

IROPCO An Iranian exploration company whose shareholders are NIOC, Getty, Skelly, Superior, Sun, Kerr McGee, Cities Service and Atlantic Richfield.

IROS Iranian Oil Services Ltd, the service company of the Consortium (*q.v.*).

IRVING OIL A refining and marketing company in Eastern Canada controlled by Standard Oil of California.

IS Intascale; *see* Tanker Rate Schedules.

ISRAEL For crude oil production *see* Table 1.

ITALPROTEINE A joint ANIC–BP company formed to construct a 100 000 ton p.a. protein plant in Sardinia.

ITALY The Italian state oil company is ENI. For crude oil production *see* Table 1.

J

JACOBS John I. Jacobs are London tanker brokers, the company being registered in 1927. The shares are quoted on the London Stock Exchange. Their twice annual review is the major source of published information on the world tanker fleet.

JAMBUR (Iraq) *See* Bai Hassan.

JAPAN For crude oil production *see* Table 1.

JAPAN OIL DEVELOPMENT COMPANY *See* JODCO.

JAPAN PETROLEUM DEVELOPMENT CORPORATION (JPDC) This Government agency, formed in 1967, gives support to Japanese companies engaged in oil exploration.

JAPAN PETROLEUM INSTITUTE *See* JPI.

JAPEX The Japan Petroleum Exploration Co. was formed in 1955 to carry out exploration and development of petroleum in Japan. When JPDC (*q.v.*) was formed Japex was amalgamated into the new organisation.

JATIBARONG (Indonesia) A crude oil of 29° API gravity and 0·1% sulphur exported by SBM.

JERSEY STANDARD The Standard Oil Company of New Jersey; *see* Exxon.

JET A OR A-1 This is the standard aviation kerosine fuel which is similar to (UK) domestic quality paraffin or the lightest quality burning oils.

JET B Jet B fuel is a wider cut from crude oil than Jet A-1 (kerosine) and includes material which is in the gasoline boiling range. Consequently its flashpoint is lower and so its use has been limited.

JET FUELS Theoretically jet engines can run on any fuel but in practice gasoline-type and kerosine-type fuels are favoured for aviation use. The prices of these two fuels are similar but kerosine is generally preferred to gasoline because of its lower volatility. It is of course important for aviation fuels to be free from impurities which would block filters and have only low concentrations of hydrocarbons (certain n-paraffins and aromatics) which could separate as solids under very cold conditions.
Prices of jet kerosines are usually much in line with ordinary domestic kerosines in bulk markets.

JET PETROLEUM A UK petroleum company registered in 1953 with a capital of only £600 000. It was taken over by Conoco in mid-1961. Conoco continued to use the Jet name for mogas marketing in the UK.

JOBBER A jobber may buy oil products from a refiner, hold on his own account and sell. Alternatively he may introduce a customer to a refining company and take commission on the sale without using his own capital.

JOBBER MARGIN The difference between a jobber's buying and selling prices.

JODCO The Japan Oil Development Company which was formed in 1973 in connection with the purchase of an interest in ADMA. The company is owned by the Overseas Petroleum Corporation of Japan.

JONES ACT This Act stipulates that shipments between domestic US ports, including Alaska, Hawaii and Puerto Rico be carried in US built ships operating under the US flag and manned by US seamen. The Act was named after Senator Wesley Jones who succeeded in making it part of the 1920 US Merchant Marine Act.
Waivers of the Act are permitted only in the interests of national defence. For example in 1975, although there was a gas shortage in New England, Treasury Secretary Simon refused to allow four shipments of liquefied natural gas to be made from Alaska to Boston in foreign flag vessels to ease the shortage.

JOSEPHINE A UK North Sea oil field (Block 30/31). Original shareholdings were as follows:

Phillips Petroleum Exploration UK Ltd	35%
Agip (UK) Ltd	15%
Fina Exploration Ltd	30%
Century Power & Light Ltd Plascom (1909) Ltd Oil Exploration Ltd Halkyn District United Mines Ltd	20%

JOURNALS *See* Petroleum Economist (formerly Petroleum Press Service) O & G.J, PIW, Platt's, Petroleum Times, MEES, Bulletin Mensuel, API, BOM and Petroleum Business News Index.

JP.4 JP.4 is an aviation turbine fuel also known as Jet B, (*q.v.*). The JP or jet paraffin designation of aviation fuels was used by IATA and grades JP.1 to JP.8 have all been used but superseded by other descriptions.

JPI The Japan Petroleum Institute, which was formed in 1958, is much concerned with the commercial side of the industry and is a good source of information through its regular publications.

K

KEROGEN Kerogen is the main organic component of shale (*q.v.*). It contains about 1% sulphur but is rich in nitrogen (2%) and oxygen (6%). Kerogen itself, free from mineral matter, would have a calorific value similar to that of poor grade coal.

KEROSINE The kerosine fraction of crude oil, approximate boiling range 175–250 °C is sometimes included in middle distillates. Aviation fuels, burning oil, paraffin and vaporising oil, could all be substantially kerosine material.

KERSHAW REPORT M. Kershaw was the Rapporteur for the Council of Europe Report 'European Energy Problems', CEC, Strasbourg 1962, Document 1483. This was an important oil and energy policy document. Basically it endorsed the conclusions of the Robinson Report (*q.v.*).

KHAFJI (Neutral Zone) A crude oil of 29° API gravity and 2·9 % sulphur exported from Ras Khafji.

KIRKUK (Iraq) The Kirkuk oil field is one of the world's most prolific fields. Production to end of 1973 amounted to 826 million tons. The crude is of 36° API gravity and 2 % sulphur.

KISSINGER PLAN The US policy statement made by Dr Henry Kissinger in 1974 set out a policy for coping with the sharp rise in energy prices in 1974. *See* IEA.

KLAMONO (Indonesia) A crude oil of 19° API gravity and 1 % sulphur.

KNPC The Kuwait National Petroleum Company which was formed in 1960. The Kuwait Government holds 60 % of the shares and the other 40 % are held by nationals of Kuwait.

KOC *See* Kuwait Oil Co. Ltd.

KOGEC The Kuwait Oil Gas and Energy Corporation which was set up in August 1974 to handle the Government's interests in oil and gas operations.

KOTC Kuwait Oil Tankers Company. In 1974 the Company operated six tankers totalling nearly 800 000 tons dwt.

KUWAIT For crude oil production *see* Table 1. *See also* KOGEC.

KUWAIT CRUDE This major export crude oil, shipped from Mina al Ahmadi, is of 31° API gravity and 2·5 % sulphur.
Umm Gudair crude, which is not a normal export crude, is of 24·5° API gravity and 3·7 % sulphur.

KUWAIT OIL CO. LTD This company which was formed in 1951, 50/50 BP and Gulf, carried out exploration, development and production of petroleum in Kuwait. The main fields in Kuwait operated by the company were Burgan, Magwa, Umm Gudair and Minagish. The loading terminal is Mina-al-Ahmadi. Production in Kuwait rose from 2·3 million b/d in 1967 to 3·0 million b/d in 1972 falling back to 2·3 million b/d in 1974.
The company was fully participated in 1975.

KYODO OIL A grouping of several small refining and marketing companies which pooled their facilities under Government sponsorship in 1965. This was effectively Japan's first move in forming a state backed oil group.

L

LACQ GAS A natural gas field in France which has been one of the main supply sources of gas to the French market.

LAFTA The Latin American Free Trade Association. Includes the following: Argentina, Brazil, Chile, Colombia, Ecuador, Mexico, Paraguay, Peru, Uruguay and Venezuela.

LAMBERT BROTHERS A London firm of shipping brokers which is a subsidiary of the Hill Samuel Corp. (Merchant Bankers).

LAPCO The Lavan Petroleum Company which was incorporated in 1965 in Teheran. It is an Iranian exploration and production company whose shareholders are NIOC, ARCO, Union Oil of California, Murphy Oil and Sun Oil. Oil production is from the Sassan field.

LAPIE REPORT *See* EEC of Six.

LAURÉ COMMISSION A commission set up by the French Government in 1975 to reassess the arrangements under which the French oil industry operates.

LDF Light Distillate Feedstock. These are usually materials prepared by atmospheric distillation of crude oil supplemented by small quantities from the hydro-cracking process. They will have boiling ranges within the range 15–220 °C and can have quite different qualities and values.
The prices of LPGs, PFDs, SRNs, SRGs, and SRBs are not quoted frequently but will be of the same order as LDF prices.

North West Europe, Bulk LDF Prices (*$/Tonne*)

1961	1963	1965	1968	1970	1972	1973	1974
24	18	19	26	20	27	80	100

LEAD ALKYLS Addition of lead alkyls to motor gasoline is a relatively cheap and efficient way of improving antiknock quality, i.e. increasing the octane rating as measured by all methods. The most common lead alkyl is tetra–ethyl–lead (TEL) but tetra–methyl–lead (TML) is also extensively used. TML usually costs about 10 % more than TEL and is not quite so effective as TEL for enhancing the research octane number. It is used in preference to TEL when market conditions justify improvement of the 'front end' octane quality of a motor gasoline to make the gasoline less susceptible to pinking (low speed knock).
In 1974 US fob prices of TEL and TML mixes were about \$1400/ton. Improvements in the manufacturing process reduced the price of TEL through the 1950s and 1960s. In 1973 the price was the same as in 1954 although in 1974 it increased fairly rapidly with the general inflation.

LEAD FREE GASOLINE Until the invention of tetra–ethyl–lead in 1922, motor gasoline was of course lead free. Addition of TML and other lead alkyls to hydrocarbon mixtures gave them a superior performance and engines became more efficient. In the late 1960s a lobby developed for the removal of lead from gasoline on the grounds of reducing lead pollution in the atmosphere and also allowing cars to have catalyst systems fitted to reduce the content of hydrocarbons and certain other gases in exhaust emissions.
There is no technical problem in manufacturing high octane lead free gasoline but it is costly both in monetary and energy terms. In the US additional investment in refining capacity of up to $\$20 \times 10^9$ has been quoted as being necessary, depending on how severely exhaust emission standards are enforced. Cost/US gallon would increase by 1–2 cents on the 1974 price of about 40 cents/US gallon.

LEAGUE OF ARAB STATES The Pact of the Arab League was signed in 1945 with the Economic Council and Joint Defence Councils being formed in 1950. In 1964 the Council of Arab Economic Unity was established by the Economic Council. Its aims included removal of internal tariffs, establishment of common external tariffs, freedom of movement of labour and capital and for common economic policies.
The members include Algeria, Bahrain, Egypt, Iraq, Jordan, Kuwait, Lebanon, Libya, Mauritania, Morocco, Oman, Qatar, Saudi Arabia, Somalia, Spanish Occupied Arab Territory, Sudan, Syria, Tunisia, United Arab Emirates, Yemen Arab Republic and Yemen PDR.

LEASE COMMITMENTS A lease commitment is effectively an extension of a companies debt. Commitments could include tankers, service stations, offices and other properties.
The commitments of Burmah Oil and Occidental in the tanker market proved to be severe embarrassments to these companies.

LIBYA For crude oil production *see* Table 1.

LIBYAN NATIONAL OIL CORPORATION The Libyan state company which was formed in 1970.

LIFO The method of stock valuation last in–first out as opposed to the first in–first out (Fifo) method. Under steady trading conditions it obviously makes little difference which accounting method is used. However, when the freight markets are volatile resulting in temporary 'highs' in product prices, and in periods such as 1973 when a great increase in crude oil prices occurred, it does matter which method is employed. There is no convention used by the major oil companies. In its 1972 report Chevron stated that 47 % stocks were on a Lifo basis and 53 % on a Fifo basis. In 1973 61 % of inventories were on a Fifo basis and in 1974 100 % were on a Lifo basis. BP in 1974 stated that 100 % were on a Fifo basis.
The main reason why Lifo became the favoured method was that it minimised unrealistic inflation of reported earnings during periods of rapidly inflating crude oil and product prices.

LIGHT-LOADED A light-loaded vessel is one carrying only a part cargo. Vessels could be light-loaded because the vessel is capable of carrying more oil than is required; in the case of barges because water levels on a river are not sufficient for the barges to be fully loaded. In this latter instance low water surcharges can sometimes apply.

LIGHTENING Lightening is a procedure for reducing the load in vessels so that they may enter a port which would not take the vessel fully laden. Thus oil may be shipped from the PG to NWE in a VLCC and before docking part of the load of about 250 000 tons would be transferred to another tanker. The part-laden VLCC would then dock and discharge, with the other vessel discharging at the same, or another, location. The purpose of the operation is that at full costs, a VLCC would ship crude from the PG to NWE at about half the cost of crude shipped in two vessels of the 100 000 tons dwt class.

LIGNITE In energy terms lignite (sometimes known as brown coal) ranks between hard coals and peat. Poor quality coals, such as some found in Pakistan and Nigeria, are also termed lignitious coals. The calorific value of a lignite should be specified together with the water content of the sample on which this value was determined.

Lignites or brown coals are important fuels in Germany, the Indian sub-continent and Australia.

LINDSEY OIL REFINERY A company formed by CFP in 1956 to build and operate a refinery on the Humber (UK). Formerly known as Total Oil Refineries (GB). Capacity 8 million tons p.a. and now owned 50% by CFP and 50% by Petrofina.

LIQUID PARAFFIN *See* White Oils.

LLOYDS LIST This London daily paper, dating from 1734, is mainly concerned with shipping and shipping movements but is also a useful source of information of the oil and energy industries.

LOAD-ON-TOP This is an arrangement to minimise oil pollution at sea by minimising the amount of oil discharged in tank cleaning operations. The tanks in a vessel are cleaned successively with the dirty washings (water plus oil) being pumped into one tank designated the slop tank. Maximum separation of oil from the water is achieved. The economic attraction of the system is that the oil recovered largely pays for the operation.
The phrase load-on-top comes from the loading of crude oil on top of the remaining contents of the slop tank at the loading terminal.

LOCAL EXCHANGES A local exchange is an agreement between different companies to exchange products within the same country. This allows companies to reduce transportation costs by supplying their markets from a convenient refinery owned by a competitor. Exchange agreements may be negotiated to cover period of several years. *See also* International Exchanges.

LOW GRADE HEAT Low grade heat is heat which economically is of little or no value. A typical example would be cooling water in a power station. Cooling water from an industrial process may be too warm to discharge into a river (*see* Thermal Pollution) but not warm enough to be useful for central heating purposes.

LP Linear Programming. A mathematical technique which has been much used by oil companies to optimise their supply programmes. The solution is a minimum cost solution subject to the many constraints specified in the mathematical model of the company's operations.

LPG Liquefied Petroleum Gases. It is composed of propane (C_3) and butanes (C_4s) when obtained from wells, crude oil distillation, cat. reforming or hydrocracking. Sometimes C_3 and C_4 hydrocarbons from cracking processes are also blended into LPGs. LPGs are frequently sold in small quantities which involve high distribution and packaging costs. In bulk their value will be similar to that of LDF (*q.v.*).

LR1, LR2 *See* Tankers, Large Range.

LSMO The London and Scottish Marine Oil Co. Ltd, a participant in the UK North Sea Ninian Field. The company is a member of the Ranger Group.

LTBP The London Tanker Broker Panel.

LWRS *See* Nuclear Reactors.

M

MACH Societa Mach is part of the Italian Monti Group which bought BP Italiana.

MacKINNEY REPORT The MacKinney report was published in February 1956 in Volume 2 of the Report of the Panel on the Impact of the Peaceful Uses of Atomic Energy to the Joint Committee on Atomic Energy.
In the report the views were quoted of W. E. Pratt that adequate reserves of oil existed to satisfy demand for at least the next 20 years.

MAGNUS A commercial UK North Sea oil field owned 100% by BP with reserves of 400 million barrels.

MAIN PRODUCTS Main Products are those which form the major proportion of crude oil (major share of the barrel). They include mogas, LDF, kerosines, gas oils and fuel oils. The minor products are termed specialities (*q.v.*).

MAJORS The Majors is the collective name frequently employed for the world's seven largest oil companies. They are Exxon, Socal, Gulf, Texaco, Mobil, BP and Shell. The term is also sometimes used instead of integrated, i.e. an oil company engaged in all aspects of the oil industry.

MALAYSIA The Malaysian state oil company is Petronas; for crude oil production *see* Table 1.

MALONE REPORT A US report on energy supply by the Minerals, Materials and Fuels Economic Subcommittee (Chairman G. W. Malone), *Accessibility of strategic and critical materials to the United States in time of war for our expanding economy*. US Government Printing Office, Washington DC, 1954.

MANDJI (Gabon) This crude of 29° API gravity and 1·3% sulphur is exported from the Cap Lopez terminal.

MANGANESE *See* MMT.

MANUFACTURED GAS Could be blast furnace, city, coal, coal hydrogenation plant and coke oven gases, Lurgi plant gas from oil gasification, synthesis, town and water gases.

MARATHON OIL COMPANY An integrated US oil company which was registered in 1887 as Ohio Oil Company changing its name to Marathon in 1962.

MARGIN The term is frequently used loosely and unless defined could be the difference between costs or prices, or a profit margin.

MARGINAL COST The marginal cost of a crude oil is that of a suitable crude oil bought to be refined to meet a particular market demand. It could be an arms length price negotiated with a crude oil seller or a producing company's acquisition cost. The marginal cost of freight will be the going spot market rate for the vessel available for the route. It is likely to be the rate quoted for the largest vessel that can trade on the route.
The marginal cost of a product or a 'barrel' of products will equate to the cost of crude and freight and to the marginal cost of refining. The marginal cost of refining will include the cost of fuel in the refining operation but not as a return on refinery capital.

MARGINAL CRUDE OIL As a result of there being a strong demand for a marginal product (*q.v.*) this demand will cause a certain type of crude oil to become the marginal crude oil. When fuel oil is in relatively short supply, the marginal crude oil will be a heavy crude oil such as Kuwait. When light distillate is short, the marginal crude oil will be a light crude oil such as a light North African crude.
In the sense that refineries contain crackers to manufacture more light product than is naturally available in any given crude oil, the lightest crude oil available is the marginal crude. Light crude oils are therefore at higher prices than heavy crudes.
In the early 1970s the World's marginal crude oil was Arabian Light in the sense that demand for oil was met, and could only have been met by expansion of Saudi Arabian crude production. Saudi Arabian crude oil was however the cheapest crude oil in the world with production costs of about 10 cents/bbl. (There is no relation between the cost of a crude oil and its price.)

MARGINAL FIELD AID To encourage production of oil from fields that would not be economic at the taxation rates applied to large fields, provision is sometimes made to encourage production. This was done by the UK Government in their proposals of February 1975 by giving the Secretary of State for Energy the power to refund the UK Government's $12\frac{1}{2}$% royalty should this prove necessary in the smaller North Sea fields.

MARGINAL FREIGHT RATE This is the rate paid for a single voyage charter. When rates are low, the marginal rate is that for the largest vessel that can operate on the route. When rates are high, i.e. when there is only a small surplus of shipping, rates for all size vessels will be similar. Then a payment of $30/ton may be made for shipment of crude oil from the PG to north-western Europe whether it is carried in a vessel of 35 000 or 350 000 tons/dwt. However when the markets are slack the rate for a VLCC may be about $3/ton whilst for a small vessel the minimum rate may be near $10/ton.
AFRA quotations of course are not marginal rates and neither are various composite indices published by shipbrokers.

MARGINAL MARKETS On the marginal petroleum markets, the so-called Spot Markets, there is trade daily in the major petroleum products. The important markets include New York, Rotterdam and Genoa.
See also Platt's European Prices.

MARGINAL PRICE In times of relative surplus, marginal prices will be much in line with marginal product costs (*q.v.*). However when there are relative shortages, marginal product prices may be substantially higher; double, for example, the landed cost of crude in a particular market. Marginal prices have a substantial influence on all market prices and also on oil company profits.

MARGINAL PRODUCT Crude oil is refined to a particular pattern of products depending on the refinery or group of refineries, the season and the type of crude oil input. However even with product shipments between continents, from time to time one product will be in relatively short supply and production cannot be readily increased. This will be the marginal product. Increased supplies of the marginal product may be achieved by bringing into the refining system a greater proportion of a certain quality crude oil; this will be the marginal crude oil (*q.v.*).

MARGINAL RESOURCES These may also be termed sub-economic resources. Submarginal reserves are sometimes defined as that part of the reserves which require a substantially higher price, at least 1·5 times the price at their evaluation, or a major advance in technology for them to be developed. However, the economic basis for classifying resources is poor. For example, the sharp increase in crude oil prices in 1974 did little to advance the status of tar sands and shale deposits.

MARKET PRICE As with other commodities, the market price for crude oil or any petroleum product depends on a number of factors the most important being: the location of the sale, the size of the deal and the operating price in the nearest marginal market (*q.v.*). The price of crude oil itself is much dependent on the world energy balance.

MARKET SHARES, WORLD OIL PRODUCTION Market shares data are usually concerned with domestic markets but sometimes the concept is applied to the world. The following data were quoted by ENI in their 'Sommario Statistico' for 1955–1973:

	1970		1973	
	Production (million tonnes)	*World %*	*Production (million tonnes)*	*World %*
Exxon	271·5	14·0	276·3	11·8
Texaco	161·4	8·3	226·8	9·7
Gulf	162·0	8·3	166·2	7·1
Socal	127·9	6·6	166·8	7·1
Mobil	91·3	4·7	113·5	4·9
Shell	206·6	10·6	230·4	9·8
BP	189·0	9·7	229·0	9·8
CFP	57·5	3·0	69·3	3·0
Total above	1267·2	65·1	1478·3	63·1
Total world	1945·9	100	2342·5	100

MARKETABLE NATURAL GAS Is defined as raw gas from which certain hydrocarbons and non-hydrocarbon compounds have been removed or partially removed by processing. Marketable natural gas is sometimes referred to in the US as pipeline, residue, or sales gas.

MARTHA A marketing company of OMV (*q.v.*).

MAUREEN A UK North Sea oil field (Block 16/29).
Original shareholdings were as follows:

Phillips Petroleum Exploration UK Ltd	33·78%
Agip (UK) Ltd	17·26%
Fina Exploration Ltd	28·96%
Century Power & Light Ltd	9·0%
Ultramar Exploration Ltd	6·0%
British Electric Traction Ltd	5·0%

MBD Million barrels daily also expressed as mb/d and million b/d.

MEDIUM RANGE TANKER A tanker of 25 000 to 44 999 tons dwt as defined in the AFRA index of 1964. Such a vessel may be used as crude or a products carrier.
For cost of operation *see* Tankers, Large Range.

MEDRECO The Mediterranean Refining Company; the company is owned 50% Caltex/50% Mobil.

MEES *The Middle East Economic Survey;* an influential weekly journal published in Beirut. An index has been published since 1973.

MEGA The symbol is M and the multiple, 10^6.

MEK Methyl–Ethyl–Ketone. An organic solvent which is employed in the dewaxing of lubricating oil basestocks.

MER The maximum efficient rate of production of crude oil. Considerable judgement is exercised in setting this rate and there need not always be agreement on what is the MER to within a factor of two.

METHANOL Methanol can be made from natural gas and transportation of natural gas as methanol as an alternative to liquefaction is sometimes considered. For consumption methanol could be reconverted to gas (methane), used as a fuel, or as a motor fuel either alone or mixed with petroleum motor spirit. However methanol suffers two disadvantages when used as a motor fuel; it is leached out by water bottoms in storage tanks and it has a low calorific value. Fuels with low calorific values give low mpg but the relationship is not linear.

Fuel	*Calorific value (gross Btus/gallon)*
Methanol	76 000
Ethanol	101 500
Motor Spirit	148 000
Motor Benzol	158 000

See also Development Costs, OECD Estimates.

METHMIX These are methanol–water mixtures used in aviation to lower the temperature of the combustion mixture and maximise engine power for take off.

MEXICO The Mexican state oil company is Pemex; the Mexican oil industry being nationalised in 1938. For crude oil production *see* Table 1.

MIDDLE DISTILLATES The term is used to describe those petroleum products in the boiling range approximately 160 to 360 °C. The products include domestic kerosine, jet kerosine and all types of gas oils and distillate diesel fuels.

MIDDLE EAST Shipments of oil from the 'Middle East' or 'Gulf' (which may be confused with the US Gulf) are sometimes quoted to avoid using the terms 'PG', 'Persian Gulf' or 'Arabian Gulf'.

MIDDLE EAST EMERGENCY COMMITTEE This committee, set up at the time of the first Suez Crisis in 1956, comprised fourteen leading US oil companies and was formed under the auspices of the US Government.
In London the comparable organisation was the Oil Emergency (London) Advisory Committee (*q.v.*).

MIL The abbreviation MIL is used in the US for Military and is followed by a letter and numbers to denote product specifications.

MINAS (Indonesia) This crude oil, also known as Sumatran Light, is of 35° API gravity and 0·1% sulphur. It is exported from the Dumai terminal.

MINERAL FUELS These include crude oil, condensates, natural gas, bituminous coals, anthracites, lignites, tar solids and shales. Excluded are nuclear power, hydropower, fuelwood and baggasse.

MINERAL OILS A collective term describing crude oil products and also those from shale, tars and coal to distinguish these hydrocarbon oils from vegetable oils.

MINERALOLSTEUERGESETZ This was a German Mineral Oil Tax introduced on 20th December 1963. It was levied under Article 17 of the Treaty of Rome whereby member countries were authorised to make good the loss to their budgets incurred by elimination of custom tariffs. However the tax was not supposed to have a protective effect on any trade or industry.

MMT Methylcyclopentadienyl–manganese–tricarbonyl is a gasoline antiknock or octane improvement additive. MMT is more expensive than lead compounds such as TEL and TML and has been little used.

MOBIL OIL CORPORATION One of the world's seven giant integrated oil companies and a descendant of the Standard Oil Trust.
The company was incorporated in 1882 in New York as Standard Oil Company of New York (SOCONY), split from the Standard Oil Trust in 1911, amalgamated with the Vacuum Oil Company in 1931 to form the Socony–Vacuum Corporation. The name was changed to Socony–Mobil in 1955 and to Mobil in 1966.

MOGAS *See* PMS.

MOND REPORT A UK Government White Paper *Report of the National Fuel and Power Committee*, Chairman Sir Alfred Mond.
HMSO Cmd 3201, 1928.

MONOPOLY The closest approach to a theoretical monopoly in the oil industry was achieved by Rockefeller in 1879 when he controlled about 90% of US refining capacity. The early 1870s had seen poor trading conditions due to excess capacity in refining and more particularly production. Rockefeller realised that control of product was practically impossible and set out to control the US oil industry through control of the refining stage. In 1911 the Standard Oil Trust was split into over thirty companies the main being:

1. Standard Oil (New Jersey), frequently known as Jersey Standard, then as Esso and finally Exxon.
2. Standard Oil Company of New York (Socony); in 1931 merged to become Socony Vacuum, in 1955 Socony Mobil and in 1966 Mobil Oil.
3. Standard Oil Company of California (Chevron).
4. Standard Oil Company of Indiana (Amoco).
5. Standard Oil Company of Ohio (Sohio).

MONSANTO One of the world's largest chemical companies with substantial interests in petrochemicals and lubricating oil additives.

MONTHLY DIGEST The UK Government publication of monthly energy statistics. *See* Digest of UK Energy Statistics.

MONTROSE A commercial UK North Sea oil field 130 miles east of Aberdeen. Peak production rate estimated to be 50 000 b/d, with estimated recoverable reserves at 200 million barrels. Original share holdings were:

Amoco UK Petroleum Ltd	30·77%
Gas Council (Exploration) Ltd	30·77%
Amerada Exploration Ltd	23·08%
Texas Eastern (UK) Ltd	15·38%

The crude is of 42° API gravity and 0·2% sulphur. Offtake is by SBM.

MOP The UK Ministry of Power which was superseded by the Ministry of Technology, the Department of Trade and Industry and in 1974 by the Department of Energy.

MOROCCO For crude oil production *see* Table 1.

MOT The UK Ministry of Transport World War II freight rate schedule. *See* Tanker Rate Schedules.

MOTOR SPIRIT *See* PMS *and* Star Rating System.

MOVEMENT MATRIX A movement matrix equates the consumption of oil with the supplies. Along the top of the matrix are the consuming countries whilst down the side are the supply areas. A column therefore indicates how the supplies of oil to one consuming country or area are made up. A row indicates the disposal pattern of a supply country or area.
Annual movement matrices are published in the *BP Statistical Review* (time lag of about 4 months) and by the OECD with a longer time lag.

MPC The Mosul Petroleum Company of Iraq. MPC was officially incorporated with ICOO (*q.v*) on 1st March 1973. It owns the Butmah and Ain Zalah oilfields. MPC was a member company of IPC.

	Production to 31*st Dec.* 1973 (*long tons*)
Ain Zalah	21 171 513
Butmah	4 664 926

MSA The Mutual Security Agency.

MUBARRAS *See* ADMA and ADMA Crude.

MULLION London tanker brokers which for many years have provided an index of spot market tanker freight rates for the *Petroleum Economist*. The index is a worldwide index and consequently is influenced by the rates obtained for all categories of vessels on all major routes.

MULTIGRADE OILS *See* SAE Viscosity Ratings.

MURBAN (Abu Dhabi) An onshore crude of 39° API gravity and 0·7% sulphur exported from Jebel Dhanna.

MURPHY OIL CORPORATION An integrated US oil company. Murco Petroleum, which operates in the UK, is a wholly owned subsidiary.

MWV Mineralolwirtschaftsverband—the trade organisation for the West German Oil Industry.

N

NAFTA A Russian oil company operating in Europe usually with small market shares.

NAM Nederlandse Aardolie Maatschappij; the Dutch gas exploration company owned jointly by Exxon and Shell. First significant gas strike was at Coevorden in 1948 followed by the substantial Slochteren or Groningen field in 1960. (Slochteren is a village in the province of Groningen.)

NAPHTHA Naphtha, not specified as virgin naphtha, could contain some olefinic materials. Its value is likely to be the same or fractionally lower than virgin naphtha. (*q.v.*).

NAPHTHENES Naphthenes and compounds containing naphthenic rings are present in all crude oils. The most important individual naphthene is cyclohexane which is used as a chemical feedstock.

NAPHTHENIC ACIDS Naphthenic acids are extracted from certain crude oils which have relatively high concentrations (about 1%). The price of naphthenic acids is little influenced by the price of crude oil. When the price of crude oil was about $15/ton, the price of naphthenic acids was about $800/ton.

NATIONAL BENZOLE COMPANY A UK petroleum products marketing company registered in 1919. It was formed initially as a trade association controlling quotas and output of UK benzol producers.
The benzol was produced by distillation of coal and could consist of more than 99% benzene or be a mixture of benzene with up to 20% toluene. Motor spirits with a high benzol content would have high gravities (0·78). But as the compression ratios of engines rose from 7:1 to as high as 9:1 more attention was paid to the composition of the lower boiling components of motor spirits to minimise low speed knock (pinking). Until 1975 the company was part of the Shellmex and BP marketing organisation. In 1976 it became part of BP Oil. *See also* Retail Outlets, UK, Benzol, Aromatics.

NATIONALISATION The term participation is more frequently found in petroleum literature than nationalisation. Participation has been used for the progressive nationalisation of crude oil production companies whilst nationalisation has tended to be used for government takeover or other assets. The motivation for nationalisation or governments taking financial interests in companies has been varied; e.g.

the British Government's interest in BP in 1914, the Iranian nationalisation of BP's assets in Iran in 1951 and the establishment of State Oil Companies (*q.v.*). Recent examples of nationalisation or expropriation of subsidiaries of international oil groups include Cuba (1959), Egypt (1961 and 1964), Ceylon (1962), Syria (1965), Peru (1968/69), Bolivia (1969), South Yemen (1969), Somalia (1970) and Ethiopia (1975).

NATO The North Atlantic Treaty Organisation which issues quality specifications for many petroleum products. NATO comprises; Belgium, Canada, Denmark, Greece, Iceland, Italy, Luxembourg, Netherlands, Norway, Portugal, Turkey, UK, US and West Germany.

NATURAL GAS Natural gas is composed of methane and ethane. From some areas, e.g. the North Sea, the gas is almost entirely methane. Methane is largely used as a fuel but ethane may be used to produce ethylene. In many parts of the world natural gas has had zero value and when produced with oil (associated gas), has been flared. This is because transportation costs for natural gas have been high and it has been unable to compete with other fuels. Because natural gas is a convenient sulphur free fuel, it should command a premium over the thermal equivalent of oil in particular locations. In the US in the 1960s and 1970s control of the price of natural gas by the FPC led to potential demand exceeding supply. The US Natural Gas Act was passed in 1938 but the FPC did not interpret it to apply to independent producers until 1954. From 1954 onwards the FPC encouraged demand and discouraged supply. In 1960 the 'second Phillips case' arose, where Phillips and others maintained that costs exceeded gas revenues. The FPC did not agree. In 1965 the FPC maintained that a 3% increase in the price of gas would suppress demand increases. By the early 1970s potential demand for gas exceeded supply by an enormous amount.
This example is probably the most pertinent in the energy market where price control has led to acute imbalance between potential demand and supply.

NATURAL GASOLINE This is composed of pentanes and heavier hydrocarbons in the gasoline boiling range. *See also* SRG.

NCA The US National Coal Association. *See* Bituminous Coal Annual.

NCB The UK National Coal Board, formed in 1947, which produces most of the UK's coal.
From time to time the NCB produces forecasts of West Europe's energy and oil demand and puts forward, sometimes with the WECP, policy desirable from the coal producer's point of view. *See* Energy Policy.

NCW The non-communist world; but the term really requires further definition. Yugoslavia, due to its association with the OECD, is likely to be included in the OECD grouping.

NEB National Energy Board of Canada. Controls the oil market by issue of licences for imports/exports.

NEDC The Near East Development Corporation, formed in 1928, jointly owned by Exxon and Mobil. Held a $23\frac{3}{4}$% interest in IPC.

NEDO The UK National Economic Development Office which is a forum for the UK Government, Industry and the Trade Unions. NEDO issues reports on aspects of UK economic policy and other reports of interest to the oil industry, e.g. on conservation and on chemicals. The committees which report on individual topics producing development plans for the industry are known as Little Neddies.

NEEDLE COKE This is produced from highly aromatic petroleum feedstocks such as the heaviest oil from cat. cracking (decant oil) or the residue from ethylene manufacture (the steam cracking process). Needle coke is high quality coke and is used in electric arc furnaces for steel manufacture. About 5·5 kg coke is used in the manufacture of 1 tonne steel.

NELSON INDICES Nelson Indices are cost indices of refinery plant and engineering supplies, etc. published at frequent intervals in the *Oil and Gas Journal* (*q.v.*).

NEPCO The New England Petroleum Company which was incorporated in New York in 1947. NEPCO is an oil trading company and has a 65% interest in BORCO (the Bahamas Refining Company which has a 500 000 b/d refinery in the Bahamas). The other 35% is owned by Socal.

NESTE OY The Finnish state oil company formed in 1948, the state holding nearly 98% of the shares.

NETHERLANDS For crude oil production *see* Table 1.

NETWORK PURIFICATION This term is a euphemism for eliminating relatively unprofitable outlets.
For an example *see* Retail Outlets, UK.

NEUTRAL ZONE This is territory between Saudi Arabia and Kuwait which produces heavy crude oils, Burgan (Wafra), Hout, Eocene and Ratawi, now known as the Partition Zone. Crude oil production statistics frequently include 50% of the Neutral

Zone's production with that of Saudi Arabia and 50% with that of Kuwait.

NEW BUILDINGS This term is used to describe the orderbook for tankers to distinguish them from the present fleet.

NEW OIL A special meaning of the term existed in the US in the post 1973 period when new oil was defined as oil production in excess of the base production control level (*q.v.*) of a property.

NEW ZEALAND For crude oil production *see* Table 1.

NEWFOUNDLAND REFINING CO. The Newfoundland Refining Co. was formed in 1967 to operate the 100 000 b/d Come by Chance refinery in Newfoundland which came on stream in late 1973. It is wholly owned by Shaheen Natural Resources Co., the New York company of J. M. Shaheen.

NGL Natural Gas Liquids. These are the hydrocarbons propane, butanes and pentanes. They are recovered from raw gas liquids by processing in field separators, scrubbers, gas processing and reprocessing plants, or cycling plants.
The value of NGLs is similar to that of LPG (*q.v.*) and condensates (*q.v.*).

Chemically: LPGs are composed of C_3 and C_4 hydrocarbons.
NGLs are composed of C_3, C_4 and C_5 hydrocarbons.
Condensates are composed of C_4 and higher carbon number hydrocarbons.

NHC The National Hydrocarbons Corporation of the UK which was first proposed at the annual conference of the Labour Party in 1967 when the National Executive Committee presented a policy statement on North Sea gas. A report by the Labour Party's Fuel Study Group was published in 1968. The report published organigrams of proposed Transport and Supply Exploration and Production and Gas Council Production and Supplies Divisions and detailed budgets for twelve years ahead. On setting up 150 staff were proposed. The report was not well received; the (UK State) Gas Council did not exactly welcome the proposal.
Talk of a UK NHC was revived in 1974 in the North Sea participation discussions but the original Labour Party report was not mentioned. With the formation of BNOC (*q.v.*) the NHC ceased to be an issue.

NICKEL *See* Vanadium.

NIGC The National Iranian Gas Company was formed in 1965 and is a subsidiary of NIOC (*q.v.*).

NIGERIAN CRUDE OILS These include Bonny, Brass River, Forcados, Pennington and Qua Iboe. For Nigeria's crude oil production *see* Table 1.

NIGERIAN LIGHT This export crude, also known as Bonny Light, is of 37·6° API gravity and 0·13% sulphur. Bonny is the export terminal.

NIGERIAN MEDIUM This export crude, also known as Bonny Medium, is of 25·7° API gravity and 0·24% sulphur. Bonny is the export terminal.

NIGERIAN NATIONAL OIL CORPORATION The Nigerian state company formed in 1971.

NIGERIAN PETROLEUM REFINING CO. The company was formed in 1960 and operated the 60 000 b/d Port Harcourt refinery which came on stream in 1965. In July 1967 operations were halted with the start of the Civil war. The refinery was recommissioned in May 1970.
The original shareholding was 60% Government, 20% BP and 20% Shell.

NINIAN A commercial UK North Sea oil field 280 miles NNE of Aberdeen. Initially peak production rate estimated to be 400 000 b/d with estimated recoverable reserves at 1700 million barrels. The reserves estimate was later downrated to 1200 million barrels. Production planned for 1978.
In 1975 the shareholdings were as follows:

	(%)
Chevron (operator)	15·6
Burmah	19·5
ICI	16·9
BP	17·5
Ranger	7·0
Ocean Drilling	6·5
Murphy	6·5
LAMSO	5·4
SCOTT	2·5
Cawoods	1·3
National Carbonising	1·3

NIOC The National Iranian Oil Company formed in 1951 by the Iranian Government. NIOC operates in Iran including many joint operations with other companies. Overseas it operates in India, South Africa and the North Sea.

NIOGAS The Austrian state owned gas development company.

NITC National Iranian Tanker Company. In 1974 it operated four tankers totalling about 185 000 tons dwt.

NITROGEN Nitrogen compounds occur extensively in petroleum but are not of economic significance. The presence of nitrogen compounds is probably essential in lubricating oil basestocks. A refining treatment which removed all nitrogen would probably produce a poor quality oil.

NLGI The US National Lubricating Grease Institute. It sets standards for greases and publishes a technical journal, the *NLGI Spokesman*.

NOC The National Oil Corporation of Libya.

NOJC The US National Oil Jobbers Council.

NON-COMMERCIAL FUELS These include peat, bagasse (sugarcane waste), waste wood, fuelwood and especially in India and Pakistan, animal refuse. *See also* Calorific Value.

NOR National Oil Refinery. In recent years many refineries have been termed National Refineries. One of the earliest was built by the Anglo–Persian Oil Company in South Wales in 1916 coming on stream in 1922. The refinery is now known as BP Refinery (Llandarcy) Ltd.

NORSK HYDRO A Norwegian company dating from 1905 with 50·75% of the shares being owned by the Norwegian Government. The company has industrial and chemical interests and is active in the Norwegian North Sea.

NORTH SEA Exploration in the North Sea began in earnest in the 1960s, partly promoted by the discovery of large quantities of gas in Holland. Gas was found in the southern sector of the North Sea in the mid-1960s and by the early 1970s there were many discoveries of oil and gas in the UK and Norwegian sectors.
In the early 1970s there was much controversy about the peak production level. Estimates varied from 2 to 3 million b/d to double and treble that amount. A UK Government estimate published in November 1974 suggested that UK reserves were worth $93 000 million at the then current prices.

NORTH–SOUTH DIALOGUE The Conference on International Economic Co-operation held in Paris; first session 1975.

NORWAY The Norwegian state oil company is Statoil (*q.v.*). For crude oil production *see* Table 1.

NUCLEAR FISSION The splitting of atomic nuclei whereby very large amounts of heat are released which may be used to generate electricity or used for other purposes. There are many commercial reactors fuelled by uranium oxide. The sodium cooled reactor is fuelled by a mixture of uranium metal and plutonium oxide.
For a comparison of technical detail of nine reactors *see* G. R. Bainbridge, Nuclear Options, in *Nature* **249,** (1974), p. 733.
However, at present supplies of uranium do not allow for generation of considerable quantities of power without concern for uranium reserves. The future may be with the Breeder Reactor which produces additional quantities of radioactive fuels as it produces power.

NUCLEAR FRACTURING Exploding nuclear devices in oil or gas fields to break up the structure in the hope of releasing quantities of hydrocarbons that could not be otherwise extracted. The technique has not been of great commercial significance.

NUCLEAR FUSION Nuclear fusion has enormous potential for commercial energy production but technical problems need to be solved; the most ambitious target date announced for a commercial fusion reactor is 1995. Controlled fusion needs a temperature of about 200 million °C.
If and when nuclear fusion becomes a commercial reality, the radioactive waste problems associated with fission reactors will be bypassed.
Uncontrolled thermonuclear reactions, such as atomic bombs, release enormous amounts of energy and at the theoretical limit, where matter could be converted entirely into energy, one pound of matter would yield as much energy as one megaton of oil.

NUCLEAR POWER COSTS In the late 1940s many estimates were made of the cost of electricity generated from nuclear power. One of the lowest estimates (the California Report, *q.v.*) gave a cost increase of $1\frac{1}{4}$ times thermal whilst the Thomas Report (*q.v.*) stated the relative cost would be $5\frac{1}{2}$ times thermal generating costs. Since these early estimates many more have followed invariably being higher than the then current thermal costs.

NUCLEAR REACTORS By mid-1975 there were three commercial nuclear reactors operating with development and capital costs written off or down. The bare operating costs made all three reactors attractive compared with mid-1975 fossil fuel prices. The reactors were: Magnox (UK designed gas cooled), Light Water Reactors (LWRs), Candu (Canadian designed heavy water reactor).
Five AGRs, Advanced Gas Cooled Reactors, were under construction in the UK but have been delayed due to technical problems. High Temperature Reactors (HTRs) and Fast Breeder Reactors (FBRs) were assessed as reactors with considerable future potential.

O

O AND G J *The Oil and Gas Journal.* An American weekly publication covering all aspects of the oil industry. Frequently publishes forecasts. It is one of oldest oil journals being founded at Beaumont in the Spindle Top boom of 1902. *See also* Nelson Indices.

OAPEC The organisation of Arab Petroleum Exporting Countries which was formed in 1968 by Kuwait, Libya and Saudi Arabia. Algeria, Bahrain, Egypt, Iraq, Qatar, Syria and the United Arab Emirates subsequently joined the organisation. Saudi Arabia is the driving force behind OAPEC and its main object is the co-ordination of oil policies within the Arab countries.

OASIS The Oasis Oil Company of Libya was formed in 1960 and was owned by Amerada, Conoco, Marathon and Shell. It was responsible for much of Libya's production, exporting through the Es Sider terminal. In 1970 production reached 946 000 b/d.

OBO An Ore-Bulk-Oil carrier also known as a combined carrier. These vessels may trade in oil or otherwise in bulk trade such as grain, coal or iron ore. The OBO designation arises because of the possibility of trading on a triangular route with two legs in bulk and one in oil. Many OBO vessels were about 80 000 tons dwt.

OCAW The US Oil, Chemical and Atomic Workers Association.

OCCIDENTAL PETROLEUM CORPORATION An integrated US oil company incorporated in California in 1920. The company expanded considerably following success in finding and rapidly exploiting oil in Libya.

OCEAN LEAKAGE This term describes the losses incurred in ocean transport of crude oil and products. A round figure of 0·5 % is frequently used to allow for this loss.

OCIMF Oil Companies' International Marine Forum.

OCRS Organisation Commune des Recherches Saharienne. An organisation concerned with French oil policy overseas. With the overthrow of the Ben Bella regime and subsequent Algerian independence, a new agreement was signed between France and Algeria in July 1965 covering the future French participation in Algeria with the holdings existing in July 1965 remaining subject to the OCRS code. However, tax rates were increased to a maximum of 55 %. *See also* Franc Zone Crude Oil.

OCTANE RATING The octane rating is the most important property of a motor gasoline and may be determined by one of several methods. The actual test procedure is not important but what is, is the limit set, to which a gasoline is manufactured. The most frequently quoted octane measurement is the Research Octane Number as measured on a standard (CFR) engine by the F-1 procedure. The Motor Octane Number or F-2 method, carried out on the same standard engine, gives ratings that may be about 10 to 20 numbers lower than the Research rating. The greater the amount of cracked spirit (olefinic material) in a motor gasoline blend, the greater will be the difference between the Research and Motor ratings. However, the price difference between gasolines of low and high olefin content need not be great; it could be zero or of the order of one per cent of the price; the lower olefin content gasoline having the high value. *See* PMS.

OECD The Organisation for Economic Co-operation and Development (OECD) which superseded the OEEC, was set up under a Convention signed in Paris on 14th December 1960, which provides that the OECD shall promote policies designed:

(i) to achieve the highest sustainable economic growth and employment and a rising standard of living in Member countries, while maintaining financial stability, and thus to contribute to the development of the world economy;
(ii) to contribute to sound economic expansion in Member as well as non-member countries in the process of economic development;
(iii) to contribute to the expansion of world trade on a multi-lateral, non-discriminatory basis in accordance with international obligations.

The Members of OECD at the end of 1976 were Australia, Austria, Belgium, Canada, Denmark, Finland, France, the Federal Republic of Germany, Greece, Iceland, Ireland, Italy, Japan, Luxembourg, The Netherlands, New Zealand, Norway, Portugal, Spain, Sweden, Switzerland, Turkey, the United Kingdom and the United States. Yugoslavia was an associate member.

OEEC The Organisation for European Economic Co-operation which was formed by the countries participating in the Marshall Plan in 1948. The organisation was responsible for commissioning energy and oil forecasts and influencing Energy Policy (*q.v.*). The OEEC was superseded by the OECD (*q.v*) in 1960.

The members of the OEEC were Belgium, Luxembourg, Netherlands, France, West Germany, Italy, UK, Norway, Denmark, Sweden, Iceland, Greece, Turkey, Ireland, Austria, Switzerland and Portugal. The US and Canada were associate members.

OEP The US Office of Emergency Planning. Security of supply of US oil imports has been one of its concerns. *See also* OSAC.

OETB The Offshore Energy Technology Board of the UK announced by the Secretary of State for Energy in May 1975 to advise on R & D programmes to ensure efficiency and safety of offshore operations.

OFF BALANCE SHEET FINANCING Methods of obtaining additional gearing which need not necessarily be indicated in the accounts. Such methods include hire purchase, forms of leasing and operating through companies the accounts of which are not consolidated. *See also* Lease Commitment.

OFF-PEAK Off-peak periods are times during the day when consumption of gas or electricity is relatively low. Cut Price tariffs are quoted to encourage consumption in these periods.

OFP (Omnium Francais de Petroles) OFP was set up in 1938 and marketed under the OZO brand. It was integrated into CFP in 1960 and the OZO brand name was changed to TOTAL.

OGJ *See* O and G J.

OIL EMERGENCY (LONDON) ADVISORY COMMITTEE This committee was set up at the time of the 1956 Suez Crisis and comprised Shell, CFP and BP. *See also* Middle East Emergency Committee.

OIL SEARCH LTD The company was incorporated in Papua in 1929 and has carried out mostly unsuccessful exploration in Papua.

OIL SPILLS Oil spills have accompanied oil production through the last 100 years but it is only comparatively recently that they have been a direct cost to the industry. Routine oil spills from cleaning crude oil tankers have been largely eliminated by the 'load on top' arrangements. The industry in the 1970s also took major steps for insurance of oil spills by tankers. A strong environmental lobby certainly cost the US oil industry substantial sums through the limiting of drilling in the Santa Barbara Channel in California. One of the largest direct costs to the industry was the $33 million claim to compensate fishermen in Japan following the Mizushima refinery spill in December 1974 (Mitsubishi Oil Company).

OIL STATISTICS *See* Energy Statistics.

OIL TAXATION BILL (UK) *See* Petroleum Revenue Tax (UK).

OIL WELL PRODUCTIVITY The large onshore wells in the Middle East are amongst the world's most productive with flows of 10 000 b/d or more at a cost of about 10 cents/bbl. Areas such as Libya and Nigeria have typical wells producing about 2 000 b/d at a cost of the order of 50 cents/bbl. Small wells, such as many in the US, have production of the order of 20 b/d and in these wells the oil has to be pumped up and the cost may be $1·00/bbl or more. Production from offshore wells is necessarily more expensive; even in favourable locations offshore oil cost might be $1·00/bbl and in unfavourable locations, such as the North Sea, of the order of $5·00/bbl.

OILER A naval term for a small fuel oil carrier. Loosely used for dirty category product carriers up to about 45 000 tons dwt., but could also include larger vessels. *See* Tankers, Categories for classification system.

OK Oljekonsumenternas Riksfoerbund is the Swedish oil consumers co-operative founded in 1946. It is the central organisation of 30 co-operative associations which had altogether about 700 000 members at the end of 1974. It held just over 20 % of the Swedish mogas market in 1974. It operates the 8·3 million ton/year refinery at Scanraff with Texaco as minor partner.

OLADE Organizacion de Latinoamerica de Energia formed by 22 Latin American countries in 1972. The members are Argentina, Bolivia, Brazil, Chile, Colombia, Costa Rica, Cuba, Dominican Republic, Ecuador, El Salvador, Guatemala, Guyana, Honduras, Jamaica, Mexico, Nicaragua, Panama, Paraguay, Peru, Trinidad, Uruguay and Venezuela. Its objectives were higher prices for oil shipped outside Latin America, uniform policy and better co-operation within Latin America.

OLD OIL A special meaning of the term existed in the US in the post-1973 period when it was the same as controlled crude oil (*q.v.*).

OLEFINS Olefins are introduced into petroleum products by the various cracking processes. The most important olefins are ethylene, propylene and the di-olefins isoprene and butadiene which are all important chemical feedstocks. These olefins are manufactured mostly from naphtha by the steam cracking process. The price of the olefins is firmly linked to the price of naphtha.

OMAN A crude of 35°API gravity and 1% sulphur. The export terminal is Mina al Fahal. For crude oil production *see* Table 1.

OMNI PETROL A cut price mogas marketer in Austria owned by BPAG.

OMNIREX The Omnium de Recherches et d'Exploitations Pétrolieres formed in September 1957. Omnium Française de Pétroles owns 35%.

OMV Oesterreichische Mineraloelverwaltung AG; the Austrian state oil company which owns the main refinery in Austria at Schwechat. It was commissioned in 1961.

ONGC The state Oil and National Gas Commission of India.

OPB Oil Policy Branch of the Canadian Department of Energy, Mines and Resources. Administered the oil import subsidy programmes until transferred to ESAB (Energy Supplies Association Board) on 1st November 1974.

OPEC The Organisation of Petroleum Exporting Countries was formed in September 1960, the founder members being Iran, Iraq, Kuwait, Saudi Arabia and Venezuela. By 1974 there were twelve members; Algeria, Eçuador, Indonesia, Libya, Nigeria, Qatar and the United Arab Emirates having joined. Each of the five founding members has veto power over new applicants. OPEC, which is sometimes known as a producers' cartel, was formed because of the concern over cuts in oil prices. In the 1960s it was not effective in stopping price erosion but maintained revenue through frozen posted prices.

OPEC BULLETIN A vehicle for expressing OPEC views, for example, OPEC has always been greatly concerned with its relatively small take compared especially with consumer government taxes on oil. The September/October 1969 number published the following:

Breakdown of Cost of one average barrel of oil in Western Europe in 1967

	$	%
Cost of Production	0·285	2·7
Cost of Refining	0·350	3·3
Tanker Freight	0·680	6·3
Storage, dealer's margin, etc.	2·790	26·0
Oil company's net profits	0·681	6·3
Oil taxes in consuming country	5·100	47·5
Revenue of producing countries	0·853	7·9
	10·739	100·0

The bulletin is published irregularly by OPEC Information Dept, 1010 Vienna, Austria.

OPEG The OEEC Petroleum Emergency Group composed of BP, CFP and Shell. This was the oil industry group which advised the OEEC Oil Committee. London was the headquarters of the group but a small permanent group in Paris was formed for liaison with the OEEC.

OPERATING COSTS The operating costs of an oil field are the variable costs of production, i.e. the costs exclusive of those associated with the capital required to develop an oil field. They could range from less than ten cents to a few dollars per barrel. *See also* Production Costs.

OPERATION CONTRACT The operation of an oil field by a company whilst the ownership of the oil field remains with another party. *See also* Service Contract.

OPPORTUNITY COST This is the real cost of satisfying a need or want expressed in terms of the cost of the alternative choice. Thus because energy cost is usually a small part of total manufacturing cost, an industrialist can afford to pay a very high price to avoid closing down his factory. This factor is undoubtedly responsible for sharp increases in marginal market prices of oil products in times of relatively short or expected short, supply.

ORGANIC TRADE This is the ongoing business of a company which may be subject to positive or negative growth, but has continuity. It is essentially the long-term business contracts; sales of mogas and gas oil through most types of service stations and for other products through owned facilities or wholesalers which have some tie with the company. Most sales classified as organic trade are to consumers. *See also* Inorganic Trade.

ORGANISATION OF AMERICAN STATES Comprised of the following: Argentina, Bolivia, Brazil, Chile, Colombia, Costa Rica, Cuba, Dominican Republic, Ecuador, El Salvador, Guatemala, Haiti, Honduras, Mexico, Nicaragua, Panama, Paraguay, Peru, Uruguay, US and Venezuela.

ORIGINAL IN PLACE RESERVES These are defined as the total quantity of crude oil or raw gas initially in place within the estimated area of an oil or gas field from which production has been obtained or for which reserves have been credited. This term represents the sum of (1) the anticipated ultimate crude oil or raw gas production and (2) the crude oil or raw gas estimated to be non-recoverable under existing economic and operating conditions.

ORINOCO TAR BELT *See* Venezuela.

OSAC The UK Oil Supplies Advisory Committee. A body which advises on the supply of oil to the UK, assessed on a worldwide basis, in times of emergency or crisis. It would deal with questions such as; if oil supplies were stopped from one or more areas, would the supply of tankers be able to move supplies from alternative areas? The committee is essentially an industry advisory service to government. *See also* PIAC.

OTC The Overseas Tankship Corporation which is owned by Standard Oil of California. The company was incorporated in 1946 in Panama.

OVERSEAS PETROLEUM CORPORATION This was formed in 1970 in Tokyo as the North Slope Oil Co., the name being changed in 1972. In association with US companies it entered into exploration in Alaska. In December 1972 the company bought a 30% interest in ADMA from BP. *See also* JODCO.

OZO *See* OFP.

P

PACIFIC PETROLEUMS LTD An integrated Canadian oil company incorporated in 1939.

PAD The US Petroleum Administration for Defense. The PAD Districts of the US, 1 to 5, are the same as the API Regions (*q.v.*).

PAJ The Petroleum Association of Japan.

PAKISTAN For crude oil production *see* Table 1.

PAKTANK An oil storage company which is a subsidiary of the Dutch company Pakhoed Holding NV. It has storage facilities in the UK, Belgium, Holland and Scandinavia.

PALEY REPORT A US report on energy supply and demand; President's Materials Policy Commission, *Resources for Freedom*, US Government Printing Office, Washington DC, 1952.

PANAMA CANAL The Panama Canal was completed in September 1913. It has been of far less importance to the international oil industry than the Suez Canal and relatively little oil is moved through it. In the period 1970 to 1975 about 14 000 vessels each year used the canal. Revenues in the year 1974/5 amounted to $143 million. The toll for the canal was constant from its opening to July 7th 1974 when there was a 20% increase. *See also* Canal Tolls.

PARAFFIN In the UK paraffin is a word in common use for kerosine. It usually refers to the special domestic grade which has a high smoke point (i.e. gives off relatively little smoke when burning in lamps and room heaters) and has little odour.

PARAFFIN WAX There are several grades of paraffin wax; 120–125 °F, 130–135 °F and 140–145 °F, the numbers referring to the approximate melting point. The name is due to the fact that they are composed almost entirely of n-paraffins. Paraffin waxes are really a by-product of lubricating oil production and it is never a straightforward exercise to allocate how much of the cost of the combined lubricating oil/wax production operation should be borne by the lub oil and how much by wax. In the US in 1974 the price of wax in large bulk was about 33 cents/lb (about ten times the price of crude oil).

PARAFFINS Paraffins occur extensively in crude oils. High concentrations of certain paraffins are desirable in some petroleum products but undesirable in others. Modern refineries deal adequately with crude oils regardless of their paraffin content or paraffinicity. Consequently the paraffinicity of a crude oil has little economic significance other than that indirectly represented in the gravity differentials associated with crude oil prices.

PARAMARGINAL RESERVES OR RESOURCES That portion of sub-economic reserves or resources that might soon be worthwhile producing if the selling price of the resource increases.

PARTICIPATION In the late 1960s the oil producing countries, in their desire to obtain greater revenues for their crude oils, discussed participating in the activities of the operating companies. This participation, taking an equity stake in the companies, was seen as preferable to outright nationalisation, particularly in the light of Iran's experience in the early 1950s.

The idea that host governments should participate in old as well as new concession agreements was discussed at the OPEC conference of 1968. In the late 1960s there was much discussion of the form that participation should take. Participation at the producing end only was considered by some as merely partial nationalisation carrying with it the same dangers as total nationalisation. These dangers were associated with the oversupply of crude oil in world markets which thought might depress crude oil prices. However in 1970 Libya, following its September 1969 revolution, took a much harder line in negotiating with the producer companies eventually nationalising

BP's production assets in Libya in December 1971. Algeria was also successful in nationalising all production assets in 1970 and 1971 and on 1st June 1972, Iraq nationalised the Iraq Petroleum Company.
Meanwhile the producing companies in Abu Dhabi, Kuwait, Qatar and Saudi Arabia were participated to the extent of 25% by January 1973 following lengthy negotiations.
In 1973 the climate for World energy supply changed considerably with the possibility of reduced crude oil prices becoming remote. The Arab oil embargo of 1974 and the unilateral posting of high crude oil prices on 1st January 1974, signalled complete control of crude oil by the host governments in nearly all areas even though discussions on payments to the producing companies over the value of equity stakes continued.
The participation by OPEC countries produced three main outcomes; complete nationalisation of the producer companies with their staff expelled but with negotiated compensation for the assets, 100% participation with the original producer companies retained as virtually service contractors, and thirdly, participation to a limited extent where it was felt by the host government worthwhile to retain the producing company as a partner to provide expertise and capital.
In 1976 participation was again prominent with BNOC (*q.v.*) negotiating 51% equity shares in the North Sea oil fields.

PARTICIPATION CRUDE Crude oil that belongs to the producing country government; i.e. that proportion of the crude oil production which is in proportion to the government's equity stake in the producing company. Such crude oil could be sold by the government to a third party or sold to the producing company as buyback crude oil (*q.v.*).

PEAK LOAD The period or periods during a day in which consumption is relatively high. *See also* Peak Shaving.

PEAK SHAVING The practice of introducing additional supplies of gas or electricity during peak periods, for example, extra gas from special storage or, additional hydro electricity from pumped storage.

PEAT Peat is usually classified as a non-commercial fuel and its heating value varies with the water content. *See* Calorific Value.

PEMEX Petroleos Mexicanos; the Mexican state oil company.

PENNINGTON (Nigeria) A crude oil of 38° API gravity, 0·1% sulphur exported from the Apoi offshore terminal.

PERTAMINA Perusahan Negara Pertambangan Minjak Dan Gas Bumi Nasional, the state oil and mining company of Indonesia. A feature of this company is its massive debt. Even in late 1974, after the rise in oil prices, its debt was estimated to be about $2000 million. The bulk of Indonesia's production has been from the onshore Sumatra fields run by Caltex and Stanvac. Pertamina's own production of about 150 000 bbl/day in 1974 was tied in part to repayment of development costs.

PERU For crude oil production *see* Table 1.

PETRA A Shell oil trading company formed in 1975.

PETROBRAS Petroleo Brasileiro; the Brazilian state company engaged in all aspects of the oil industry in Brazil. The company was incorporated in 1954.

PETROCAN The Canadian state oil company which was officially inaugurated in January 1976. On its formation Petrocan took over about $500 million Government investments in Canadian oil plus a 20% equity of Syncrude participation, the Polar Gas project and other interests.

PETROCHEMICALS Petrochemical prices are dependent on the price of feedstocks (light distillates, gases, aromatics) and also on petrochemical plant capacity availability and trading conditions. In the plastics industry hydrocarbon feedstocks account for only approximately 5% of selling prices (US 1974). In 1974 only about 5% of total US oil and gas was used for petrochemicals and about 1% for plastics. In Western Europe 6% of oil was used for petrochemicals and in Japan 8%.

PETRODOLLARS A word which first came into prominence in 1974 and is used to designate a dollar which is owned by a resident of an oil exporting country. *See also* Recycling (Money).

PETROGAL The state oil company of Portugal (*q.v.*).

PETROL The term used, outside the oil industry, for motor gasoline in the UK.

PETROLEUM BUSINESS NEWS INDEX A monthly index service established by the API in January 1975; subscription $100/year. The index covered *Platt's Oilgram*, *PIW*, *Oil Daily*, and the *Petroleum Economist*.

PETROLEUM COKE *See* Sponge Coke, Honeycomb Coke, Needle Coke.

PETROLEUM ECONOMIST A long established monthly journal owned by a Trust. It is written in London and published in several languages. It has an annual index and is the most useful single source for tracing items in the field of oil economics.

PETROLEUM INTELLIGENCE WEEKLY (PIW) This is an influential weekly newsletter published by Petroleum and Energy Intelligence Weekly, Inc. of New York. It is a good source of economic information and up-to-date statistics. In 1976 the annual subscription was $540 airmail.

PETROLEUM PRESS SERVICE In 1974 this was named the *Petroleum Economist* (*q.v.*).

PETROLEUM REVENUE TAX (UK) A bill introduced in the UK Parliament in November 1974 to levy a special tax on North Sea oil production profits. The proposed tax (rate not specified) was to be levied on production before applying Corporation Tax (which stood at 52 % at the time). The oil industry was concerned with how the tax was to be levied in view of a claimed average $4·50 cost per daily barrel of production and the possibility of a falling market price of crude oil.

PETROLEUM TIMES A weekly general petroleum journal published by IPC, London. Quotes oil prices being particularly detailed on UK product prices by company and area.

PETROLIBER The Campania Iberia Refinadora de Petroleos SA. The company operates a refinery at La Coruna in North Spain and has a 30 % holding in Hispanoil.

PETROMIN Petromin is the Saudi Arabian Government Agency (General Petroleum and Mineral Organisation). The Chairman is the Saudi Arabian Minister for Petroleum, H. E. Ahmed Zaki Yamani.

PETRONAS The Malaysian state oil company which was created by the Petroleum Development Act of July 1974.

PETROPAR The Sté. des Participations Pétrolieres formed in December 1957. BRP 72 %.

PETROREP The Sté. Pétroliere des Recherches dans la Region Parisienne. Formed in October 1955. Nahmias Brothers 51 %; Rothschilds 14 %. No French state participation.

PETROSAREP Formed in October 1957. PETROFRANCE Group (Nahmias Bros.) 60 %, finance companies 40 %.

PETROSUL The Portuguese company was 34 % Government owned until March 1975 when, with the nationalisation of Portuguese banks, the Government stake rose to 75 %. At the time of the increase in Government stake the company was building a 180 000 b/d refinery at Sines.

PETROVEN The state oil company of Venezuela; the company is Petroleos de Venezuela which in 1975 took over the interests of the 21 concessionary and 16 participating oil companies operating in Venezuela.

PG The Persian Gulf. *See* Middle East.

PHILIRAN An Iranian exploration company owned by NIOC and Phillips Petroleum.

PHILLIP REPORT A report issued in 1958 by the Energy Commission of the European Parliament, on the gas industries in the EEC of Six, comparing gas reserves in the EEC of Six with those in other countries.

PHILLIPS CASE *See* Natural Gas.

PHILLIPS PETROLEUM An integrated US oil company ranking just after the majors. The company was incorporated in 1917 and is unusual in that its head office is away from the main petroleum quarters being situated in Bartlesville, Oklahoma. The company has been particularly strong in the natural gas and petrochemical industries.

PIAC The Petroleum Industry Advisory Committee for the UK Government. It deals with many aspects of the UK petroleum industry and collates statistics for government including forecasts of oil demand.

PIB The Petroleum Information Bureau of the UK which was sponsored by the UK Petroleum Industry and was active in the 1950s and 1960s publishing UK and World Oil Statistics. Some of its activities were taken over by the UK Institute of Petroleum in the 1970s.

PIB Polyisobutylene; a polymer which is manufactured from a four-carbon olefin (polyethylene or polythene is from a two-carbon olefin), the olefin also being obtained from an ethylene cracker. The hydrocarbon polymer has been used as a viscosity index improver in lub oils and, when other elements such as nitrogen are introduced into the PIB, as a detergent additive. The concentration of a PIB additive in lub oil may be as high as 6 % and the unit cost of the additive is about 10 times the cost of the lub oil base stock.

PINK BOOK The UK Government annual publication covering the UK Balance of Payments.

PIPER A commercial UK North Sea oil field 120 miles north-east of Aberdeen. Peak production estimated to be 220 000 b/d with estimated recoverable reserves at 800 million barrels. Original ownership; Occidental Petroleum (UK) Ltd 36·5 %, Getty Oil International (England) Ltd 23·5 %, Allied Chemical (GB) Ltd 20·0 %, Thompson Scottish Associates Ltd 20·0 %. The crude is of 36·5° API gravity.

PIW *See* Petroleum Intelligence Weekly.

PLANNING The methods used are numerous. A survey of the state of the art in 1974, with 28 references, is a useful starting point. *See* Cost of Capital.

PLASTICS *See* Petrochemicals.

PLATT'S COAL WEEK A weekly news and price service first published on the 31st March 1975.

PLATT'S EUROPEAN PRICES Platt's European Bulk Prices are quoted for cargo and barge lots at Rotterdam and cargoes at Genoa. For each centre a price range is quoted for each trading day and monthly averages are quoted for the previous month of 'lows', 'highs' and 'highs and lows'. The prices quoted reflect deals done in the markets or the prices at which trade might be done. Deals can also be done using Platt's Prices as reference, e.g. Platt's low or Platt's high plus $2, in which oil does not pass through the market. The high and low prices quoted are *not* equivalent to jobbers' selling and buying prices. The prices quoted are European as opposed to EEC. Product quality also influences price. Thus the lower part of the range quoted could be for poor quality or non-EEC materials whilst the maximum price would be for a material meeting an exacting product specification, e.g. a low lead premium motor spirit with good antiknock characteristics. Platt's Rotterdam Prices have been quoted since the late 1960s. *See also* Prices—AFM; Barge Prices.

PLATT'S OIL PRICE HANDBOOK & OILMANAC Complete record of the oil industry's prices available in a single volume. Published annually, 50th edition in 1973.

PLATT'S OILGRAM LEGISLATIVE SERVICE A reference bulletin providing comprehensive coverage of Congressional activity on energy matters. Published bi-monthly from October 1974.

PLATT'S OILGRAM NEWS SERVICE A daily international oil and gas news service established 1934, published by McGraw-Hill, New York.

PLATT'S OILGRAM PRICE SERVICE A daily US and international oil price reporting service issued simultaneously from New York and Houston.

PMA The Petroleum and Minerals Authority of the Australian Government.

PMA Polymethacrylate; the polymer finds a wide variety of uses including 'perspex' and as a viscosity index improver in lub oils. PMA additives can be used at a concentration of about 6 % in lub oils and the unit cost of additive is about 10 times that of the base oil.

PMS Premium Motor Spirit. This material has a Research Octane Number of approximately 97 to 99; best grade motor spirit has a research octane number of approximately 100 to 102. The main process for manufacturing PMS is cat. reforming. A light boiling fraction from crude oil, a typical boiling range would be 65 to 170 °C, is processed to obtain about 75 % catalytic reformate and 25 % gas (methane, ethane, propane and butane). Butane is the predominant gas and up to 10 % may be blended into motor spirit depending upon local volatility specifications. Thus about 15 % of the feedstock is converted to gas which may well have only fuel oil value. The value of PMS is thus dependent on the price of the naphtha feedstock and on the price of fuel oil. The greater the price difference between naphtha and fuel oil, the greater will be the price difference between PMS and naphtha. When there is surplus refining capacity, the price of PMS above naphtha reflects only the naphtha–fuel oil price difference and the amount of fuel used in the reforming process. Thus a depressed distillate price structure could be PMS = 100, RMS = 91, naphtha = 90, FO = 70. RMS need not have a price greater than naphtha; when straight run naphtha is in relatively short supply and cracked spirits are relatively in surplus, then the price of straight run naphtha (LDF) may be slightly greater than RMS. A more typical price structure would be PMS = 170, RMS = naphtha = 150, FO = 70.

When cat. reforming capacity is tight, as it was in Europe in the second quarter of 1973, then PMS can be at a substantial increment over naphtha and also RMS will be priced higher than naphtha. Prices, in $/tonne, at Rotterdam in the second quarter of 1973 averaged as follows: PMS, 95; RMS, 85; Naphtha, 50 and HFO, 17.

In this period there were reports of cat. reforming capacity being offered at $50/ton.

In 1964 the average difference between PMS and RMS was $7/ton, in 1968 $5/ton, in 1972 $8/ton and in 1974 $10/ton; all these differentials corresponded to about eight octane numbers. So over this period, the value of one octane has been about $1/ton in the range 90 to 98 Research rating. Taking the octane number of cat. reformer feedstock as being 60 and RMS as 90, over much of the period octane appreciation below 90 has been worth little. *See also* Product Giveaway.

PNOC The Philippine National Oil Company.

POLAND For crude oil production *see* Table 1.

POLENG (Indonesia) A crude oil of 43° API gravity and 0·2% sulphur, exported from the Surabaja terminal.

POOL AGREEMENT (OF 1928) *See* Achnacarry Agreement.

POOL PETROL In the UK brand names for motor spirits were suspended during World War II and the product was distributed and sold as 'Pool'.

PORTUGAL The state oil company is Petrogal which was formed from Sacor (*q.v.*), Sonap, Petrolsul and Cidla.

POSTED PRICES, CRUDE OILS The posted price system originated with the Standard Oil Trust when it 'posted' prices at which it was prepared to buy crude at the well head. After the breakup of the Standard Trust in 1911 the system continued where the larger crude oil buyers would announce the price at which they were prepared to purchase crude oil. Subsequently prices were posted by the oil companies controlling production in particular fields. The oil companies jealously guarded their right to post crude prices without necessarily having to consult with producer country governments.
See Equity Cost Oil and Crude Oil Prices, Unitary Crude, Prices, Crude Oil.

POSTED PRICES, PRODUCTS Product prices, wholesale prices, are quoted for many of the main towns and terminals in the US. For example Mid-west wholesale prices are quoted at St Louis and Chicago. Comprehensive ranges of product prices are quoted at lake terminals and marine fuel oils are, of course, quoted at all the major ports. The major oil companies quote product prices, usually similar, at each location in the Caribbean, Middle East and Far East. For quotations *see* Platt's Oilgram Price Service. *See also* Prices, Spot Markets.

POTENTIAL RESOURCES (OF CRUDE OIL) *See* Ultimate Recoverable Reserves.

POTT–BROCHE PROCESS A coal liquefaction process developed in the 1920s and used by Germany in World War II to produce liquid fuels.

POUR POINT This is the temperature at which an oil solidifies plus 5°F, i.e. the temperature below which it may not be possible to use an oil without some heating to maintain flow from storage. Although widely used, the test is relatively imprecise and is of little economic importance. On residual fuel oils the viscosity and pumpability are the important parameters, on gas oils and kerosines it is the cloud point (*q.v.*) which is important.

PPS *Petroleum Press Service;* now named the *Petroleum Economist* (*q.v.*).

PRECISION As with all other data, there are errors associated with the routine measurements made in the oil and energy industries. The following estimates of the reproducibility* are guesses and are not based on strict statistical appraisal.

Crude Oil Production Measurements	±0·01%
Crude Oil Transportation Measurements	±0·1%
Refining of Crude Oil	±1%

(Refinery Fuel and Loss in a modern simple refinery is 4%; in a complex refinery 7%.)

Petroleum Products Within countries	±0·1%
Petroleum Products Between countries	±0·2%
Energy Measurements Within countries	±0·5%
Energy Measurements Between countries	±1·0%

* The reproducibility is defined as the limits within which estimates made by different sources should agree most of the time (i.e. 95% of the time).

PREISSPIEGEL A bi-monthly report containing quotations of independent German importers of gas oils and fuel oils.

PREPA The Sté. de Prospection et d'Exploitation Pétrolieres en Alsace. Formed in April 1953. BRP and SNIP together hold 65%.

PRICE INFLATION Price inflation occurs when there is an increase in the price of a goods or service without a change in the quantity or quality. Strictly, inflation in oil prices occurs when prices rise in line with general inflation (*see* Deflators). However sharp rises in oil prices are sometimes due to an increase in economic rent (*q.v.*).

PRICE RINGS *See* Club Agreements.

PRICES In the marginal markets (*q.v.*), daily reference prices may be quoted as 'highs' and 'lows'. Also quoted in market reports are 'offers' and 'enquiries' which may differ by a small amount from the high–low range. 'Bids' and 'awards' usually refer to tenders. Many 'bids' and 'awards' are published in the US trade press covering all petroleum products.

See also BLS Prices for US Wholesale Prices, SOEC for European Energy Prices.

PRICES—AFM These prices are similar to Rotterdam spot market prices and are usually quoted in $/tonne. Where the materials originated outside EEC countries, the EEC tariff is added before quoting the price. The prices originated in the early 1960s when a group of German oil importers decided to establish Rotterdam quotations. Each Monday averages of the previous week's business were calculated and published; the prices being quoted in DM fob Rhine barge. Although similar to Platt's Prices, they are based only on business actually done, not on feeling in the market. AFM is Aussenhandelsverband fur Mineralöl EV.

PRICES, CRUDE OIL Until the producer governments obtained control of their oil around 1974, the producing companies guarded their right to 'post' prices without necessarily consulting the government. The posted price of a crude oil usually set the maximum level at which oil was sold but the open market or arms length price of crude oil could be a lot less; perhaps 75% of the posted price during times of crude oil surplus. *See also* Profitability, Posted Prices, Crude Oils, Equity Cost Oil.

PRICES, NON-DISCRIMINATORY Following participation of crude oil production companies, the former owners continued to lift crude under preferential terms. A non-discriminatory price is one in which all buyers of crude oil are treated equally, i.e. the former owners of the field negotiate a genuine arms length deal.

PRICES, PLATT'S *See* Platt's European Prices.

PRICES, PRODUCTS *See* Posted Prices, Products *and* Prices, Spot Markets.

PRICES, SPOT MARKETS The main Spot Markets for oil products' trading are the US Gulf Coast, New York, Rotterdam and Genoa. The prices at Rotterdam are usually for barge lots; at the other three centres cargo lots. *See* Platt's European Prices.

PRIVATE BRAND DISTRIBUTOR In the US, he undertakes the same function as a Branded Distributor (*q.v.*) except that he uses his own private brand name on the gasoline.

PROBABLE RESERVES A realistic assessment of the reserves that will be recovered from known oil or gas fields based on the estimated ultimate size and reservoir characteristics of such fields. Probable reserves *include* those reserves shown in the 'proved' category.

PRODUCT QUALITY GIVEAWAY The quality of petroleum products is strictly controlled. They are blended to manufacturing specifications which may cover one or more product or brand specification. Most specification clauses are readily met without there being any economic incentive to make a blend nearer the specification limit. However there may be a clause on which failure to blend near the limit does incur a cost penalty examples being the sulphur content of a fuel oil or the octane number of a gasoline. The most important example of product giveaway is associated with octane numbers. To market a mogas of 99 research octane, a target level of 99·4 may be used to ensure that 99% of blends have octane numbers greater than 99. This product giveaway of 0·4 octane number would cost a refiner at least 50 cents/ton.

PRODUCT SPECTRUM A term sometimes used to describe the range of products carried. The major oil companies carry a range of more than 5000 products.

PRODUCT STOCKS *See* Stocks.

PRODUCTION COSTS For crude oil; these should include the contribution from Development Costs (*q.v.*) and Operating Costs (*q.v.*). The prolific Middle East wells had historic production costs of the order of 10 cents/barrel but offshore wells, such as those in the North Sea, could have development costs of a few dollars/barrel and operating costs of the order of a dollar/barrel. Small oil fields are usually uneconomic because of high development costs and not because of high operating costs.
The following production costs (cents/bbl) were provided by *MEES in* **XVIII**, 15th November 1974.

Arabian Light, Medium, Heavy, and Light Berri	10
Abu Dhabi Murban	15
Abu Dhabi Marine (Umm Shaif and Zakum)	30
Qatar (Dukhan)	17
Qatar Marine	22
Iran Light and Heavy	12
Kuwait	6
Iraq (Basrah)	12
Oman	29
Khafji and Hout	45

Great varieties of costs are quoted but it is not always clear how they have been calculated and expressed. Facilities installed during previous years at costs lower than prevailing levels, are sometimes mixed with relatively new facilities to quote average costs. Thus in end 1974 $ and end 1974 equipment prices, production costs for Middle East land concessions were probably about 30 cents/bbl. Details of production costs in Egypt, Iran, Libya, Trinidad and on Amoco crude oil imports to the US were revealed in a US Congress investigation in 1974. The data covered the period 1971–1973 and was published in several sources. *See for example*, *PIW* 27th June 1974.

PRODUCTION WELL A well that is drilled in a proved field area and through which oil and/or gas will be produced. The process of installing production wells is known as development drilling or sometimes step out drilling.

PROFITABILITY Oil company profitability depends on many factors; companies can of course trade at a loss whilst in early 1974 profits of $1/bbl were possible. At the end of 1974 profits were estimated at $0·22/bbl compared with about $0·35 cents/bbl average in the 1960s and early 1970s. (*See MEES* 20th December 1974.)

PROFITS In the petroleum industry profits have been more dependent on the degree of excess capacity than on the state of individual countries or the World's economy. The two most important elements have been the crude oil production surplus capacity (*q.v.*) and the spot market freight rate (*q.v.*). Refinery capacity has been less important. *See also* Crude Oil Sales Profits.

PROJECT INDEPENDENCE Following the big rise in energy prices in 1973/74, the US adopted *Project independence* the aim of which was to make the US independent of energy imports in a decade or so. *See Project independence, background paper*, US FEO, 1974.

PROPANE *See* LPG.

PRORATIONING This is a measure to limit the production of crude oil to eliminate drastic competition amongst crude oil sellers. In the US, voluntary prorationing was first introduced in 1914. In 1934 the Interstate Oil Compact Commission was formed to co-ordinate state activities. Texas was the state which gained most from prorationing. Many US wells were not subject to prorationing including stripper and marginal wells. *See also* Spacing.

PROVED RESERVES The estimated quantity of crude oil, natural gas or natural gas liquids which analyses of geological and engineering data demonstrate with reasonable certainty to be recoverable from known oil or gas fields under existing economic and operating conditions.

PRT The Petroleum Revenue Tax of the UK (*q.v.*).

PURCHASING POWER OF OIL The concept was of particular interest with the high inflation rates from 1972 onwards. Oil producing countries naturally wished for a constant volume of oil to be able to purchase a constant volume of manufactured goods.

PURE OIL COMPANY As a result of a meeting of independent oil producers at Butler, Pennsylvania on 24th January 1895 to discuss the problems posed by Standard Oil, a new oil company was proposed. The Pure Oil Co.'s organisation was completed by November 1895 and it represented the consolidation of independents' interests in production, refining, pipelines and marketing.
The formation of the Pure Oil Co. was well received by dealers who were reportedly bitter at the 'adultered' oil marketed by Standard Oil. David Kirk was the first President of the company but to some extent the formation of Pure Oil was due to Roger Sherman in seeking to limit the monopolistic activities of Standard Oil.

PUTNAM An early quite comprehensive source-book on energy supply and demand—P. C. Putnam, *Energy in the future*, Van Nostrand, New York, 1953.

Q

QATAR CRUDE OILS Qatar land crude, also known as Dukhan, is of 41° API gravity, 1·3% sulphur and is exported from Umm Said.
Qatar marine crude is of 37° API gravity, 1·5% sulphur exported by SBM from Halul Island.
For crude oil production, *see* Table 1.

QATAR GENERAL PETROLEUM CORPORATION A company established by the Qatar Government in 1975 to engage in all aspects of the petroleum industry.

QATAR PETROLEUM CO. LTD The company was originally registered as Petroleum Development (Qatar) Ltd in 1936, the name being changed to the above in 1953. Amongst the original shareholders were BP, CFP, Exxon and Shell. Initial participation was in 1973.

QUA IBOE (Nigeria) This crude oil is of 37° API gravity, 0·1% sulphur and is exported by SBM from Qua Iboe.

QUAD A Quad is 10^{15} Btu or roughly about 25 million tonnes oil. In 1974 US energy consumption was 72 Quads and total World energy consumption about 152 Quads.

QUALITY CONTROL *See* Product Quality Giveaway.

QUOIN ISLAND A small island at the entrance to the Persian Gulf which is used as a convenient point for calculating freight rates from the PG. It is the practice to calculate the rate from Quoin Island to the discharge port and add to this the rate from the loading terminal in the PG to Quoin Island. Similarly Europa Point is a fixed point (in the Straits of Gibraltar) from which distances to North European ports are tabulated.

R

RACK PRICE This is the price quoted by a refiner to jobbers for truck loads picked up at the refinery or major terminal.

RAG Rohoel-Gewinnungs AG. An Austrian exploration company, formed by Socony Vacuum and Shell, which found oil in 1937 and the development of the Gaiselberg field took place in 1938. Another discovery was made by RAG at Voitsdorf in 1963.

RAMBOUILLET DECLARATION This was made at Rambouillet on 17th November 1975 setting out 15 principles to restore economic growth. The signatories were France, West Germany, Italy, Japan, UK and the US. Item 13 mentioned continued co-operation to develop alternative energy sources and the steady and harmonious development of the world energy market.

R & D Following the increase in energy prices in 1974, most nations expanded their R & D budgets with nuclear energy continuing to get the major share.

	1974 (million US $)	
	Total Budget	*Nuclear*
US	1 000	630
Canada	105	74
France	352	224
West Germany	457	299
*UK**	228	164
Japan	187	158

* Year 1973/4

The figures however do not reflect the total R & D position in the countries because nearly all the work on oil is carried out by commercial concerns. In the early 1970s, major oil companies spent about 20 pence (50 cents) per ton of product sales on total research and development including technical services. Such costs were written off in the year in which they were incurred.

RANGER OIL The Ranger Oil Company of Canada was registered as the Maygill Petroleum Co. in 1950 taking the name Ranger in 1958.
Ranger is a production and development company with interests in the North Sea, US, Australia and Spain.

RANGER (UK) LTD Ranger has interests in Block 3/8 of the Ninian North Sea oil field.
Original shareholdings were as follows:

BP	50·0%
Ranger Oil (UK) Ltd	20·0%
Scottish Canadian Oil Transportation Co. Ltd	7·0%
London and Scottish Marine Oil Co. Ltd	15·5%
Cawood Holdings Ltd	3·75%
National Carbonising Co. Ltd	3·75%

RAP Régie Autonome des Pétroles formed in July 1939, a public company. RAP controls SOGERAP. In 1966 BRP and RAP were merged to form ERAP (*q.v.*).

RASMUSSEN REPORT A USAEC report; *An assessment of accidental risks in US commercial nuclear power plants*, August 1974. A report which had considerable publicity which provided calculations of the probabilities of serious nuclear accidents happening.

RATAWI (Neutral Zone) A crude oil of 23·5° API gravity 4·1% sulphur exported from Mina Saud.

RAW GAS Raw gas is hydrocarbon gas produced from a non-associated gas field. The higher boiling components and perhaps undesirable components are removed before marketing. *See also* Marketable Natural Gas.

RECON CRUDE Reconstituted crude. A crude oil which has been blended, usually in a producing country, to meet the needs of a refinery in a consuming country. A reconstituted crude may have a lower sulphur content or higher distillate content than the natural crude oil. *See also* Spiked Crude.

RECOVERY FACTOR The amount of crude oil that may be recovered from the total oil in a reservoir. Published estimates are usually conservative and do not take into account the full potential of secondary and tertiary recovery (*q.v.*) and the probability of higher energy prices.

RECYCLING (MONEY) The word applied, mainly from 1974 onward, to the arrangements for enabling consumer countries to pay for oil imports whilst running large trade deficits. The main problem in recycling is to re-allocate the funds from the countries where the OPEC countries prefer to hold their money, such as the US, to the countries which need the funds to pay for further oil imports.

RED LINE AGREEMENT Named after a red line drawn on a map of the Middle East which enclosed Saudi Arabia, Turkey and Iraq. This agreement, made in October 1927, defined the area within which American, British and French oil companies agreed not to compete for concessions without first consulting their partners. The agreement influenced Middle East concessions for many years after and was one of the most significant agreements ever made to curtail competition in the oil industry.

REDDAWAY ESTIMATES These were estimates provided by W. B. Reddaway and colleagues of the Department of Applied Economics of Cambridge University, on the foreign earnings of the three major UK oil companies. The work was done at the request of the CBI and covered the period 1956–1964.

REFINER DISTRIBUTOR In the US one who acts as a wholesaler in gasoline, may also retail and may buy additional gasoline to supplement that which is refined in his own refinery.

REFINERY COST *See* Refinery Fuel, Refining Cost.

REFINERY FUEL Refinery fuel can be any petroleum product or waste (unmarketable gas or oil) and may include purchased electricity and steam which may have been generated from non-petroleum fuels. In a simple refinery, refinery fuel is about 3 % of the crude run (and loss about 1 %) but in a complicated refinery with lots of secondary processing and specialities, it could be 10 % or more. The figure of 3 % refinery fuel was representative for 1975; before 1950 the figure was about 6 % for a simple refinery.

REFINERY GAS This is gas obtained from the operation of a refinery. Its composition is variable and it can contain hydrogen, the hydrocarbons present in natural gas, unsaturated hydrocarbons and possibly nitrogen and carbon dioxide. It is usually consumed as refinery fuel with any temporary surplus being flared.

REFINING COST In 1975 $, the average cost of refining a barrel of oil was about $4·50. This included a return on all capital investment associated with refining.

REGENT OIL CO. LTD The Regent Oil Co. of UK was the distributor of Caltex products in the UK using the brand names of Regent and Caltex. With the break up of the Caltex arrangement, the Regent Company of the UK went to Texaco and by the 1970s Texaco was marketing under its own name.

RENT In economic terms the rent is the payment made to a supplier or producer over and above the fully built up cost of production. Economic rent is a function of supply and demand for oil and also of the supply and demand for alternative fuels. There have been instances when the price of oil has increased sharply in times of general inflation but there is no relationship between rent and price inflation.
Strictly, the production cost would be a fully built up cost including exploration, development, production and overheads and a return on capital. The situation is relatively straightforward if inflation is ignored. Thus for a crude oil selling at $10/bbl with total costs $1/bbl commands a rent of $9. In many instances with large prolific fields, much of the historic cost has been considerably belittled by inflation and the total cost is little above current day to day running costs.

REPGA The Sté. de Recherches et d'Exploitations de Pétrole et de Gaz which was formed in August 1956. Schneider Group 51 %, finance companies 41 % and BRP 8 %.

RESALE AGENT There is little if any practical difference between the activities of an oil jobber (*q.v.*) and a resale agent. Both terms however sometimes appear together in trade reports, e.g. 'as a result of consumer stockdraft, consumption is not showing up in refilling by jobbers and resale agents'.

RESEARCH AND DEVELOPMENT COSTS *See* R & D.

RESERVES REMAINING These are the quantities of crude oil, natural gas, natural gas liquids as estimated under proved or probable reserves after deducting those quantities produced up to the date of the estimate.

RESERVES, TERTIARY RECOVERY The quantity of oil in place and the proportion that may be obtained by various recovery methods are subjective matters. Gulf Oil Co. in 1975 published some figures that gave considerable future potential in the US. They estimated that enhanced or tertiary techniques could raise reservoir yields from 32 % to an average level of 45 %; the increased quantity of oil amounting to 58 billion barrels. The debate on oil reserves and the extent to which they may be produced influences governments' policies and so can influence energy prices.

RESERVES, ULTIMATE RECOVERY Allocation of oil reserves in the ground as to; that proportion that can be produced under existing pressure, the additional amount by techniques such as gas and water injection, the proportion that might be yielded by tertiary recovery, and that proportion which will always remain in the ground has always involved much subjective judgement. Typical figures which form a basis for discussion are:

30% readily produced.
30% tertiary recovery including future technical developments.
40% ultimately unrecoverable.

For further definition *see* Ultimate Recoverable Reserves.

RESOURCE In geological terms a concentration of naturally occurring solid, liquid or gaseous materials in or on the Earth's crust in such form that economic extraction is potentially feasible. *See also* Identified Resources.

RETAIL OUTLETS, UK Retail outlets are also sometimes termed filling stations, service stations, petrol stations, garages.

	1968	1974
Shell	8 000	7 450
BP	5 000	3 475
National Benzole	4 650	2 200
Exxon (as Esso)	9 756	7 200
Texaco	4 405	2 900
Mobil	1 500	1 400
Fina	1 400	1 190
Total	600	1 210
Burmah	365	861
Jet	785	806
VIP	700	545
Murco	448	370
Gulf	200	376
ICI	100	360
Amoco	172	336
Total UK outlets	39 958	32 662

REUTERS Reuters News Service issues a daily petroleum report which contains news items and facts covering the energy industry.

RHINE–DONAU PIPELINE This runs from Karlsruhe to Ingolstadt to Neustadt. It is 270 km long and 26 in. in diameter. Original shareholdings were:

	%
Exxon	20
Shell	15
BP	11
Mobil	11
ENI	10
Marathon	7
Gelsenberg	6
Texaco	6
DEA	3
Scholven	3
Continental Oil	3
Wintershall	3
CFP	2

RICHFIELD OIL The Richfield Oil Corporation of the US which was absorbed in the build up of ARCO (*q.v.*).

RIDLEY COMMITTEE A White Paper on UK Energy, *Report of the Committee on National Policy for the use of fuel and power resources*, HMSO, Cmd 8647, 1952.

RMS Regular Motor Spirit. In the late 1960s and 1970s, this had an octane rating of about 90 by the research method but in earlier periods could have been considerably lower. For economics *see* PMS.

ROAD OIL This is a petroleum residue material intended for cold application to road surfaces—Bitumen (*q.v.*) requires heating before application. Road oil is a very minor product but the cost of production and market prices should be much in line with those for bitumen.

ROBINSON REPORT The last major energy study made by the OEEC: *Towards a new energy pattern in Europe*, OEEC, Paris, 1960.

ROMASHKINO (USSR) A crude of 32° API gravity 1·6% sulphur exported from Ventspils (Baltic).

ROP Russian Oil Products. An extensive UK 'independent' marketer and price cutter in the period up to 1939. ROP was superseded by NAFTA.

ROSTAM (Iran) An offshore crude oil of 36° API gravity 1·6% sulphur exported from the Lavan Island terminal.

ROTTERDAM PRICES *See* Platt's European Prices.

ROYAL DUTCH SHELL *See* Shell.

ROYALTY A royalty is a sum paid to use another person's property. The royalty paid by concession holders to host governments was frequently $12\frac{1}{2}\%$. Governments' take was also increased by taxation and, post 1973, by participation.
The general economic definition of royalty is the payment for concessions of any kind or payment for the use of patents or copyrights.

RUMANIA For crude oil production *see* Table 1.

RUNDOWN TANKS Tanks which receive the distillate fraction or product direct from the refinery process unit. From these the material could go into intermediate storage or blending tanks and then into final product storage.

S

S The chemical symbol for sulphur; *see* Sulphur *and* Fuel Oil.

SACOR Sociedade Anonima Concessionaria Da Refinacao De Petroleos Em Portugal. The company operates refineries at Lisbon and Oporto.

SAE The US Society for Automotive Engineers. It publishes a volume 'SAE Standards' which describes technical standards for lubricants.

SAE VISCOSITY RATINGS The commonly used ratings for internal combustion engine or crankcase oils. Thin oils, for use in cold conditions, would be rated 5W or 10W, other oils 20W, 30W, 40W or 50W. Multigrade oil, e.g. 10W/40 means an oil which is quite thin (10W) when cold but 40W (i.e. quite thick rating) when hot. This is achieved through the addition of a chemical polymer (polymethacrylates or polyisobutenes) to a mineral oil, the polymer serving to increase the viscosity as the temperature of the oil is increased. There is no fundamental reason why different viscosity oils should command different prices.

SAFREP The Sté. Anon. Française de Recherches et d'Exploitations de Pétrole, formed in February 1956 with BRP having a 64% holding. Linked with SPAFE (*q.v.*).

SAFRIPOL A South African chemical company owned by Sentrachem and Hoechst. In 1974 Safripol undertook with BP to build a huge coal based acetylene complex for $225 million; a significant move for BP, moving away from oil as a chemical feedstock.

SAIPEM An engineering subsidiary of the Italian state oil company ENI (*q.v.*).

SALARIED BULK AGENT In the US a wholesaler in gasoline who is an employee of the wholesaling company.

SALT DOME A deposit of salt which forms a dome in hard rock. When the salt has been removed by pumping out as brine, the cavern may be used for storage of hydrocarbons. Unit storage costs may be a fraction of conventional storage costs. *See also* Stocks, Types of Storage.

SAMARCO A shipping company owned by the Saudi Arabian Government, Mobil and Fairfield Maxwell Ltd. The company is based in Jeddah and managed by Atlas Maritime Co.

SAN Styrene–acrylonitrile resins which are important petrochemically derived resins.

SAP The Sté. Africaine des Pétroles which was formed February 1955. BRP and SNIP together held 73%.

SARAS S.p.A. Raffinerie Sarde. The company, formed in 1962, operates a refinery in Sardinia and also has an interest in petrochemicals.

SARIR (Libya) A crude oil 37° API gravity 0·15% sulphur exported from Marsa al Harigo.

SAROM Sarom Raffinazione S.p.A. The company operates a refinery at Ravenna.

SASOL The South African Coal Oil and Gas Corporation which was incorporated in 1950. Its main business was the manufacture of liquid fuels and chemicals from coal. The business was vital to the economy of South Africa since the country lacked oil but had substantial and cheap coal production.

SASSAN (Iran) An offshore crude oil 34° API gravity 1·9% sulphur exported from the Lavan Island terminal.

SAUDI ARABIA The Saudi Arabia state oil company is Petromin. For crude oil production *see* Table 1.

SAUDI ARABIA, CRUDE OIL RESERVES The Ghawar field in Saudi Arabia is the biggest oil field in the world. An interesting statistic is that with production at the 9 million b/d level, the natural gas flared each day approximates to UK daily consumption of gas.

Saudi Arabian reserves have been estimated as near 100 billion barrels which is about one seventh of the World's proven reserves. However there is speculation that Saudi Arabia's reserves might allow for ultimate production of 200 to 300 billion barrels out of 400 to 600 billion barrels in place. Peak production levels may ultimately prove to be about 20 million b/d and there has been speculation of even higher levels.

SBM Single Buoy Mooring System (*q.v.*).

SBP Special Boiling Point spirits. These products are used mainly as solvents, e.g. turps substitute, and can be of a narrow boiling range, e.g. 100–120 °C within the light distillate boiling range although they can have a wide boiling range.
SBPs cost about the same to manufacture as light distillate chemical feedstocks and less than regular grade motor spirit but the relatively small quantities sold involve high handling and transport charges. *See also* LDF.

SBR Styrene–butadiene rubber which is the general purpose synthetic rubber manufactured from petrochemicals. About three-quarters of butadiene usage is for SBR manufacture.

SCALE NOS. 1, 2 AND 3 *See* Tanker Rate Schedules.

SCHNEIDER REPORT The OECD Report *Energy Prospects to* 1985, OECD, Paris 1974 which was developed under the guidance of Professor H. K. Schneider of the Institute of Energy Economics, Cologne.

SCHUMAN PLAN The Schuman Plan originated in France and was supported by the French Government. Initially six countries formed the European Coal and Steel Community (ECSC) which was a common market for coal and steel. West Germany, Belgium, The Netherlands, Luxembourg, Italy and France signed an agreement on 19th March 1951 which was ratified on 25th July 1952. The ECSC agreement was a forerunner of post World War II energy policy in Europe.

SCOT The Scottish Canadian Oil and Transportation Co. Ltd, a participant in the UK North Sea Ninian Field. The company is a member of the Ranger Group.

SCP The Sté. Cherifienne des Pétroles, formed in April 1929. Kingdom of Morocco 50%; the rest mainly French state.

SDRs Special Drawing Rights of the IMF. As an alternative to using one currency or gold to express prices, prices may be expressed in SDRs which were established by the IMF in 1968. A SDR is a weighted index of 16 major currencies. In June 1975, OPEC took a decision to express oil prices in SDRs as an alternative to US dollars but the decision was not implemented. When the Suez Canal was reopened in June 1975 the rates were also expressed in SDRs.

SDW The summer dead weight of oil tankers; a measurement usually quoted in long tons as opposed to tonnes.

SEASONAL RATIOS Seasonal ratios express the relative amounts of petroleum products used in the different seasons. They are usually expressed in the quarters January to March, April to June and so on and sometimes also as monthly ratios. Seasonal ratios for consumption, supply and refinery runs are all employed in short term forecasting when they are used to break down annual forecasts into quarterly data. Typical ratios are quoted below, in each instance the annual rate is 400 units;

	Jan.– Mar.	*April– June*	*July– Sept.*	*Oct.– Dec.*
Refinery runs, USA	101	99	99	101
Refinery runs, Europe	103	97	97	103
Total petroleum consumption, Japan	109	93	91	107
Total petroleum consumption, Europe	110	91	90	109
Mogas consumption Europe	87	104	112	96
Fuel oil consumption, Europe	118	86	82	114

SEASONAL SPECIFICATIONS These are changes in product specifications with the season. The most important change is with motor gasoline specifications. Low vapour pressure specifications in the summer may permit the use of little or no butanes in the gasoline, whereas, winter specifications may permit about 10% butanes to be blended. Butanes which cannot be blended into gasolines might otherwise have to be used for fuel at much reduced values.

SEASONALITY Consumption of most petroleum products is seasonal. Mogas consumption is high in summer with the holiday season but lubricating oil peaks before the main holiday months. Aviation fuel and asphalt or bitumen also peak in the summer. Gas and fuel oils peak in the winter due to space heating and relatively high industrial production. *See also* Seasonal Ratios.

SEATO The South East Asia Treaty which was signed in September 1954 by Australia, New Zealand, Pakistan, The Philippines, Thailand, France, the US and the UK. Essentially a defence treaty. Article 3 of the treaty also specified technical assistance and promotion of economic progress.

SECONDARY PRODUCTION Also known as secondary recovery; a production technique of much greater interest with the substantial rise in energy prices in 1974. Oil reserves are porous rock formations which hold the oil under pressure and it is this pressure which forces the oil to the surface when the reservoir is tapped. As the reservoir pressure diminishes, so does the flow of oil from the well. However pressure in the reservoir can be enhanced by injecting gas or water into the reservoir; gas injection and water injection are the two main methods of secondary recovery. Not all production can be stimulated by injection and obviously availability of cheap water or gas is necessary to practise secondary recovery.

In really prolific fields, such as some in the Middle East, historic production costs have been less than 10 cents/bbl and although cheap, even valueless gas was available, it was not re-injected but flared. As countries became aware of conservation some stipulated that the gas separated from the crude oil must not be flared but be re-injected or used in some useful way. Whereas in pre-1973 conditions secondary recovery costs of cents a barrel might have been considered uneconomic in the Middle East, costs of the order of a dollar or so a barrel might have been considered worthwhile in the Western Hemisphere. In post-1973 conditions much higher total production costs, up to $10/bbl, made sense in countries importing oil and with balance of payments problems.

Venezuela is a country with many fields which have been in production for decades and which practises secondary recovery on a substantial scale. In 1974, 54 gas injection plants and 60 water injection plants were in operation. This scale of operation meant that about 50 % of gas produced was re-injected, 30 % used as fuel and only 20 % wasted with the aim of cutting the waste to a few per cent within a short period.

SEISMIC SURVEY A survey carried out to obtain geological information of subterranean rocks. A small explosion generates energy waves sending them down through the strata. The waves travel faster through hard rocks than through soft. A receiver situated some distance from the explosion, a seismograph, records the rebounding waves and from the records the subterranean formations may be deduced. *See also* Geophysical Costs.

SENEGAL For crude oil production *see* Table 1.

SENTRACHEM The South African Chemical Company.

SEPL The South European Pipeline which runs from Lavera (near Marseilles) to Strasbourg. It is 780 km long and 34 ins in diameter. Original shareholdings were:

	(%)
Exxon	28
Shell	23·2
CFP	15
BP	8
Mobil	5·6
DEA	4
Antar	4
Gelsenberg	3·6
ERAP	2·8
Scholven	2·4
Wintershall	2·4
Petrofina	1

SEREPCA The Sté. de Recherches et d'Exploitation des Pétroles du Cameroun, formed in September 1951. Cameroun Republic 11 %, BRP 55 %.

SEREPT The Sté. de Recherches et d'Exploitation des Pétroles en Tunisie, formed in February 1949. Tunisian State 24 %, CFP 11 %, the remainder was mostly BRP.

SERIA The export grade of this offshore crude oil is of 39° API gravity, 0·05 % sulphur and is shipped by SBM.

SERVICE CONTRACT Exploitation of an oil field by a company in which the ownership of the field remains with the state.

SERVICE STATIONS Establishments selling gasoline and perhaps operating other facilities according to the whims of the oil companies and the fashion of the day. A service station may be company owned and operated, company leased and operated (land not owned), employee operated with the employee carrying the financial risks of operation. At the other end of the spectrum service stations can be owned by the dealer who may buy gasoline from a variety of suppliers as he wishes. In the US, a service station is defined as a retail operation which derives the majority of the income from the sale of motor spirit and lubricants. *See also* Retail Outlets, UK.

SETTLING TANK A settling tank is an oil storage tank with a heating surface of not less than two square feet/ton of oil capacity.

SEVEN MAJORS A term which has frequently been used for BP, Shell, Exxon, Gulf, Mobil, Socal and Texaco. Mattei, when Chairman of ENI, termed these companies the Seven Sisters.

SEVERANCE TAX A tax introduced by US states on all crude oil and natural gas production.

SHALE Oil shales are clays impregnated with organic materials of which kerogen (*q.v.*) is the main component. Yields of hydrocarbons can vary appreciably depending on the quality of the oil shale; a yield of 20 gallons/ton would be average, 40 gallons/ton would be good. The yield of oil products will depend on the refining technique but 20% naphtha, 50% kerosine/gas oil and 30% fuel oil may be considered typical values. These yields are similar to an average crude oil. On this basis, with running costs alone being about $10/ton (1975), a ton of shale yielding half a barrel of oil is equivalent to an oil price of about $20/barrel. *See also* Development Costs, Shale.

SHARJAH For crude oil production *see* Table 1.

SHELL The Shell Group of Companies is the group controlled by two holding companies which are quoted separately on stock exchanges. Royal Dutch Shell owns 60% of Shell operations and Shell Transport and Trading 40%. There are also other companies in the Shell Group which have independent stock exchange quotations, the biggest is Shell Oil of the US. The Royal Dutch company was formed in The Hague in 1890 and Shell Transport and Trading in London in 1897. The London company took over the interests of the firm founded by Samuel Samuel in the 1830s which had general trade with the Far East including ornamental shells from which the group took its name in 1946. Before the Shell Group ended up in its present form several regroupings were involved. Earlier companies included the Asiatic Petroleum Company (1903), The Anglo Saxon Petroleum Company (London, 1907) and NV De Bataafsche Petroleum Maatschappij (The Hague, 1907).

Shell is second only to Exxon and operates worldwide through several hundred main companies. It has always been relatively strong on the marketing as opposed to the production of petroleum. It has large chemical interests and in the 1970s took two important steps of diversification in acquiring large interests in the metals field and in partnership with Gulf in nuclear energy. In the UK its long standing marketing organisation with BP (Shell-Mex and BP) was dissolved in 1975 as were some of the other joint interests in Africa and the Middle East.

SHERMAN ACT The Act was passed in the US in 1889 and declared it unlawful for companies to exercise unreasonable restraint of trade, i.e. those companies that had too much power in a market. It was used to break up the Standard Oil Trust in 1911. *See also* Pure Oil Co.

SHUT-IN FIELD This is where an oil or gas field has been discovered and has never been or is not being exploited.

SIGNAL OIL AND GAS CO. Signal, which was incorporated in 1925, was mainly an exploration and development company. In 1974 it was bought by Burmah Oil just before Burmah's tanker problems came to a head. By 1975 Signal's interests, excluding its North Sea holdings, had been placed on the market to raise money to enable Burmah to survive and they were sold in 1976 to R. J. Reynolds Industries, for $20 million.

SIMON COMMITTEE A White Paper on UK Energy; *Domestic Fuel Policy*, Cmd. 6762, 1946.

SINCLAIR OIL The Sinclair Oil Corporation was a US integrated oil company incorporated in 1919. It had extensive interests outside the US. In 1958 it signed an agreement with BP for joint operation in certain areas but this proved not to be long lasting. Following a take over bid by Gulf and Western (a US conglomerate) in the late 1960s, Sinclair was absorbed largely by ARCO but US Antitrust activities forced ARCO to divest much of the Sinclair company to BP. These interests were then later taken over by Sohio.

SINGLE BUOY MOORING SYSTEM The single buoy mooring system is an alternative to building deep water discharge or loading terminals, thus eliminating the need to lighten vessels or use relatively high cost small tankers. The cost of such systems can obviously vary considerably with location. As a guide, a system built by Shell for operation in 1976 to serve its Stanlow (UK) refinery, cost about £70 million.

SINGLE POINT MOORING *See* Single buoy mooring system.

SIRIP Societé Irano–Italienne Des Pétroles. An Iranian exploration and production company formed in 1957 whose shareholders are NIOC and ENI. The company was successful with offshore exploration and production from the Bahrgansar field, commenced in 1962, and from Nowruz in 1971. A third oil field discovered by the group was Hendijan. SIRIP was of great importance to ENI being one of the few bright spots in its exploration and production activities.

SKINNER'S Walter R. Skinner's *Oil and Gas International Year Book* which was first published in 1911. This annual publication is now published by the London *Financial Times* and contains brief financial details of many of the important oil and gas companies together with information on their production interests.

SLURRY OIL This is an oil which contains catalyst fines after the cat. cracking operation. When the catalyst is removed the product becomes known as decant oil (*q.v.*).

SMMT The UK Society of Motor Manufacturers and Traders. Its annual statistical publication *The Motor Industry of Great Britain* has been published since 1946 and is a detailed source of UK and World statistics on many aspects of the motor industry.

SNAM PROGETTI An engineering subsidiary of the Italian state oil company ENI (*q.v.*).

SNAP Sté. Nationale des Pétroles d'Aquitaine. *See* Aquitaine.

SNEA The Societé Nationale Elf Aquitaine formed by the merger of Elf–Erap and SNPA which was approved in January 1976.
The combination of Elf–Erap and SNPA gave France a second major integrated oil company, the other being CFP. At the time of the merger Elf–Erap was 100% state owned whilst SNPA was 51% state owned.

SNPA The Sté. Nationale des Pétroles d'Aquitaine. Formed in December 1941 with BRP and SNIP having 53%. The company is more usually referred to as Aquitaine. It was merged with ELF in 1976.

SN REPAL The Sté. Nationale de Recherche et d'Exploitation des Pétroles en Algerie, formed in November 1946. Algerian State $40\frac{1}{2}\%$.

SOCAL *See* Standard Oil Company of California.

SOCIETA ITT–INDUSTRIA ITALIANA PETROLI The name given to the business run by Shell Italiana following the sale to ENI in December 1973.

SOCONY–VACUUM OIL CO. *See* Mobil Oil Corporation.

SODIUM *See* Vanadium.

SOEC The Statistical Office of the European Community which publishes much energy data on the EEC of Nine. The regular Eurostat publications contain some of the most comprehensive tables on European energy prices (*q.v.*).

SOFIRAN An Iranian exploration company whose shareholders are NIOC, Elf–Erap and Aquitaine.

SOFT LOAN A loan made at a rate of interest that is well below commercial rates to help to make a deal. Also used by oil producing countries after participation to allow other countries to buy oil.

SOGERAP A French investment company incorporated in 1957 with about 75% of the shares being held by ERAP (*q.v.*).

SOHIO The Standard Oil Company of Ohio was one of the smaller companies originating from the break up of the Standard Oil Trust. Its main activity was marketing in Ohio and some adjacent states. Following BP's discovery of oil in Alaska, a merger was agreed between BP's North American interests in 1970, when BP took a 25% stake in Sohio which would rise to a controlling interest when Alaskan oil supplies rose to 600 000 b/d. *See also* Sinclair Oil.

SONATRACH The Algerian state oil company which in 1974 controlled about 80% of Algerian production. It was formed on 31st December 1963 and participates in all parts of the oil industry.

SOUR CRUDE This is crude oil which contains appreciable amounts of sulphur in the form of hydrogen sulphide and mercaptans. Conversely a crude which is lacking in these materials is termed 'sweet'.

SOUR OILS Sour oils contain unacceptably high concentrations of hydrogen sulphide gas or mercaptans. A sour oil is not an oil with a high total sulphur content. Sour oils are further treated to make marketable products. The process might be termed 'sweetening' and a sour oil is sometimes termed 'unsweetened'.
A modern process for sweetening is the DHT (distillate hydrotreating) in which all distillates from the crude distillation unit are subject to a mild hydrogenation.

SOSIM An Italian company set up in 1960 to study the laying of a Tunisian–Sicily pipeline for North African natural gas.

SOUTH AND EAST AFRICA The following countries are included: Botswana, Ethiopia, French Somaliland, Kenya, Lesotho, Malagasy, Malawi, Mozambique, Rhodesia, Somali Republic, South Africa, South-West Africa, Sudan, Swaziland, Tanzania, Uganda and Zambia.

SOUTH EAST ASIA The countries in SE Asia include: Cambodia, Hong Kong, Laos, Macao, Malaysia, Pacific Islands, Singapore, South Korea, South Vietnam, Taiwan and Thailand.

SOJUZNEFTEEXPORT This is the USSR oil export organisation.

SPACING The minimum spacing of crude oil wells was one of the devices used in prorationing (*q.v.*) in the US. Thus a state with a well density of 1/20 acres might legislate for a maximum density of 1/40 acres.

SPAFE The Sté. des Pétroles d'Afrique Equatoriale, formed in July 1949. BRP and SNIP had a 58 % holding.

SPAIN The Spanish state oil company is Hispanoil. For crude oil production *see* Table 1.

SPC The Singapore Petroleum Company which posts product prices at Singapore.

SPC PRICES Singapore Petroleum prices which are representative of trade in that area and are quoted in Platt's Oilgram Price Service.

SPECIALITIES These products include special solvents, lubricants, waxes, extracts and bitumens (asphalts). Total production of products in this class usually comprises only a few per cent of total petroleum production. There are however special refineries which manufacture bitumens or lubricants as the greater proportion of their output.

SPIKED CRUDE A crude oil to which a component has been added, e.g. butane or light distillate. This is a cheap way of transporting the added component to the consuming area. *See also* Recon Crude.

SPONGE COKE (OR GREEN COKE) This is produced from high asphaltenes petroleum feedstocks such as vacuum residues. The name is derived from its appearance. Sponge coke has low electrical conductivity, usually high impurities (e.g. sulphur, nitrogen, nickel, vanadium) and so is not suitable, without further treatment, for manufacturing anodes. It may be used for blending with coal-coke for use in blast furnaces or as a smokeless fuel. The coke, as it comes from the coking unit, may contain about 5 % water and also perhaps 10 %, or more, volatile materials. The term 'green coke' may be applied to all cokes before calcining. Unless qualified, green coke usually refers to low grade coke. *See also* Needle Coke, Honeycomb Coke.

SPOT MARKET FREIGHT RATES The important spot market for oil tankers is a substantial market with typically 10 to 20 % of the World tanker fleet engaged on single voyage charter supplemented of course by whatever surplus exists as being potentially available for single voyage charter. Spot freight rates have been a major influence in spot market oil product prices. *See also* Marginal Markets.

Spot market rates are quoted for clean and dirty vessels; the clean market rates usually being higher because smaller size categories are used. The most important market is the dirty market which may be 10 times larger in tonnage terms than the clean market although in vessel terms fixed only about three times as large.

Tanker fixtures 4th quarter 1974 *and* 1*st quarter* 1975

	Clean	*Dirty*
Number of vessels	400	1 350
Total Tonnage (dwt)	8 600 000	86 000 000

SPOT MARKETS *See* Marginal Markets.

SPS The Sté. des Pétroles du Senegal, formed in July 1956; BP 50 %, BRP 26 %.

SPV The Sté. des Pétroles de Valence, formed in December 1955; BP 50 %, BRP and SNIP 20 %.

SRB Straight Run Benzine. A crude oil distillate usually within the boiling range 95–175 °C. Premium feeds for cat. reformers for xylenes manufacture have end points of approximately 140 °C. For prices *see* LDF.

SRG Straight run gasoline. A crude oil distillate usually within the boiling range 65–95 °C. It is sometimes defined as C_5 (hydrocarbons) to 95 °C. For prices *see* LDF.

SRN Straight run naphtha. A term used in the US to denote catalytic reformer feedstock of boiling point range 65–200 °C. In the US there is more pressure on motor gasoline production than there is in Europe. Consequently it is the practice to put high end point light distillates into a cat. reformer because it increases the yield of the relatively high value gasoline that may be obtained from a given crude oil. *See also* Virgin Naphtha. For prices *see* LDF.

SRP Suggested Retail Price. This is the price recommended by suppliers to dealers for retail customers.

SSRP The Sté. Saharienne des Recherches Pétrolieres, formed in June 1960. BP 50 %, BRP and SNIP 20 %.

STAHMER *See* German Oil Information Service.

STANDARD OIL *See* Standard Oil Companies of: California (Chevron), New Jersey (Exxon), Indiana (Amoco), New York (Mobil) and Ohio (Sohio).

STANDARD OIL COMPANY The original Standard Oil Company, built up by Rockefeller. By

1911, it controlled 90% of US pipelines, 85% of refining and marketing and about 11% of crude oil production. In 1911 the US Supreme Court ordered the break up of the company and 33 or more units were formed into independent concerns. Most of the units flourished and are still a force in the industry. They include Exxon, Mobil, Socal, Standard Indiana and Sohio. *See also* Monopoly.

STANDARD OIL COMPANY OF CALIFORNIA One of the World's big seven major oil companies. The company was formed in 1911 with the break-up of the original Standard Oil Trust. It is one of the Aramco partners. Indonesia and the US are its other major sources of crude oil.
Amongst the names used are Socal, Chevron, Irving Oil, OTC and Vectra. The Caltex company is owned 50% by Socal and 50% by Texaco. In 1975 Socal made a significant diversification move with the purchase of a 20% stake in Amax, the US coal and mineral mining group.

STANDARD OIL OF INDIANA After the big seven, one of the largest integrated oil companies. It operates worldwide, the main brand name being Amoco.

STANDARD OIL TRUST The Standard Oil Trust of Rockefeller was dissolved in 1911. *See* Monopoly; Standard Oil Company.

STANDARD TANKSHIP There have been many standard tankships in the industry and it is likely that there will be many more. The nearest concept to a generally accepted standard is based on the World War II tanker, a T-2. A standard shipping unit is employed to compute tanker demand from oil consumption and demand forecasts, oil production and tankship movements. The demand for tankers may then be compared with supply, also expressed in the standard tanker unit.

Some Standard Tankship Units

	Tons dwt	*Speed (Knots)*
MOT	12 000	9·5
USMC	12 551	12·5
Scale 1	12 000	11
Scale 2	12 000	11
Scale 3	19 500	14
ATRS	16 600	14·25
Jacobs T-2	16 600	14
Esso T-2	16 600	14·6
BP T-2	16 600	14·5
Shell T-2	16 600	14·5
Sun Oil T-2	16 765	14·5
Intascale	19 500	14
Worldscale	19 500	14

Port times and other factors also vary.
See also Tanker Supply and Demand.

STANIC The two Italian refineries at Bari and Leghorn are owned by Stanic Industria Petrolifera S.p.A., a company formed in 1949 which is owned 50/50 by ENI and Exxon.

STAR RATING SYSTEM From March 1967, motor gasolines (petrols) in the UK have had their quality specified by a star rating system (BS 4040: Part I 1971). The system was designed to enable the public to better appreciate the effect of octane quality of motor spirits on engine performance and hopefully not to waste money buying a better grade than a particular car needed. There were four grades specified.

Star Grading	*Minimum Research Octane Number*
5	100
4	97
3	94
2	90

Extensive use of blender pumps in the UK however enabled motorists to purchase to any specified whole number between 90 and 100.

STATE OIL COMPANIES Apart from the USA, most countries now have state oil companies. If BP is classified as a state oil company it would be one of the first (1914). Other prominent European state oil companies include ENI, Italy; ERAP, France; OMV, Austria; Hispanoil, Spain; Petrogal, Portugal and Statoil, Norway.

STATFJORD (Norway) A North Sea field developed by Mobil with Statoil. The crude is 38° API gravity with 0·3% sulphur.

STATISTICS *See* Energy Statistics.

STATOIL Statoil is the Norwegian state oil company and the decision to form the company was made by a unanimous vote in the Storting on 14th June 1972. Statoil is automatically brought in as a partner in all Norwegian North Sea interests through state participation.

STEAM CRACKED FUEL OIL OR TAR This material is obtained from the bottom of the fractionating column which separates the products from the ethylene cracking operation. It is a highly aromatic oil and, when the feed to the ethylene cracker is LDF, contains little sulphur and nitrogen. It is used as a feed for needle coke production (*q.v.*) or as carbon black feedstock or may be blended into fuel oils for burning.

STEP OUT WELL *See* Production Well.

STIR Societé Tuniso–Italienne de Raffinage. STIR, owned 50% by the Tunisian state and 50% by ENI, operates a small refinery at Bizerta.

STOCK LEVELS Stock levels vary considerably from one country to another not only in total quantities but also as measured in days supply. Since World War II there has been a tendency to increase stocks in nearly all areas. In the early 1970s normal stock levels in several European countries were about 90 days and ideally stocks of each product would be approximately equal. A reduction in stock levels from about 90 days to 70 days would indicate a tight supply (and a high price) and 100 to 110 days stock would be an easy supply. From the point of view of security of supply, high stock levels are desirable. However oil companies or governments have to pay for holding stocks. At a consumption rate of 50 million b/d in the Free World, about $500 million (end 1974 $) needed to be found to buy the oil. The cost of freight and tanks will be an additional financial burden.

STOCKBUILD A building up of stocks of any kind for any purpose. The term is not usually applied to a single domestic consumer but the terms stockbuild and stockdraft are used for classes of consumers. In the petroleum industry there is a seasonal pattern of stockbuild and stockdraft. Fuels used for space heating are stored as winter approaches whilst mogas stocks are increased for the summer season. In the US, for example, distillate fuel oil stocks may be as high as 200 million barrels in October and, depending on the severity of the winter, fall to near 100 million barrels by the end of the first quarter.
See also Seasonality *and* Seasonal Ratios.

STOCKDRAFT A running down of stocks of any kind—from crude oil stocks to product stocks held in domestic consumer tanks. The opposite of stockbuild (*q.v.*).

STOCKS Stocks of oil may have different definitions in different areas. The following definitions were published in the agreement on an International Energy Program between ten European Governments, the US, Canada, and Japan in Brussels on 21st September 1974:

1. Total oil stocks are measured according to the OECD and EEC definitions revised as follows:

A. Stocks included: crude oil, major products and unfinished oils held
- in refinery tanks
- in bulk terminals
- in pipeline tankage
- in barges
- in intercoastal tankers
- in oil tankers in port
- in inland ship bunkers
- in storage tank bottoms
- in working stocks
- by large consumers as required by law or otherwise controlled by Governments.

B. Stocks excluded:
(a) crude oil not yet produced
(b) crude oil, major products and unfinished oils held
- in pipelines
- in rail tank cars
- in truck tank cars
- in seagoing ships' bunkers
- in service stations and retail stores
- by other consumers
- in tankers at sea
- as military stocks.

Stocks are frequently expressed in terms of days supply using the annual rate of consumption observed in the previous calendar year.

STOCKS, COMPOSITION Stocks are usually quoted in total but under normal conditions not all stocks are accessible, the quantity of stock above the minimum operating level is the important parameter. The minimum operating level is not necessarily the same for each product or for crude oil because of different storage and transportation requirements.

STOCKS, SPOT MARKET Spot market stock levels are not necessarily related to total market stocks in the area. It is the total stocks in the area; stocks potentially over-hanging the market coupled with expectation, that influence prices. If traders take a bearish view they may seek to run down stocks. Thus, in late April 1975, guestimates were made that gas oil stocks at Rotterdam were very low; in the range 200 000 to 400 000 tonnes which was about one quarter of the stock level of six weeks earlier. An additional indicator of the low stocks position was that little gas oil was in transit (in tankers) unsold. Gas oil prices had been depressed throughout the first four months of 1975 and the bearish position on stocks was being taken on an expectation of even lower price levels to follow.

STOCKS, TYPES OF STORAGE A great deal of oil is needed to fill pipelines and tank bottoms and this level is referred to as the minimum operating level. The minimum operating level could be half of the total oil in stock or even more. Unconventional storage has come more into use as consumer governments have specified higher minimum stock levels. Unconventional storage could be salt caverns but these are likely to contain only 10 % or less of total oil stored. One of the highest figures was in West Germany where at the end of 1974, 5·6 million out of 47 million m^3 total, was cavern storage.

STOCKS, UK UK oil stocks are published monthly by the UK Government (Department of Energy). Traditionally UK stocks have been at high levels compared with other European countries. Taken alone, UK stocks are not an important factor in influencing product price levels.

STOCKS, US Stock levels in the US are reported weekly by the API for motor gasoline, kerosine, distillate fuel oils, residual fuel oils and unfinished oils. Variations in these stock levels are not in general prime influences on prices.

STRIPPER WELLS Stripper wells are usually defined as wells in the last stages of their life and production is usually 20 b/d or less. Many strippers' production is hardly economic but since this has accounted for about 20% of US production, if the wells were closed the crude would have to be replaced by depleting the larger fields at a faster rate, or by imports. Closure of strippers would therefore benefit the economy in the short term but security of supply considerations suggest that strippers should be kept in production.

SUEZ CANAL The Suez Canal was inaugurated in 1869 and in its early years was important for the Europe to India and Far East trade. From World War I to 1967 the canal was of importance to the oil trade from the Middle East to Europe. When it was closed in the June 1967 war, it could handle ships of 60 000 tons dwt loaded or 150 000 tons dwt in ballast. In 1966 revenue totalled about £95 million and estimates of the lost revenue due to the 1967–1974 closure have been as high as £1 200 million It was formally re-opened by President Sadat on 5th June 1975.

SULPHUR Elemental sulphur does not occur in crude petroleum but sulphur compounds do extensively. In most crude oils about 25% of the compounds present contain at least one sulphur atom and in fuel oils about 50% of the material is composed of sulphur compounds. Sulphur compounds are removed from petroleum distillates by hydrogen processes; hydrotreating (a mild process) and hydrofining (a more severe process).
The presence of sulphur compounds is beneficial and indeed essential in lubricating oil basestocks and bitumens. The sale of elemental sulphur is not economically important to the petroleum industry.

SULPHUR DIOXIDE Sulphur dioxide (SO_2) is a major cause of atmospheric pollution arising from the burning of sulphur containing fossil fuels. Thus fuel oils and coals have price variations in accordance with their sulphur contents. *See also* Sulphur, Fuel Oils, Crude Oil Quality and Egerton Report.

SULPHUR EMISSION DATA Sulphur emission data (sulphur dioxide) from the burning of solid fuels and oil, are subject to more variation from one report to another than most statistics. A Swedish Royal Commission working in 1975 reported the following sulphur emissions in 1975:

	Million tonnes
Sweden	0·8
Western Europe	22
Eastern Europe	38

SUMATRAN LIGHT CRUDE OIL *See* Minas.

SUMED The 207 mile, 42 inch diameter pipeline from Suez to the Mediterranean which was inaugurated in mid-December 1976. It is owned by Egypt 50%, Saudi Arabia, Kuwait and UAE 15% each and Qatar 5%.

SUPERTANKERS The word supertanker has always been used to describe the largest size categories of the day. Thus in the late 1950s vessels above 24 000 tons dwt were described as supertankers. In the late 1960 the term fell into relative disuse when the largest vessels (more than 200 000 tons dwt) were termed VLCCs (very large crude carriers).

SUN OIL COMPANY An integrated US oil company which was founded in 1886 and has worldwide operations. For many years the company has published a detailed analysis of the world tanker fleet.

SUNNINGDALE OILS LTD A Canadian exploration and development company incorporated in 1969 which has interests in many parts of the world including the UK North Sea.

SUNRAY A company and trade name used by Sun Oil Co. (*q.v.*).

SUPERIOR OIL COMPANY The Superior Oil Co. of the US was incorporated in 1936. It is a production and marketing company in the US. Outside the US it has interests in the UK North Sea with a 25% holding in Arpet Group (Hewett Gas field etc).

SURPLUS CAPACITY, CRUDE OIL *See* Crude Oil Production, Surplus Capacity.

SURPLUS CAPACITY, FREIGHT In the post World War II period substantial surpluses of tankers have been the rule. Tight freight markets (i.e. few surplus ships) existed in the winters of 1950/1, 1951/2 (Korean War) and 1956/57 (First Suez Crisis); late 1967 (Suez Canal closure); the winter of 1970/71 (economic boom conditions) and in 1973 (Arab oil embargo). Otherwise there has existed a surplus, frequently substantial, of tankers. The relationship between tanker surplus and spot market freight rates may be approximately expressed as follows. At 10% surplus tankers, spot market rates are in accordance with a fully built up cost of operation including adequate return on capital. As the surplus declines to zero, the spot market rate rises to about five times the cost rate. With a tanker surplus of 20 % or more there are usually many vessels in lay-up and spot rates reflect only variable operating costs and sometimes not even those. Thus in 1975 $ and conditions, distress rates from the Middle East to North-western Europe using a VLCC would be about $3/ton, 'normal' rates about $7/ton and rates in a hypothetical tight market situation about $35/ton.

SURPLUS CAPACITY, REFINING It is seldom that the petroleum industry does not have available surplus refining capacity somewhere in the World. Outside the US in the post World War II period there have been many times when the refining of crude oil has not been in itself an economic proposition. This is because product prices in the marginal markets have not allowed a margin over the cost of crude and freight. In Europe the large building programme of the 1960s and early 1970s saw a surplus emerge before the break in the growth of oil consumption in 1974. In the summer of 1975 European refineries were operating only at about 60 % capacity with the likelihood of being in substantial surplus for a long period.

SVARLBARD TREATY The Svarlbard Treaty was signed in 1920 by 36 nations specifying the terms under which Norway was given jurisdiction of the Svarlbard (Spitzbergen) Islands. The treaty is of importance for the development of oil and gas fields on the continental shelf.

SVENSKA PETROLEUM A Swedish state owned oil company formed in 1976.

SWEDEN, MARKET SHARES The following market shares (%) in 1974 were estimated by Veckans Affaerer:

	Mogas	*Light Heating Oil*
OK	20	15·5
BP	14	14·5
Shell	18	14·5
Nynas	5	10·5
Texaco	10	10·5
Ara-Bolagen	—	8·0
Gulf	11	7·5
Exxon	14	7·5
Mobil	3·5	2·5

SWEDISH CARTEL A cartel which operated from 1930 to 1947 when it was investigated by the Riksdag's Oil Investigation Committee. The cartel included four of the majors, the Swedish Nynas company and the Russian Nafta.

SWING PRODUCTION Swing production of crude oil is the production of the country, zone or area that acts as a balancing region. The swing producer needs adequate reserve installed capacity to meet surges in demand and can tolerate relatively low production when demand is slack. From the late 1960s, Saudi Arabia became regarded as the swing producer for both OPEC and the Freeworld.

SWISS OIL CORP. *See* Ashland.

SYRIA For crude oil production *see* Table 1.

T

T-2 *See* Standard Tankship.

TAIWAN For crude oil production *see* Table 1.

TAL The Trans-Alpine pipeline runs from Trieste to Ingolstadt. A 480 km, 40 inch diameter line built at a cost of about $270 million, it was commissioned in 1967. There are 15 storage tanks at Trieste of total capacity of 810 000 tonnes. The original principal shareholdings were as follows:

	%
Exxon	20
Shell	15
BP	11
Mobil	11
ENI	10

TANKER RATE SCHEDULES Prior to 1939 charter rates were negotiated in some detail for each voyage. At the outbreak of war, the UK Government requisitioned all British Flag vessels. Standard schedules were introduced, the last of these issued in December 1945 becoming known as the MOT (Ministry of Transport) Schedule. The rate under this schedule from the Caribbean to North-western Europe was about £1·60/ton. Vessels requisitioned by the US Government operated under a similar schedule the USMC (US Maritime Commission). At the instigation of BP and Shell, the London Tanker Brokers Panel introduced a schedule which superseded the MOT Schedule on 1st November 1952 and lasted until 30th June 1954. This was known as Scale No. 1. The Scale 1 rate for the Caribbean to North-western Europe route was the same as in the MOT. On 1st July 1954 the London Tanker Brokers Panel introduced a new schedule, Scale No. 2 which lasted until 14th December 1958. Scale No. 3 was introduced on 15th December 1958 and lasted until

14th May 1962. In 1956 the Tanker Committee of the Association of Ship Brokers and Agents, New York, introduced the American Tanker Rate Schedule (ATRS). In London Scale 3 was replaced by Intascale which lasted from 15th May 1962 to 14th September 1969 when it was replaced by Worldscale.
There are no general rules for changing rates in one scale to another because technical factors change, e.g. changes in definition of the standard tanker (*q.v.*), fuel and port costs. In the case of Intascale and Worldscale an approximate conversion is possible, Worldscale 100 = Intascale flat, and deviations from these points are calculated on the basis of Worldscale = Intascale + 10%.
Hence: Intascale − 50 = Worldscale 55
Intascale + 50 = Worldscale 165
Intascale + 80 = Worldscale 200
The ATRS scale is approximately half the Worldscale index.

TANKER SUPPLY AND DEMAND The detailed methodology which may be employed to quantify the demand for oil tankers is described in a report published by the Institute of Petroleum, London, in March 1976. In outline, the method consists of quantifying World oil movements (World oil trade) and expressing demand for tankers in standard units. The supply of tankers is evaluated by a count of the World tankship fleet expressed in the same standard units.

TANKERS, CATEGORIES The following are size category notations used in the oil industry but the tonnage ranges are subject to some variation: Specialist—small vessels up to about 17 000 tons dwt. Included in this would be all bitumen carriers. The lower weight limits used may be 6000 or 10 000 tons dwt.

		(tons dwt)
GP	(General Purpose)	About 17 000–24 999
MR	(Medium Range)	About 25 000–44 999
LR	(Large Range)	About 45 000–159 999
VLCC	(Very Large Crude Carrier)	Above 160 000

See also Tankers, Large Range.

TANKERS, CHEMICAL Major petrochemicals such as benzene and xylenes may be carried in clean product carriers on the same basis as other clean petroleum products. Petrochemicals that are shipped in smaller quantities may be shipped in Parcel Carriers. *See* Tankers, Parcel Carriers. Specialised chemical tankers may be as small as 2000 tons dwt.

TANKERS, LARGE RANGE Also referred to as an LR tanker and in the 1964 AFRA revision, as a tanker of 45 000 to 69 999 tons dwt. Subsequently LR1, LR2 and LR3 classes were specified:

LR1	45 000 to 79 999 tons dwt
LR2	80 000 to 159 999 tons dwt
LR3	160 000 to 319 999 tons dwt

The LR3 is the VLCC class. Ships 320 000 tons dwt or more are known as ULCCs (*q.v.*).

Relative operating costs between categories

GP	150
MR	120
LR1	80
LR2	60
LR3	40
ULCC	35

These numbers give an idea of the economy of scale in tanker operation. *See also* Tankers, Categories.

TANKERS, PARCEL CARRIERS Parcel carriers are tankers which have a relatively large number of separate tanks. The larger vessels may be about 25 000 tons dwt and the lower limit is about 6000 tons dwt. A large carrier may have up to 40 separate tanks of capacity between 200 and 1500 tons. Some tanks may be rubber coated or have glass or stainless steel linings. There is nothing to prevent parcel carriers from being used to ship petroleum products such as lubricants or extracts or alkylates. The rates for parcel carriers should be in line with those quoted for small tankers trading in clean products.

TANZER REPORT A report drawn up by Tanzer Economic Associates of New York which was used as a basis for tax proposals by Alaska State.
The report gave companies' costs in 1980 values on a North Slope production of 1·7 million b/d as follows.

	\$/bbl
Development and operating	1·00
Pipeline tariff	3·50
Ocean freight Valdez–California	0·50
State Royalty/Severance Taxes	1·80
	6·80

At an estimated selling price of \$13/bbl, this would give a profit of \$6/bbl.

TAPLINE A 30–31″ crude oil pipeline running the 1000 miles from Saudi Arabia's oil fields to Sidon in the Mediterranean. It has a capacity of 25 million tonnes/year and was built in 1950 at a cost of £80 million.

TAPS The Trans Alaska Pipeline operated by Alyeska Pipeline Service is owned by eight of the large oil companies including Exxon, Arco and BP. The construction of the line was subject to much delay and cost escalation. In June 1975 the companies took \$6375 million as a control estimate as the cost of construction with the intention of completing by late 1977 with an initial capacity of 1·2 million b/d.

TARAKAN (Indonesia) A crude oil of 19·5° API gravity and 0·15% sulphur exported from the Tarakan Island terminal.

TARIFF, EEC The EEC Commission has not always been successful in setting common external energy tariffs. Neither has there been unrestricted trade in energy or oil products within the Community. The French, for example, have kept a tight control on their own market through licencing systems. Rotterdam price reports frequently refer to EEC qualified material which may be allowed in tax free due to quotas allocated to countries such as Rumania. In 1975 tariffs ranged from 3·5–6% for petroleum products. *See also* Bulk Supply Tariffs.

TAR SANDS The extensive Athabasca tar sands in Canada have received considerable attention for decades but costs have always proved a deterrent for commercial exploitation on a large scale (*See* Development Costs). The word tar is a correct description because the tar or heavy oil separated from the sand is a highly aromatic, high sulphur material more closely related to a tar derived from coal than to crude oils. Since the tar is deficient in hydrogen compared with crude oil, it needs considerable addition of hydrogen (hydrogenation) in addition to the removal of sulphur and nitrogen to make an acceptable synthetic crude oil. Extensive tar sands also occur in Venezuela.

TAR SANDS, PRICE Tar sands are one of the most expensive forms of hydrocarbon reserves to produce. There is substantial cost in mining the tar sand, in extracting the oil from it and then in treating the oil to make it of acceptable quality for marketing. The oil is so high in sulphur and so low in hydrogen that it cannot be marketed without treatment. It must therefore be hydrogenated and desulphurised to manufacture a 'synthetic' crude oil. Forecasts of production costs have always run ahead of the price of conventional crude oils and the large increase in the price of crude oils in 1974 did not change the situation. In 1975 Canadian Government estimates of the price of a synthetic crude from Athabasca Tar Sands were \$13·70/bbl in 1979, rising to \$15·55 by 1985. *See also* Development Costs.

TARS Heavy crude oils, ca. 10° API gravity, are sometimes known as tars. The term is also used for a product with similar properties obtained from the distillation of coal.

TAX ALLOWANCES Tax allowances in the oil industry covered many activities but two items were of major international importance. The first is that for many years certain countries allowed overseas taxation on crude oil production to be offset against taxes that would have been incurred on domestic market activities. For many years BP's tax bill in the UK benefited enormously from this. In the US, the depletion allowance on crude oil production did much to lower the tax bill for US crude oil producers.

TAX PAID COST The tax paid cost of crude oil is the total of the cost of production and producer government taxes. The tax paid cost is not directly related to the selling price. In 1970 the posted price of a Middle East crude oil might have been about \$1·80/bbl, production costs perhaps less than 10 cents/bbl and total government taxes about \$1·00 so the tax paid cost was about \$1·10. The market price for the oil could have been about \$1·45/bbl giving the producing company a profit of \$0·35/bbl. In 1974 'Equity Crude' (*q.v.*) had a posted price \$11 to \$12/bbl.; production costs were still about 10 cents and government taxes about \$7·00/bbl. *See also* Equity Cost Oil.

TAXATION *See* Fabrication Tax, Marginal Field Aid, Royalty, Petroleum Revenue Tax (UK).

TAXATION OF PETROLEUM PRODUCTS Traditionally, petroleum products have been relatively lightly taxed in producing areas (such as North America and the Middle East) but heavily taxed in the European Markets. Consumer country taxes were quite low in the period 1930 to 1950, in the UK for example tax on motor spirit was only about 5 to 10 cents/US gallon. Taxes increased quite sharply after 1950: in 1955 approx. 25 cents/US gallon, in 1968 approx. 40 cents/US gallon, and in 1975 70 cents/US gallon. In 1975 consumer government taxes in European countries varied from about 65 cents/US gallon in Austria and Sweden to 80 cents in West Germany and 115 cents in Italy. The total price varied from about 125–170 cents/US gallon. In the US the price was about 55 cents of which only 12 cents was tax.
Taxation of diesel fuel for road vehicles was generally in line with mogas tax. Industrial gas oils in 1975 bore taxes from about 2 cents/US gallon in the US to 14 cents/US gallon in Austria. Heavy fuel oil in 1975 was taxed in the UK and in Germany.

TAXATION, US In the US Federal taxation on gasoline dates from June 1932 when it was introduced at 1 cent/gallon. Lubricating oil was taxed at 4 cents/gallon and pipeline taxes and import taxes were also introduced in 1932. In addition to the Federal taxation, State taxation was gradually introduced although by 1946 only two States had introduced taxes on lubricating oil and 23 had introduced the crude oil severance tax (*q.v.*).

TBA Tertiary Butyl Alcohol, a high octane gasoline component which also functions as an anti-icing additive. It can be manufactured as a by-product from the ethylene oxide process using isobutane feedstock.

TBA Tyres, Batteries, Accessories. Sales of these items are sometimes evaluated in filling station economic studies.

TDTW *See* DTW.

TEHERAN AGREEMENT The Teheran Agreement of 1971 was an agreement in which the 'take' levels of the producer countries were to rise steadily until 1975. It was supplemented and superseded by the Geneva Agreement, participation and the hefty increases in oil prices which took place in October 1973 and in January 1974.

TEL *See* Lead Alkyls.

TEMBUNGO (Malaysia) This offshore crude is of 37° API gravity and 0·05% sulphur. It is exported by SBM from the field.

TEMPERATURE EFFECT ON OIL AND ENERGY CONSUMPTION In a cold winter the quantity of energy used for space heating is obviously higher than the quantity which would have been used for normal weather, i.e. average temperatures. Consumption can be 20% or more up or down on that expected for normal temperatures in one month but through a year or heating season the percentage effects are usually much less. The following sensitivity data apply to all energies used for space heating (the proportion of a fuel used for space heating varies appreciably from fuel to fuel and from area to area);

There is a 1 in 5 chance of energy demand being ±4 per cent from normal in a year.
There is a 1 in 10 chance of energy demand being ±6 per cent from normal in a year.
There is a 1 in 100 chance of energy demand being ±10 per cent from normal in a year.

One of the best analyses published, based on data from 1931 to 1973, is that by the US National Oceanic and Atmospheric Administration.
Variability of seasonal total heating fuel demand in the United States September 1973. Office of Public Affairs NOAA, 6010 Executive Blvd, Rockville, Maryland 20852, US.

TENNECO The Tennessee Gas and Transmission Co. was incorporated in 1947, and became the Tenneco Oil Company in 1960. Tenneco is a US gas and oil company with relatively small interests outside the US.

TERA The symbol is T and the multiple 10^{12}.

TERMINAL PRICE This is the price quoted at a specified terminal at which dealers may buy fob.

TERTIARY RECOVERY Four methods of tertiary recovery which have been tested in oil fields are;

Miscible Flooding; the use of first hydrocarbon solvents such as LPG, naphtha or kerosine followed by water injection.
Micellar-Polymer Flooding; this is the injection of an aqueous solution of detergent polymer.
Carbon Dioxide Flooding; injection of high pressure gas.
In situ Combustion; igniting part of the oil in the reservoir to generate steam and pressure to promote oil flow.

TEST DRILLING When an oil field has been discovered, the extent of the field is estimated by geophysical and seismic methods. However test drilling is used to verify the geophysical work and provide information from which the reserves of hydrocarbons may be estimated. For example in Venezuela in the period 1969–1974 there were 374 test drillings of which 232 gave productive results.

TEXACO Texaco, one of the World's big seven integrated oil companies, took its present name in 1959. The Texas Company was incorporated in 1902 succeeded by the Texas Corporation in 1926. As with all the majors, it operates through several hundred subsidiary companies many of which use the name Texaco. In 50% partnership with Chevron it operated Caltex companies in many parts of the World. In the UK Texaco operated under the brand name of Regent before changing to Texaco.

TEXAS RAILROAD COMMISSION The most influential of the US regulatory bodies for setting state crude oil production levels. The state regulatory bodies set the production levels for the oil fields in their territories to avoid wasteful or over-production and low prices. However following the large rise in crude oil prices in 1974, there was pressure to allow higher production rates from some fields. For example, in 1974 the East Texas field was producing at 80–90% of the MER which was assessed at 240 000 b/d. However there was some debate as to whether the MER should have been 400 000 b/d.

THAILAND For crude oil production *see* Table 1.

THERMAL EFFICIENCY The amount of useful heat extracted from a fuel, expressed as a percentage of the total calorific value of the fuel consumed, is termed the thermal efficiency. With electricity generation the thermal efficiency is the total calorific value of the electricity supplied expressed as a percentage of the calorific value of the fuel consumed in the power station.

THERMAL POLLUTION This is the discharge of waste heat from industrial processes which has an adverse effect on the environment; e.g. discharge of power station cooling water into a river.

THERMAL RECOVERY The recovery of crude oil beyond that amount which can be produced by the natural pressure, by the ignition and burning of part of the reserves to generate additional pressure.

THERMAL TAR This is a heavy aromatic oil obtained from the thermal cracking operation. It is a good petroleum coke feedstock.

THISTLE A commercial UK North Sea oil field 320 miles NNE of Aberdeen. Peak production rate estimated to be 200 000 b/d, with estimated recoverable reserves at 400 million barrels. Original shareholdings (%):

Signal Oil and Gas Co. Ltd	24*
Champlin Petroleum Co. Ltd	22·5
United Canso Oil and Gas UK Ltd	20
Santa Fe Oilfield Trading Ltd	22·5
Charterhouse Securities Ltd	1
Tricentrol Ltd	10

* Later acquired by Burmah.

The crude oil is of 37° API gravity and 0·3% sulphur.

THOMAS REPORT An early report discussing the cost of nuclear power concluding that electricity from nuclear power would cost $5\frac{1}{2}$ times as much as electricity generated by thermal methods. In retrospect this, which was the highest of the early estimates, was the most realistic. *Nuclear power* in *The international control of atomic energy*, US Dept of State Publication, 2661.

THRIFT ANALYSIS *See* Energy Accounting.

TIED OIL In the UK this is oil (tax unpaid) which has been permitted to be delivered to a bonded user or bonded distributor.

TIME CHARTER A charter of a tanker for a fixed period of time, usually at least two years. A time charter could be a 'bare boat' charter in which the crew, etc. are supplied by the charterer or a time charter which could involve the owner of the vessel supplying crew and services. Time charter rates could be attractive to an oil company compared with the effective rate for a company owned vessel because of different methods of financing, building allowances, tax differentials and different operating standards.

TIME CREDITS Crude oil sale contracts frequently specify credit in terms of delayed payment. When there was pressure to sell crude oil through the 1960s, 90- and 120-day credits were common. In 1974 credits were reduced, in some instances to 30 days, but in 1975 buyers renewed pressure for extension of credit and 45 and 60 days were obtained by early 1975. At early 1975 prices each 30 days' credit was worth about 10 cents/barrel.

TIPCO This is the Thai International Petroleum Co. Ltd, the state company of Thailand.

TLP Term Limit Pricing, an agreement in which the term of the contract is specified.

TML *See* Lead Alkyls.

TONNES Metric tons. *See also* Tons.

TONS Unless further qualified could be short tons, long tons or metric tons. Short tons are commonly used in the US and long tons in the UK prior to 1974, but both units are giving way to the metric ton. When used with tankers the ton is most likely to be a long ton. A short ton contains 2000 pounds, a long ton 2240 pounds.

TOPPED CRUDE OIL A crude oil which has had its lighter components removed in a topping plant (*q.v.*).

TOPPING PLANT A simple crude oil distillation plant which removes the lighter components of crude oil such as gases, gasolines and kerosine.

TOTAL Total is the brand name of CFP (*q.v.*) and many of CFP's subsidiary companies.
In 1968 Total colours were used for the following companies by their merger into the financial structure of Desmarais Freres:

Total, Compagnie Francais de Distribution
Desmarais Freres
Ozo, SAF
Societé Champenoise des Carburants
Total Est
Societé Total CIN

This 1968 grouping became the market leader in France.

TOVALOP The Tanker Owners Voluntary Agreement concerning Liability for Oil Pollution. In 1969 about half the tanker owners signed this agreement and by August 1971 about 95% had enrolled. The liability accepted was $100/ton subject to a maximum of $10 million/tanker/incident. *See also* CRISTAL.

TOWN GAS Town gas is that which is piped to consumers from a gas plant. Town gas may be coal gas or a gas of similar composition made from naphtha. Town gas has been used in areas which did not have natural gas supplies but relatively cheap coal supplies.

TPC The Tax Paid Cost of a crude oil. It is the total government take in taxes plus the production cost (*q.v.*). The rates for royalty and income tax may vary from area to area and from time to time. For example on 1st October 1974, Middle East crudes were charged 16·67% royalty and 65·75% income tax. Up to the end of June 1974 royalty had been $12\frac{1}{2}$% and income tax 55% of the posted price.

TRAMP The term is used to describe ships which are not owned by oil companies. Such vessels may trade in the spot (single voyage) market or be on term or period charter. The term 'private owners' is now used more frequently than 'tramp'.

TRANSFER PRICE The price at which a crude oil or product is passed from one company to another within the same group of companies. Transfer prices apply to both crude oils and products.

TRANSPORTATION COSTS—PRODUCT DISTRIBUTION A detailed analysis of the cost of transporting oil was made by M. Hubbard, The Comparative Costs of Transporting Oil to and Within Europe, IP, 7th December 1966. The paper has been a useful reference document. Comparative costs at the time for product distribution were quoted in pence ≡ US cents per mile.

	*Fixed Cost**	*Mileage Cost†*
Road Vehicle	70	4
Canal Barge	45	1
Coastal Tanker	140	0·5
Pipeline 1 million ton/year	4	1
Pipeline 10 million ton/year	2	0·2
Pipeline 50 million ton/year	1	0·07

* per ton
† per ton mile

TRANS-SHIPMENT TERMINAL A terminal used for interim storage of crude oil receiving crude in VLCCs and shipping in smaller size vessels to ports which cannot take the large vessels. Examples are the Gulf terminal at Bantry Bay in Eire and the BOPEC (*q.v.*) terminal in the Caribbean.

TRIBUTARY REFINERIES The term is sometimes used to describe refineries which are capable of supplying the main product trading centres such as Rotterdam. Under certain conditions refineries in the Caribbean, Mediterranean and PG could be considered tributary refineries of Rotterdam and product prices at these other refineries could be equivalent to those at Rotterdam less the spot market freight adjustment.

TRICENTROL Trinidad Central Oil fields was registered in 1917. The name was changed to Trinidad Canadian Oil fields in 1962 and to Tricentrol in 1971. The company is engaged in oil and gas exploration and production in several areas including North America, Trinidad and the North Sea. The shares are quoted on the London Stock Exchange.

TRINIDAD AND TOBAGO For crude oil production *see* Table 1. The state oil company is Trintoc (*q.v.*).

TRINTOC This is the Trinidad and Tobago Oil Company; the state company which was formerly Shell Trinidad. The company came under state ownership in August 1974.

TRIPARTITE DECLARATION This declaration, issued on 25th May 1950 by France, the UK and the US, recognised that the Arab States and Israel needed to maintain a certain level of armed forces for their internal security and legitimate self defence. It was an attempt to limit an arms race in the Middle East and introduce stability.

TRUCIAL SHEIKDOMS *See* United Arab Emirates.

TUJMAZA (USSR) A crude oil of 34° API gravity, 1·5% sulphur exported from Novorossijsk.

TUNISIA For crude oil production *see* Table 1.

TURKEY For crude oil production *see* Table 1.

TURKISH PETROLEUM CORPORATION The Turkish state oil concern which was formed in 1954. Its major activities are confined to Turkey.

TURPS SUBSTITUTE *See* SBP.

TWO-STROKE FUEL This is a mixture of motor spirit with 5–10% lub oil which may or may not contain an additive.

U

UCPTE The Union for the Co-ordination of Production and Transmission of Electricity.

UGP The French state owned Union Génèrale des Pétroles.

UK For crude oil production *see* Table 1. The state company is BNOC.

UK ENERGY STATISTICS *See Digest of UK Energy Statistics and Energy Trends.*

ULCC An Ultra Large Crude Carrier; a vessel of 320 000 tons dwt or more. Such vessels are most suitable for operation on specified routes such as the Kuwait–Bantry Bay route. ULCCs are unable to pass through the Dover Straits or the Straits of Malacca.

ULLAGE The space in a tank not occupied by the oil. Ullage figures are of course quoted as a measure of the ability of a tank or tanks to hold further stocks.

ULTIMATE RECOVERABLE RESERVES These are defined as the total quantity of crude oil, natural gas or natural gas liquids estimated to be ultimately producible from an oil or gas field as determined by an analysis of geological and engineering data.

ULTRAMAR The British company Ultramar was formed in 1935 for oil exploration. It became an integrated oil company in the 1960s and it operates in many areas of the World. 'Golden Eagle' is the name used by some of its subsidiary companies. The shares are quoted on the London Stock Exchange.

UMM SHAIF (Abu Dhabi) An Abu Dhabi onshore crude of 37° API gravity, 1·4% sulphur exported from Das Island.

UN SERIES J One of the most comprehensive series of worldwide energy statistics; published annually by the UN, New York.

UNES The United Nations Environment Secretariat.

UNFINISHED OILS A term, used mostly in the US, describing oils that need further blending or processing before being classified as (finished) products. Prices are not quoted; approximate values could be assigned as the value of crude oil at the specific location.

UNICE Union des Industries de la Communauté Europeene; the organisation of European industrialists officed in Brussels.

UNIFIED PRICE Following participation and the emergence of two prices for oil, the posted price and the equity price for the former owners, dissatisfaction with two prices led to the proposal for a single 'unified' price. *See* Crude Oil Prices, Unitary Crude.

UNION OIL OF CALIFORNIA The company was incorporated in 1890 and is a medium size integrated US oil company. It is, as the name suggests, strong in California but also operates in Canada and South Korea.

UNION PETROLIERE The Swiss oil industry's trade association.

UNIONE PETROLIFERA The Italian oil trades Organisation which is one of the main sources of statistics on the Italian oil industry.

UNIT OF ACCOUNT A proposed 'unit of account' for denominating oil transactions was discussed at OAPEC's Baghdad meeting of May 1976.
See also EUR.

UNITARY CRUDE *See* Crude Oil Prices, Unitary Crude.

UNITED ARAB EMIRATES This state was formed on 2nd December 1971 and comprises Abu Dhabi, Dubai, Sharjah, Ajman, Umm al Quwain and Fujairah. The seventh Trucial Sheikhdom, Ras al Khaimah, did not join at the outset but had applied for membership before the Emirate was formed. *See also* League of Arab States.

UNLEADED GASOLINE *See* Lead Free Gasoline.

UNSWEETENED OILS *See* Sour Oils.

UPSTREAM Activities stated to be upstream are with reference to the refinery, i.e. crude oil production and crude oil transportation operations.

URUGUAY The state oil company is ANCAP.

US For crude oil production *see* Table 1.

US FUEL OIL GRADES There are six grades of fuel oil specified in ASTM fuel oil specifications.

No. 1 Kerosine type } distillates
No. 2 Gas oil type }

No. 3 Not now used
No. 4 Fuel oil for installations without preheating
No. 5 Fuel oil for installations with preheating
No. 6 Fuel oil for installations with facilities for thickest grades.

USAC The US Atlantic Coast; the abbreviation is used in the tanker industry.

USAEC The United States Atomic Energy Commission.

USMC The US Maritime Commission. *See* Freight Rate Schedules.

USNH The United States North of Cape Hatteras. Tanker rates are quoted on a daily basis in many trade journals for key routes such as PG–USNH, Rotterdam–USNH and Mediterranean–USNH.

USSR For crude oil production *see* Table 1. Although most sources agree on 1863 as the first year of production, there are reports which give earlier dates. For example, production in 1860 is quoted by Réclus, (*Universal Geography*, Virtue, 1880) as: Petroleum 568 100 centals, Coal 7 387 700 centals (a cental = 100 pounds).
The USSR does a substantial trade in crude oil and oil products, with the rest of the world, on balance being an importer.
Russia has substantial reserves of oil and gas but no reliable estimate of them has yet been produced. A major problem in developing Russia's reserves is their remoteness. Two Russian-owned oil companies are ROP (now defunct) and NAFTA.

USWC The US West Coast; the abbreviation is used in the tanker industry.

V

VACUUM OIL COMPANY *See* Mobil Oil Corporation.

VALVOLINE *See* Ashland.

VANADIUM Vanadium compounds occur extensively in crude oils in low concentrations and their presence, as with nickel and sodium compounds, is undesirable since they can cause corrosion and poison catalysts. However it is not usual to see price differentials between crude oils or products on account of these metals being present but maximum permitted levels are stated in many quality specifications.

VEBA It is 40% owned by West German Government. Originally formed in 1935 as Scholven-Chemie AG. *See* Aral.

VECTRA The Vectra Corporation, whose principal business is synthetic fibres in the US, is owned by Standard Oil of California.

VENEZUELA Venezuela has produced oil since 1917 and is a founder member of OPEC. One of the world's largest oil reserves is the Orinoco Tar Belt which measures approximately 370 miles from east to west and 35 miles from north to south. Although the first exploration well was drilled in 1935, by 1961 only 42 wells had been drilled. From these wells it was estimated that reserves were 700 000 million barrels of 8–12% API gravity; these reserves in 1961 were three times all other Venezuelan reserves. For crude oil production *see* Table 1.

VENEZUELAN OIL CORPORATION (CVP) This was formed on 19th April 1960 and was the first Venezuelan state owned oil company. By 1974 the Corporation was responsible for about 3% of the hydrocarbon output in Venezuela.

VENPROTEINAS A company formed in 1975 to study protein manufacture at Puerto la Cruz, Venezuela. The full name is Bio Proteinas de Venezuela and is owned 60% by the Venezuelan Government, 20% by BP and 20% by private Venezuelan interests.

VIGZOL A small UK oil company registered in 1928. The name continued in use until 1964 when it was changed to Amoco (UK) Ltd, a wholly owned subsidiary of Standard Oil of Indiana (*q.v.*).

VIRGIN NAPHTHA A straight run distillate from crude oil; the naphtha boiling range is about 170–240 °C. It may be used for blending jet kerosine and gasoline fuels. A virgin naphtha will not include materials from a cracker, i.e. olefin content is zero. In bulk its value will be similar to jet gasoline or kerosine values but special naphthas used for solvents will be priced according to the package size and transport costs.

VIRGIN STOCK Mineral oils which have not been processed, or used and subsequently recovered (e.g. lub oils), are sometimes termed 'virgin'.

VLCC A very large crude carrier. Defined in AFRA as an LR3 (160 000–319 999 tons dwt). A handy sized VLCC, applied to the smaller vessels in the category approximately 220 000 tons dwt. For cost of operation *see* Large Range Tanker.

W

WAFRA CRUDE OIL *See* Burgan.

WALIO (Indonesia) A crude oil of 35° API gravity, 0·7% sulphur exported from the Kasim terminal, West Irian.

WATER GAS A gas made by alternately blowing air and water over an incandescent bed of coke. It is composed mainly of hydrogen and carbon monoxide. The calorific value of water gas is about 30% that of natural gas. *See also* Carburetted Water Gas.
Water gas is not marketed and is of relatively little importance.

WATER INJECTION *See* Secondary Production.

WEATHER EFFECTS *See* Temperature Effect on Oil and Energy Consumption.

WECPA West European Coal Producers Association. *See also* CEPCEO. For publications of oil and energy forecasts *see* Energy Policy.

WEPCO The Western Desert Petroleum Company, an exploration company formed in 1967, 50% EGPC and 50% Phillips Petroleum.

WEST AFRICA The following countries are included: Africa, Angola, Burundi, Cabinda, Cameroon Republic, Canary Islands, Cape Verde Islands, Central African Republic, Chad, Congo, Dahomey, Gabon, Gambia, Ghana, Guinea, Ivory Coast, Liberia, Mali, Mauritania, Niger, Nigeria, Portuguese Guinea, Rwanda, Sao Tome and Principe, Senegal, Sierra Leone, Spanish Guinea, Spanish Sahara, Toga, Upper Volta and Zaire.

WESTERN EUROPE The following countries are included: Austria, Belgium, Cyprus, Denmark, Eire, Finland, France, Gibraltar, Greece, Iceland, Italy, Liechtenstein, Luxembourg, Malta, Netherlands, Norway, Portugal, Spain, Sweden, Switzerland, Turkey, UK, West Germany and Yugoslavia.

WHITEHALL PETROLEUM A UK company registered in 1964 with interests in the North Sea. The company is controlled by the Cowdray family which half a century before was responsible for the oil industry in Mexico.

WHITE OILS Light crude oils with a very high percentage of light distillates. Such a crude oil might be condensate from a gas field. White oils may be the class of oils which includes all gasolines and light distillates and kerosines, i.e. oils which are water white in colour. Alternatively they may be highly refined oils in the lubricating oil class which have been highly refined to remove all coloured materials; the aromatics and heterocyclics. The resultant white oil, composed of isoparaffins and naphthenes, is odourless and tasteless. Liquid paraffin is one of the white oils prepared from stock in the lubricating oil range.

WHITE PAPER A British Government document laying down policy as opposed to a Green Paper which is for discussion. A Brown Paper is a rather more colloquial description of a UK Department of Energy document. The following White Papers have been concerned with Energy Policy:

Cmd. 3201, 1928 (Ward)
Cmd. 6762, 1946 (Simon)
Cmd. 7664, 1948 (Clow)
Cmd. 8647, 1952 (Ridley)

WHITE SPIRIT A straight run distillate in the naphtha boiling range (150–200 °C) which is used as a thinner and solvent in paints. Frequently known as 'Turps substitute'. In the US it is also sometimes referred to as 'Petroleum Spirit' but in the UK this is motor gasoline. Price in bulk should be approximately that of jet gasoline or kerosine for a low aromatics content (ca 15% aromatics), or similar to that for premium motor spirits for a high aromatics content (ca 45%) material.

WILD CATTING OR WILDCAT WELL The drilling of a well into a structure which may or may not contain hydrocarbons. When hydrocarbons have been proved to be present by drilling, subsequent wells are known as development or step out wells.

WINDFALL PROFITS Profits brought about by increased prices which are not accompanied by increased costs. Such profits are usually looked on as unjustified and are sometimes subject to additional taxation.

WINTERSHALL Wintershall Aktiengesellschaft was incorporated in Germany in 1921. Wintershall holds a 15% interest in Aral and an 18·5% interest in Deminex. Wintershall has small exploration interests outside Germany.

WOCANA The World Outside the Communist Bloc And North America.

WOOD Most quoted energy statistics are for mineral fuels (*q.v.*) but unless there is a specific disclaimer, fuelwood may quite possibly be included in the solids category. *See also* Calorific Value *and* Non-Commercial Fuels.

WORLD PETROLEUM CONGRESS The papers presented at the WPC represent contemporary industry views and are sometimes a useful source of economic information. Congresses have been held in London 1933, Paris 1937, The Hague 1951, Rome 1955, New York 1959, Frankfurt 1963, Mexico City 1967, Moscow 1971 and Japan 1975.

WORLD TRADE Petroleum has long been a major component of world trade. In 1974 there was a massive increase in the value of world trade by 44% to $825 billion and about half of this increase was due to the much higher price of oil in 1974 over that of 1973. The volume increase however was only 5%. In 1973 oil producers spent about $20 billion whilst in 1974 expenditure increased to $31 billion. Even with this increased expenditure, OPEC countries were left with a surplus of more than $50 billion in 1974.

WS Worldscale; *see* Tanker Rate Schedules.

Y

YAKUTSK PACT A pact concluded in November 1974 by the USSR, Japan and the US for a two year development programme in the Yakutsk area of Eastern Siberia. The estimated cost was $400 million.

YAOUNDE CONVENTION A convention whereby the EEC of Six entered into a free trade area type of arrangement with Associated Overseas Territories. These were: Burundi, Cameroon Republic, Central African Republic, Congo (Brazzaville), Congo (Leopoldville), Dahomey, Gabon, Ivory Coast, Malagasy, Mali, Mauritania, Niger, Rwanda, Senegal, Somalia, Tehad, Toga and Upper Volta.

YIELD ANALYSIS A technique by which the value of a barrel of crude oil is calculated using yield data and product data applicable to a particular market. Thus the value of a particular crude oil will change from one market to another and yield analysis demonstrates which is the most attractive purchase to a refiner/marketer. The value calculated from a yield analysis of course depends not only on the refining pattern and product value but also on the data used for refining cost and freight rate.

YIELD OF PRODUCTS A typical crude oil would have the following yield structure

	% *Weight*
Gases, Light Distillates (Gasolines)	20
Kerosines, Gas Oils	34
Fuel Oils, Lubs, Bitumens	40
Refinery Fuel and Loss	6
	100

Such a yield would be satisfactory for marketing in Western Europe but in the US a much higher yield of gasoline is required so the yield structure would be modified to 50% gasolines and only 10% fuel oils. Such a yield shift is achieved by cracking; catalytic cracking and/or thermal cracking and/or hydrocracking. The higher prices obtained for gasolines compared with fuel oils justify the expense of installing and operating cracking capacity.

YIELD SHIFT *See* Yield of Products.

YPFA Yacimientos Petroliferos Fiscales (Argentine), the Argentine Government Company which is engaged in all aspects of the oil industry in the Argentine.

YPFB Yacimientos Petroliferos Fiscales Bolivianos. This is the state oil company of Bolivia.

YUGOSLAVIA For crude oil production *see* Table 1.

Z

ZAKUM (Abu Dhabi) An offshore crude of 40° API gravity, 1·0% sulphur exported from Das Island.

ZARZAITINE (Algeria) A crude oil of 42° API gravity, 0·1% sulphur exported from the La Skhirra terminal in Tunisia.

ZUEITINA (Libya) A crude oil of 40° API gravity, 0·25% sulphur exported by SBM at Zueitina.

ZURICH GROUP This term has been used to describe the state backed/owned oil agencies of Austria, Japan, Spain and West Germany.

TABLES

Table 1 CRUDE OIL PRODUCTION WORLDWIDE (Thousands b/d)

Country	*First Year of Production*	*Production*				
		1900	1925	1950	1960	1974
Abu Dhabi (UAE)	1962	—	—	—	—	1 414
Afghanistan	1971	—	—	—	—	neg
Albania	1933	—	—	8	14	48
Algeria	1914	—	neg	neg	184	1 038
Angola	1956	—	—	—	1	20
Argentina	1907	—	17	64	175	415
Australia	1964	—	—	—	—	379
Austria	1933	—	—	28	46	44
Bahrain	1932	—	—	30	45	67
Barbados	1966	—	—	—	—	neg
Bolivia	1930	—	—	2	10	48
Brazil	1940	—	—	1	81	175
Brunei	1931	—	—	85	97	220
Bulgaria	1955	—	—	—	4	4
Burma	1889	neg	20	1	11	21
Canada	1862	2	1	80	518	1 999
Chile	1950	—	—	2	20	29
China, People's Republic	1939	—	—	2	62	1 000
Colombia	1921	—	3	93	152	171
Congo	1960	—	—	—	1	40
Cuba	1933	—	—	neg	neg	3
Czechoslovakia	1919	—	neg	1	3	4
Denmark	1972	—	—	—	—	2
Dubai (UAE)	1969	—	—	—	—	242
Ecuador	1917	—	neg	7	7	162
Egypt	1911	—	3	45	65	238
France	1918	—	1	3	39	36
Gabon	1957	—	—	—	15	187
Germany, East	1930	—	—	neg	4	4
Germany, West	1880	1	1	22	109	131
Hungary	1937	—	—	10	25	42
India	1889	3	2	5	9	149
Indonesia	1893	6	59	133	418	1 397
Iran	1913	—	93	664	1 054	6 022
Iraq	1927	—	—	136	967	1 934
Israel	1956	—	—	—	3	1
Italy	1860	neg	neg	neg	37	18
Japan	1875	2	5	6	10	14
Kuwait	1946	—	—	344	1 624	2 548
Libya	1961	—	—	—	—	1 526
Malaysia	1910	—	12	1	1	100
Mexico	1901	—	316	198	271	624
Morocco	1932	—	—	1	2	1
Netherlands	1943	—	—	13	37	29
New Zealand	1933	—	—	neg	neg	4
Nigeria	1958	—	—	—	18	2 332
Norway	1971	—	—	—	—	34
Oman	1967	—	—	—	—	295
Pakistan	1922	—	1	4	7	9
Peru	1896	1	25	41	53	73
Poland	1874	6	16	3	4	10
Qatar	1949	—	—	34	172	518

Table 1—*contd.*

Country	*First Year of Production*	*Production*				
		1900	1925	1950	1960	1974
Romania	1857	4	46	88	234	290
Saudi Arabia	1936	—	—	547	1 247	8 483
Senegal	1960	—	—	—	neg	—
Sharjah (UAE)	1974	—	—	—	—	27
Spain	1966	—	—	—	—	39
Syria	1968	—	—	—	—	127
Taiwan	1876	neg	neg	neg	neg	3
Thailand	1964	—	—	—	—	neg
Trinidad and Tobago	1908	—	12	57	116	180
Tunisia	1966	—	—	—	—	86
Turkey	1948	—	—	neg	7	65
United Kingdom	1919	—	neg	1	2	2
USA	1859	174	2 092	5 407	7 035	8 795
USSR	1863	208	144	748	2 949	9 102
Venezuela	1917	—	54	1 498	2 846	2 975
Yugoslavia	1935	—	—	2	18	71

neg—Negligible, i.e. production of less than 500 b/d.
UAE—United Arab Emirates.
A dash represents zero production.
The figures given are for crude petroleum only—shale oils and condensates are excluded.

Table 2 WORLD PRODUCTION OF CRUDE OIL (Excluding NGLs) (Million tonnes)

Year	North America		Caribbean	South America	Western Europe	USSR, E. Europe and China	Africa	Near and Middle East	Far East	Total	Year
	USA	Rest									
1860	68	—	—	—	—	2	—	—	—	70	1860
1865	343	14	—	—	—	14	—	—	—	371	1865
1870	722	33	—	—	—	41	—	—	—	796	1870
1875	1 206	28	—	—	—	132	—	—	1	1 367	1875
1880	3 609	47	—	—	1	466	—	—	4	4 127	1880
1885	3 000	33	—	—	6	2 029	—	—	4	5 072	1885
1890	6 291	102	—	—	15	4 132	—	—	23	10 563	1890
1895	7 263	93	—	—	21	6 835	—	—	251	14 463	1895
1900	8 619	124	—	38	54	11 128	—	—	582	20 545	1900
1905	18 250	125	—	61	87	9 077	—	—	1 867	29 467	1905
1910	28 389	593	20	183	156	12 927	—	—	2 649	44 917	1910
1915	38 082	5 020	106	415	108	11 820	35	463	3 350	59 390	1915
1920	60 006	23 846	224	631	97	5 381	143	1 628	3 925	95 881	1920
1925	104 627	17 668	3 652	2 167	152	10 486	179	4 162	5 067	148 610	1925
1930	123 301	6 247	24 424	3 171	257	25 307	278	6 143	8 047	197 175	1930
1935	136 637	6 277	26 393	4 612	550	35 495	173	11 465	8 891	230 713	1935
1940	183 325	7 066	34 336	4 979	1 163	37 376	911	13 004	11 072	293 232	1940
1945	231 809	7 225	53 554	5 609	1 120	37 509	1 320	25 650	2 033	365 329	1945
1950	271 109	14 089	87 559	5 800	3 668	42 834	2 380	86 076	11 649	525 134	1950
1955	333 569	30 254	122 005	7 029	9 660	84 525	1 932	180 277	13 723	768 924	1955
1960	350 331	40 250	162 569	17 452	15 386	167 200	13 790	261 836	27 055	1 055 869	1960
1965	387 553	56 479	198 060	24 182	22 071	268 800	107 209	416 041	33 133	1 513 528	1965
1970	478 565	89 624	212 124	34 859	22 744	390 600	298 875	687 601	67 492	2 282 484	1970
1973	457 266	120 740	193 613	47 536	22 600	480 200	290 061	1 045 900	110 023	2 767 939	1973
1974	436 803	118 806	173 349	45 138	22 596	520 700	272 489	1 076 214	111 704	2 777 799	1974

Table 3 WORLD OIL PRODUCTION (Million b/d)

Country/Area		1967	1968	1969	1970	1971	1972	1973	1974
TOTAL WORLD		37·0	40·3	43·7	47·8	50·3	52·9	57·6	57·9
TOTAL NON-COMMUNIST WORLD		30·7	33·5	36·5	40·0	42·0	44·1	48·0	47·4
TOTAL OPEC		17·8	18·8	20·9	23·4	25·3	27·1	30·8	30·4
MIDDLE EAST	Iran	2·6	2·8	3·4	3·8	4·6	5·0	5·9	6·0
	Iraq	1·2	1·5	1·5	1·6	1·7	1·5	2·0	1·9
	Kuwait	2·3	2·4	2·6	2·7	2·9	3·0	2·8	2·3
	Qatar	0·3	0·3	0·4	0·4	0·4	0·5	0·6	0·5
	Saudi Arabia	2·6	2·8	3·0	3·6	4·5	5·7	7·3	8·2
	Abu Dhabi	0·4	0·5	0·6	0·7	0·9	1·1	1·3	1·4
	Others	0·6	0·9	0·9	1·0	1·2	1·2	1·2	1·4
	TOTAL	10·0	11·2	12·4	13·8	16·2	18·0	21·1	21·7
of which Gulf Shipments		8·8	9·6	10·9	12·4	14·7	16·7	19·4	20·5
AFRICA	Algeria	0·8	0·9	1·0	1·0	0·8	1·1	1·1	1·0
	Libya	1·7	2·6	3·1	3·3	2·8	2·2	2·2	1·5
	Nigeria	0·3	0·1	0·5	1·1	1·5	1·8	2·1	2·3
	Gabon	0·1	0·1	0·1	0·1	0·1	0·1	0·2	0·2
	Others	0·3	0·3	0·5	0·7	0·6	0·6	0·4	0·6
	TOTAL	3·2	4·0	5·2	6·2	5·8	5·8	6·0	5·6
SOUTH EAST ASIA	Indonesia	0·5	0·6	0·8	0·9	0·9	1·1	1·3	1·4
	Others	0·1	0·1	0·1	0·1	0·2	0·2	0·3	0·3
	TOTAL	0·6	0·7	0·9	1·0	1·1	1·3	1·6	1·7
TOTAL OTHER EASTERN HEMISPHERE		0·2	0·2	0·2	0·4	0·5	0·6	0·6	0·6
TOTAL WESTERN EUROPE		0·4	0·5	0·5	0·5	0·4	0·4	0·4	0·4
NORTH AMERICA	USA (inc. NGL)	10·2	10·6	10·8	11·3	11·2	11·2	10·9	10·6
	Canada	1·1	1·2	1·3	1·5	1·6	1·8	2·1	1·9
	Mexico	0·4	0·4	0·5	0·5	0·5	0·5	0·5	0·6
	TOTAL	11·7	12·2	12·6	13·3	13·3	13·5	13·5	13·1
LATIN AMERICA	Venezuela	3·6	3·6	3·6	3·8	3·6	3·3	3·5	3·0
	Ecuador	—	—	—	—	—	0·1	0·2	0·2
	Others	1·0	1·1	1·1	1·0	1·1	1·1	1·1	1·1
	TOTAL	4·6	4·7	4·7	4·8	4·7	4·5	4·8	4·3
SOVIET BLOC	USSR	5·8	6·2	6·6	7·1	7·5	7·9	8·5	9·1
	Eastern Europe	0·3	0·3	0·3	0·3	0·3	0·3	0·3	0·4
	China	0·2	0·3	0·3	0·4	0·5	0·6	0·8	1·0
	TOTAL	6·3	6·8	7·2	7·8	8·3	8·8	9·6	10·5

Table 4 TOTAL CONSUMPTION—OIL (NET) (Million tonnes)

	1955	1956	1957	1958	1959	1960	1961	1962	1963	1964	1965	1966	1967	1968	1969	1970	1971	1972	1973	1974
WESTERN HEMISPHERE																				
U.S.A.	415.6	433.0	434.0	443.0	460.0	473.1	478.1	498.2	513.0	527.2	548.9	575.7	595.8	635.5	667.8	694.6	719.3	775.8	818.0	785.4
Canada	30.5	36.0	37.0	38.0	41.0	42.7	43.9	44.1	47.3	51.5	55.0	58.0	61.7	66.1	69.1	73.0	75.8	79.3	83.7	88.1
Mexico						13.5	14.2	13.7	15.3	16.1	17.5	18.6	18.6	19.5	21.5	22.9	24.3	27.7	29.6	31.1
Caribbean	62.0	66.0	73.0	76.0	79.0	31.3	35.3	31.9	34.7	36.3	38.6	42.4	44.3	47.2	48.1	53.2	56.6	65.0	66.7	70.4
S. America						35.4	36.5	38.1	39.9	41.6	43.1	45.8	47.7	52.2	57.2	61.1	66.7	68.9	73.6	77.4
TOTAL WESTERN HEMISPHERE	508.1	535.0	544.0	557.0	580.0	596.0	608.0	626.0	650.2	672.7	703.1	740.5	768.1	820.5	863.7	904.8	942.7	1014.7	1071.6	1052.4
EASTERN HEMISPHERE																				
W. EUROPE																				
W. Germany	13.2	16.0	17.0	21.0	27.0	34.0	41.0	50.2	59.6	69.0	79.5	89.2	92.1	103.8	117.1	128.6	133.5	140.9	149.7	134.4
Netherlands	12.2	15.0	15.0	16.0	17.0	13.0	15.0	17.2	19.4	22.4	25.3	27.4	28.0	30.1	32.9	36.5	36.0	40.1	41.3	35.4
Belgium/Luxembourg						8.1	9.1	10.9	12.6	14.7	16.9	17.5	19.5	21.9	25.1	27.9	28.4	31.1	31.5	27.5
France	20.3	23.0	23.0	25.0	25.0	28.1	31.0	35.8	42.3	47.5	53.9	57.7	66.2	71.8	83.0	94.3	102.8	114.1	127.3	120.1
Italy	12.2	14.0	15.0	17.0	19.0	24.2	28.8	35.0	40.4	47.6	52.3	57.7	63.7	70.3	77.3	87.3	93.8	98.2	103.6	100.7
United Kingdom	29.5	31.0	30.0	37.0	43.0	49.7	52.6	56.8	61.7	67.3	73.8	79.0	83.2	89.8	95.7	101.8	103.1	109.7	113.4	105.8
Eire	← Included in ≠ →					1.4	1.6	1.8	2.0	2.2	2.4	2.7	3.0	3.4	3.7	4.1	4.5	5.0	5.4	5.2
Denmark						5.1	5.7	7.0	8.0	9.0	10.3	11.8	12.4	13.8	16.5	18.6	18.6	19.5	17.9	16.1
Norway	14.2	16.0	15.0	17.0	19.0	3.8	4.0	4.0	4.6	4.8	5.2	5.8	6.1	6.7	7.4	8.3	8.2	8.5	8.6	7.7
Sweden						12.8	12.7	13.9	15.8	17.0	19.0	21.5	21.0	23.7	26.8	29.9	28.3	28.6	29.4	27.3
Finland						2.6	2.9	3.3	3.9	4.9	5.7	6.9	7.2	8.2	9.6	10.8	11.1	11.9	13.3	11.7
Portugal						1.7	1.9	2.0	2.2	2.5	2.7	2.9	3.2	3.4	3.8	4.6	5.4	5.8	6.5	6.5
Spain						5.2	5.8	6.7	8.2	9.5	11.4	13.2	16.0	17.8	20.7	24.1	27.6	29.8	36.3	38.2
Switzerland						3.8	4.1	5.0	6.6	7.1	8.0	8.6	9.3	10.3	11.2	12.5	13.3	13.6	14.6	13.0
Austria						3.0	3.3	3.9	4.6	5.2	5.6	6.1	6.5	7.5	8.2	9.3	10.3	10.8	11.5	10.7
Greece	15.2≠	18.0≠	19.0≠	20.0≠	23.0≠	2.6	2.9	3.3	3.6	4.0	4.4	4.8	5.5	5.9	6.1	6.7	7.4	8.6	10.0	9.1
Turkey						1.8	2.0	2.6	3.0	3.4	5.0	4.5	5.6	6.4	7.1	7.7	9.0	10.0	12.4	12.5
Iceland						0.4	0.3	0.4	0.4	0.4	0.5	0.5	0.5	0.5	0.5	0.6	0.5	0.6	0.7	0.6
Malta/Gibraltar/Cyprus						0.7	0.7	0.7	0.8	0.8	0.7	0.7	0.8	0.9	1.0	1.1	1.2	1.3	1.4	1.4
Yugoslavia						1.5	1.5	1.6	2.0	2.7	3.2	3.4	4.1	4.8	6.0	7.0	9.0	10.1	11.3	12.4
Canary Islands						1.2	1.2	1.5	1.6	2.5	2.8	3.7	4.4	3.8	3.9	4.0	3.3	2.8	2.9	2.9
TOTAL W. EUROPE	116.8	133.0	134.0	153.0	173.0	204.7	228.1	263.6	303.3	344.5	388.6	425.6	458.3	504.8	563.6	625.7	655.3	701.0	749.0	699.2
Middle East						26.4	29.5	29.5	31.5	34.9	35.3	39.3	39.0	41.1	47.2	49.5	54.9	59.6	64.6	69.7
Africa	49.8	57.0	63.0	64.0	71.0	22.8	24.4	24.5	26.7	27.3	28.4	31.2	34.9	36.4	39.0	41.6	44.2	44.5	47.5	47.0
S. Asia						12.1	12.7	12.9	15.0	16.4	17.6	18.7	19.8	23.0	27.1	27.1	28.0	30.8	33.2	31.5
S.E. Asia						15.3	18.4	22.7	22.1	24.2	24.9	28.3	41.4	47.3	52.0	58.3	60.6	68.8	74.7	79.2
Japan	10.2	13.0	16.0	17.0	23.0	30.0	41.0	48.8	62.6	75.2	87.9	100.0	122.9	142.7	169.0	199.1	219.7	234.3	268.3	261.1
Australia	9.1	10.0	10.0	11.0	12.0	10.9	11.8	12.2	13.7	15.4	17.6	18.1	20.1	21.7	23.2	24.8	25.9	26.2	28.1	29.2
New Zealand/SWPI						2.1	2.2	2.4	2.5	3.0	3.1	3.6	3.2	3.8	4.0	4.0	4.1	4.4	4.7	4.7
TOTAL E.HEMISPHERE (NON-COMMUNIST)	185.9	213.0	223.0	245.0	279.0	324.3	368.1	416.6	477.4	540.9	603.4	664.8	739.6	820.8	925.1	1030.1	1092.7	1169.6	1270.1	1221.6
TOTAL NON-COMMUNIST WORLD	694.0	748.0	767.0	802.0	859.0	920.3	976.1	1042.6	1127.6	1213.6	1306.5	1405.3	1507.7	1641.3	1788.8	1934.9	2035.4	2184.3	2341.7	2274.0
COMMUNIST WORLD																				
U.S.S.R.						119.1	128.4	143.5	157.6	169.1	180.4	193.2	210.7	225.7	241.4	262.1	273.8	296.1	317.7	341.8
E. Europe	82.3	94.0	108.0	120.0	132.0	25.6	27.8	31.3	24.7	27.3	30.4	34.3	38.1	43.1	49.2	54.0	59.3	65.5	74.8	78.3
China etc.									10.5	10.5	11.8	14.2	12.9	15.2	17.0	22.8	28.0	32.1	41.7	48.8
TOTAL COMMUNIST WORLD	82.3	94.0	108.0	120.0	132.0	144.7	156.2	174.8	192.8	206.9	222.6	241.7	261.7	284.0	307.6	338.9	361.1	393.7	434.2	468.9
TOTAL WORLD	776.3	842.0	875.0	922.0	991.0	1065.0	1132.3	1217.4	1320.4	1420.5	1529.1	1647.0	1769.4	1925.3	2096.4	2273.8	2396.5	2578.0	2775.9	2742.9

Table 5 TOTAL CONSUMPTION—SOLID FUELS (Million tonnes oil equivalent. Oil = 10 000 kcal/kg)

	1955	1956	1957	1958	1959	1960	1961	1962	1963	1964	1965	1966	1967	1968	1969	1970	1971	1972	1973	1974
WESTERN HEMISPHERE																				
U.S.A.	282.6	287.1	273.5	242.6	241.1	251.5	246.8	254.0	268.1	281.9	297.7	313.6	308.3	320.9	327.1	329.5	316.3	316.6	335.0	331.9
Canada	20.5	22.5	19.6	15.3	14.6	14.1	12.8	12.9	13.7	15.0	16.2	16.1	15.4	16.1	15.6	17.0	17.0	16.8	16.4	15.0
Mexico	0.9	1.1	1.1	0.6	0.7	0.8	0.7	0.8	0.9	0.9	0.7	1.1	1.2	1.4	1.5	1.4	1.8	2.7		
Caribbean	1.4	1.5	1.8	1.8	1.9	1.9	2.1	2.2	2.4	2.3	2.4	2.0	2.4	2.5	2.6	2.8	2.9	2.6	} 16.4	} 14.0
S. America	5.0	5.1	4.8	4.9	4.8	4.1	3.9	3.9	3.9	4.1	4.4	4.8	4.9	4.9	5.4	5.7	5.6	8.5		
TOTAL WESTERN HEMISPHERE	310.4	317.3	300.8	265.2	263.1	272.4	266.3	273.8	289.0	304.2	321.4	337.6	332.2	345.8	352.2	356.4	343.6	347.2	367.8	360.9
EASTERN HEMISPHERE																				
W. EUROPE																				
W. Germany	115.2	120.1	121.3	111.1	105.7	112.1	108.4	110.4	114.7	110.9	102.9	96.0	89.1	90.9	94.1	92.0	85.4	81.5	87.9	82.9
Netherlands	12.4	12.9	12.4	11.1	10.6	11.0	10.7	11.4	12.1	10.9	9.6	8.6	8.3	7.7	6.6	5.0	3.6	3.2	3.1	3.3
Belgium/Luxembourg	21.9	22.7	22.1	18.8	19.5	20.0	20.0	20.8	21.5	20.1	19.2	17.5	17.2	17.7	17.3	16.4	13.5	13.1	11.0	11.6
France	48.0	54.4	55.8	51.1	48.4	48.3	48.3	50.0	51.1	51.2	47.0	43.9	43.6	41.8	41.5	39.1	35.1	32.3	26.3	27.0
Italy	8.1	8.4	8.7	7.2	6.8	8.0	8.1	8.8	8.8	8.0	8.2	8.4	9.8	8.4	8.9	8.9	8.4	11.3	7.6	9.2
United Kingdom	153.7	152.1	147.9	140.3	130.4	137.7	135.9	134.0	137.7	132.5	129.8	123.9	114.0	112.5	111.2	106.6	101.0	75.3	78.5	68.9
Eire	1.6	1.3	2.4	2.0	2.9	2.5	2.7	2.8	2.7	2.7	2.2	2.5	2.6	3.2	2.9	2.6	2.4	1.4	1.8	1.8
Denmark	4.7	4.4	3.6	3.8	3.6	4.0	3.8	4.2	4.3	3.9	3.6	3.8	3.3	3.4	2.8	2.5	1.6	1.5	2.3	2.7
Norway	1.1	1.2	0.8	0.8	0.7	0.8	0.8	0.8	0.7	0.9	0.9	0.8	0.8	0.9	1.1	1.1	0.9	0.9	0.6	0.7
Sweden	4.0	3.8	3.6	2.4	2.3	2.6	2.4	2.4	2.3	2.5	2.1	2.1	1.8	1.9	1.7	1.8	1.8	3.3	4.1	3.9
Finland	1.9	1.9	2.1	1.7	2.0	2.2	2.1	2.1	1.7	2.2	2.0	1.9	1.8	2.2	2.6	2.5	2.3	6.3	4.4	4.1
Portugal	0.6	0.6	0.7	0.6	0.7	0.6	0.8	0.7	0.7	0.8	0.8	0.8	0.8	0.6	0.7	0.5	0.5	0.9	1.2	1.1
Spain	9.7	9.8	11.4	11.1	10.4	10.6	11.0	11.2	11.2	10.7	11.2	10.4	10.7	10.3	10.5	11.0	11.0	11.6	11.7	12.3
Switzerland	1.8	2.2	2.3	1.7	1.6	1.8	1.6	1.7	2.0	1.4	1.2	0.9	0.7	0.6	0.6	0.5	0.4	0.6	0.4	0.4
Austria	5.9	6.1	6.4	5.7	5.3	5.6	5.3	5.5	6.0	5.7	5.5	5.0	4.6	4.6	4.4	4.9	4.1	4.6	4.0	4.1
Greece	0.5	0.4	0.5	0.6	0.7	0.7	0.7	0.7	1.1	1.1	1.5	1.4	1.4	1.6	1.8	2.1	2.8	2.5	2.8	3.1
Turkey	2.8	3.0	3.3	3.3	3.2	3.2	2.9	3.3	3.6	3.9	3.8	4.2	3.7	4.1	4.4	4.4	4.4	13.2	14.2	14.8
Iceland	neg.	neg.	neg.	neg.	neg.	neg.	neg.	neg.	neg.	neg.	neg.	neg.	neg.	neg.	neg.	neg.	neg.	neg.	neg.	neg.
Malta/Gibraltar/Cyprus	neg.	neg.	neg.	neg.	neg.	neg.	neg.	neg.	neg.	neg.	neg.	neg.	neg.	neg.	neg.	neg.	neg.	neg.	-	-
Yugoslavia	6.3	7.2	7.8	7.7	8.7	9.5	10.0	10.1	11.1	12.2	12.4	12.0	10.5	10.8	10.8	11.6	12.5	9.0	9.9	10.0
Canary Islands	-	-	-	-	-	-	-	-	-	-	-	-	-	-	-	-	-	-	-	-
TOTAL W. EUROPE	400.2	412.5	413.1	381.0	363.5	381.2	375.5	380.9	393.3	381.6	363.9	344.1	324.7	323.2	323.9	313.5	291.7	272.5	271.8	261.9
Middle East	0.3	0.3	0.3	0.4	0.3	0.3	0.2	0.2	0.2	0.2	0.2	0.2	0.2	0.2	0.3	0.4	0.4	0.4	0.5	0.5
Africa	26.6	27.7	28.8	29.9	29.5	30.2	30.8	31.3	32.1	34.3	37.1	37.0	37.5	39.6	40.2	41.4	44.3	55.3	51.1	55.0
S. Asia	27.3	28.8	31.0	33.0	34.0	37.2	40.1	43.9	46.6	44.7	48.8	48.8	49.6	51.2	54.5	51.6	51.9	55.4	55.7	58.4
S.E. Asia	4.2	4.6	5.3	5.3	6.1	7.2	5.3	5.8	6.8	7.4	7.8	8.7	8.9	7.6	7.5	8.5	8.8	12.7	20.2	20.9
Japan	32.2	35.8	38.7	35.9	38.0	43.6	45.8	46.7	45.4	45.3	45.2	47.3	52.8	56.1	60.5	62.6	56.0	52.2	59.8	58.8
Australia	15.9	15.8	15.9	16.4	17.4	18.3	17.7	18.1	18.7	19.6	20.7	21.0	21.9	22.6	23.0	22.9	23.5	23.4	28.0	24.6
New Zealand/SWPI	1.3	1.3	1.3	1.2	1.1	1.3	1.8	1.5	1.6	1.7	1.7	1.7	1.5	1.5	1.5	1.5	1.4	1.4	1.4	1.5
TOTAL E.HEMISPHERE (NON-COMMUNIST)	508.0	526.8	534.4	503.1	489.9	519.3	517.2	528.4	544.7	534.8	525.4	508.8	497.1	502.0	511.4	502.4	478.0	473.3	488.5	481.6
TOTAL NON-COMMUNIST WORLD	818.4	844.1	835.2	768.3	753.0	791.7	783.5	802.2	833.7	839.0	846.8	846.4	829.3	847.8	863.6	858.8	821.6	820.5	856.3	842.5
COMMUNIST WORLD																				
U.S.S.R.	236.0	256.3	260.5	266.8	274.4	268.9	264.0	263.0	273.7	282.8	287.8	298.8	303.7	300.9	306.1	300.9	308.8	353.9	361.8	369.1
E. Europe	149.0	155.9	163.6	168.4	173.9	180.7	189.8	199.3	209.5	217.7	214.5	214.6	212.0	218.5	227.6	237.2	241.2	214.3	218.7	225.1
China etc.	67.6	76.3	94.3	194.0	250.2	301.8	186.3	188.0	202.4	217.5	226.0	246.2	177.1	232.1	254.5	276.8	299.4	301.9	316.3	331.5
TOTAL COMMUNIST WORLD	452.6	488.5	518.4	629.2	698.5	751.4	640.1	650.3	685.6	718.0	728.3	759.6	692.8	751.5	788.2	814.9	849.4	870.1	896.8	925.7
TOTAL WORLD	1271.0	1332.6	1353.6	1397.5	1451.5	1543.1	1423.6	1452.5	1519.3	1557.0	1575.1	1606.0	1522.1	1599.3	1651.8	1673.7	1671.0	1690.6	1753.1	1768.2

Note: A dash represents zero consumption
Neg. is less than 0.05 million tonnes oil equivalent

Table 6 TOTAL CONSUMPTION—NATURAL GAS (Million tonnes oil equivalent. Oil = 10 000 kcal/kg)

	1955	1956	1957	1958	1959	1960	1961	1962	1963	1964	1965	1966	1967	1968	1969	1970	1971	1972	1973	1974
WESTERN HEMISPHERE																				
U.S.A.	245.6	261.6	276.5	291.2	317.5	337.5	350.0	372.3	394.1	416.0	431.8	463.8	488.1	522.2	561.2	588.1	606.8	587.4	572.3	560.4
Canada	4.0	4.6	6.1	7.5	9.1	11.1	12.5	14.5	17.0	19.8	22.4	25.3	26.9	31.1	35.5	40.1	42.5	46.8	63.8	64.1
Mexico	2.4	2.7	4.2	4.8	6.2	6.6	6.8	7.2	8.0	9.8	9.9	12.9	14.1	14.2	15.1	16.7	16.7	19.7		
Caribbean	3.2	3.5	4.1	4.6	4.9	5.4	5.8	6.2	6.6	7.5	8.1	8.7	9.5	9.9	10.2	11.2	11.8	14.3	42.5	48.0
S. America	1.2	1.1	1.3	1.6	1.9	3.3	4.6	5.8	6.3	6.2	6.7	7.0	7.4	8.2	8.8	9.8	11.0	12.0		
TOTAL WESTERN HEMISPHERE	256.4	273.5	292.2	309.7	339.6	363.9	379.7	406.0	432.0	459.3	478.9	517.7	546.0	585.6	630.8	665.9	688.8	680.2	678.6	672.5
EASTERN HEMISPHERE																				
W. EUROPE																				
W. Germany	0.3	0.4	0.3	0.6	0.6	0.7	0.7	1.2	1.5	2.3	3.0	3.6	4.6	8.0	11.1	15.5	20.7	21.5	27.0	32.1
Netherlands	0.1	0.2	0.2	0.2	0.2	0.3	0.4	0.5	0.6	0.8	1.7	3.0	5.6	9.1	13.4	19.1	24.6	29.0	32.2	32.9
Belgium/Luxembourg	neg.	neg.	0.1	0.1	0.1	neg.	neg.	neg.	neg.	neg.	neg.	0.1	0.5	1.3	2.6	4.2	5.9	6.6	8.2	10.0
France	0.3	0.4	0.5	0.5	1.3	2.5	3.8	4.4	4.6	4.8	5.0	5.3	6.0	7.1	8.5	9.8	11.4	13.2	15.7	17.0
Italy	3.4	4.2	4.6	4.8	5.7	6.0	6.4	6.7	6.8	7.1	7.2	8.2	8.7	9.7	11.1	12.3	12.6	12.3	14.4	15.9
United Kingdom	-	neg.	neg.	neg.	0.1	0.1	0.1	0.1	0.1	0.2	0.7	0.7	1.4	2.9	5.9	11.2	18.1	25.2	26.1	30.8
Eire	-	-	-	-	-	-	-	-	-	-	-	-	-	-	-	-	-	-	-	-
Denmark	-	-	-	-	-	-	-	-	-	-	-	-	-	-	-	-	-	-	-	-
Norway	-	-	-	-	-	-	-	-	-	-	-	-	-	-	-	-	-	-	-	-
Sweden	-	-	-	-	-	-	-	-	-	-	-	-	-	-	-	-	-	-	-	0.4
Finland	-	-	-	-	-	-	-	-	-	-	-	-	-	-	-	-	-	-	-	-
Portugal	-	-	-	-	-	-	-	-	-	-	-	-	-	-	-	-	-	-	-	-
Spain	-	-	-	-	-	-	-	-	-	-	-	-	-	-	0.1	0.1	0.3	1.1	1.0	1.0
Switzerland	-	-	-	-	-	-	neg.	neg.	neg.	neg.	neg.	neg.	neg.	neg.	neg.	neg.	neg.	0.1	0.2	0.4
Austria	0.7	0.7	0.7	0.7	1.1	1.4	1.5	1.6	1.6	1.7	1.6	1.8	1.7	1.8	2.1	2.6	3.0	3.3	3.4	3.8
Greece	-	-	-	-	-	-	-	-	-	-	-	-	-	-	-	-	-	-	-	-
Turkey	-	-	-	-	-	-	-	-	-	-	-	-	-	-	-	-	-	-	-	-
Iceland	-	-	-	-	-	-	-	-	-	-	-	-	-	-	-	-	-	-	-	-
Malta/Gibraltar/Cyprus	-	-	-	-	-	-	-	-	-	-	-	-	-	-	-	-	-	-	-	-
Yugoslavia	neg.	neg.	neg.	neg.	neg.	neg.	0.1	0.1	0.1	0.2	0.3	0.4	0.4	0.6	0.6	0.9	1.1	1.4	1.6	1.8
Canary Islands	-	-	-	-	-	-	-	-	-	-	-	-	-	-	-	-	-	-	-	-
TOTAL W. EUROPE	4.8	5.9	6.4	6.9	9.1	11.0	13.0	14.6	15.3	17.1	19.5	23.1	28.9	40.5	55.4	75.7	97.7	113.7	129.8	146.1
Middle East	-	0.4	0.6	0.9	1.2	2.3	2.4	2.8	3.2	4.0	4.2	4.6	5.7	7.7	9.9	17.3	17.9	14.1	22.5	25.4
Africa	neg.	neg.	neg.	neg.	neg.	neg.	0.2	0.3	0.4	0.7	1.1	1.2	1.1	1.2	1.4	1.5	1.7	2.4	3.2	3.3
S. Asia	neg.	0.3	0.4	0.5	0.6	0.8	0.7	0.9	1.7	1.8	1.7	1.9	2.5	2.8	3.2	3.9	3.4	4.2	8.0	8.8
S.E. Asia	1.9	2.0	2.1	2.0	2.2	2.3	2.6	2.7	2.2	2.3	3.2	3.2	2.8	3.3	3.0	3.0	3.4	4.2	4.2	4.5
Japan	0.1	0.2	0.2	0.3	0.4	0.7	0.9	1.3	1.8	1.9	1.8	1.9	2.0	2.1	2.4	3.6	3.9	5.0	4.8	5.1
Australia	-	-	-	-	-	-	-	neg.	neg.	neg.	neg.	neg.	neg.	neg.	0.2	1.4	2.1	2.9	3.6	3.8
New Zealand/SWPI	-	-	-	-	-	-	-	-	-	-	-	-	-	-	-	0.1	0.3	0.2	0.3	0.3
TOTAL E.HEMISPHERE (NON-COMMUNIST)	6.8	8.8	9.7	10.6	13.5	17.1	19.8	22.6	24.6	27.8	31.5	35.9	43.0	57.6	75.5	106.5	130.4	146.7	176.4	197.3
TOTAL NON-COMMUNIST WORLD	263.2	282.3	301.9	320.3	353.1	381.0	399.5	428.6	456.6	487.1	510.4	553.6	589.0	643.2	706.3	772.4	819.2	826.9	855.0	869.8
COMMUNIST WORLD																				
U.S.S.R.	8.2	11.1	17.1	25.9	32.8	42.0	54.7	68.3	83.5	100.9	118.9	132.5	145.7	157.5	168.3	184.8	201.4	188.5	200.4	216.6
E. Europe	6.4	7.0	7.8	8.8	9.9	10.6	11.3	13.1	14.8	16.4	18.4	20.4	23.4	26.3	30.6	34.0	38.0	35.3	39.4	43.5
China etc.	-	-	-	-	-	-	-	-	-	-	-	-	-	-	-	-	-	3.7	4.0	4.4
TOTAL COMMUNIST WORLD	14.6	18.1	24.9	34.7	42.7	52.6	66.0	81.4	98.3	117.3	137.3	152.9	169.1	183.8	198.9	218.8	239.4	227.5	243.8	264.5
TOTAL WORLD	277.8	300.4	326.8	355.0	395.8	433.6	465.5	510.0	554.9	604.4	647.7	706.5	758.1	827.0	905.2	991.2	1058.6	1054.4	1098.8	1134.3

Note: A dash represents zero consumption
Neg. is less than 0.05 million tonnes oil equivalent

Table 7 TOTAL CONSUMPTION—HYDRO ELECTRICITY (LOW FACTOR) (Million tonnes oil equivalent. Oil = 10 000 kcal/kg)

	1955	1956	1957	1958	1959	1960	1961	1962	1963	1964	1965	1966	1967	1968	1969	1970	1971	1972	1973	1974
WESTERN HEMISPHERE																				
U.S.A.	10.2	11.0	11.7	12.6	12.4	13.1	13.7	15.0	14.8	15.8	17.2	17.3	19.7	19.8	22.3	21.9	23.6	24.3	25.2	25.6
Canada	6.7	7.1	7.2	7.9	8.5	9.2	9.1	9.1	9.1	9.9	10.3	11.3	11.7	11.8	13.0	13.8	14.1	12.4	18.5	19.8
Mexico	0.3	0.3	0.3	0.4	0.5	0.4	0.4	0.5	0.5	0.6	0.7	0.8	0.9	1.2	1.2	1.3	1.3	1.7		
Caribbean	0.2	0.3	0.2	0.3	0.3	0.4	0.4	0.4	0.5	0.6	0.6	0.6	0.7	0.9	1.1	1.2	1.2	1.4	9.5	10.4
S. America	1.4	1.6	1.8	2.0	2.1	2.2	2.3	2.5	2.5	2.7	3.0	3.4	3.6	3.7	3.9	4.6	4.9	5.4		
TOTAL WESTERN HEMISPHERE	18.8	20.3	21.2	23.2	23.8	25.3	25.9	27.5	27.4	29.6	31.8	33.4	36.6	37.4	41.5	42.8	45.1	45.2	53.2	55.8
EASTERN HEMISPHERE																				
W. EUROPE																				
W. Germany	1.1	1.2	1.1	1.2	0.9	1.2	1.2	1.1	1.1	1.1	1.4	1.5	1.5	1.5	1.3	1.6	1.3	2.0	1.1	1.7
Netherlands	*	*	*	*	*	*	*	*	*	*	*	*	*	*	*	*	*	*	*	*
Belgium/Luxembourg	neg.	neg.	neg.	neg.	neg.	neg.	neg.	neg.	0.1	0.1	0.1	0.1	0.1	0.1	0.1	0.1	0.1	0.1	neg.	neg.
France	2.2	2.3	2.2	2.8	2.8	3.6	3.4	3.2	3.8	3.0	4.1	4.5	4.0	4.4	4.6	4.9	4.3	3.7	3.6	4.2
Italy	2.7	2.7	2.8	3.2	3.4	4.2	3.9	3.7	4.2	3.7	4.0	4.1	4.0	4.0	3.9	3.9	3.8	4.1	3.4	3.7
United Kingdom	0.1	0.2	0.2	0.2	0.2	0.3	0.3	0.3	0.3	0.3	0.4	0.4	0.5	0.4	0.4	0.5	0.4	0.4	0.4	0.4
Eire	neg.	0.1	0.1	0.1	0.1	0.1	0.1	0.1	0.1	0.1	0.1	0.1	0.1	0.1	neg.	0.1	0.1	0.1	0.1	0.1
Denmark	*	*	*	*	*	*	*	*	*	*	*	*	*	*	*	*	*	*	*	*
Norway	2.0	2.1	2.2	2.4	2.5	2.7	2.9	3.3	3.5	3.8	4.3	4.2	4.6	5.3	4.9	5.0	5.5	4.9	5.7	7.1
Sweden	1.9	2.1	2.4	2.5	2.5	2.7	3.2	3.5	3.4	3.8	4.1	4.0	4.3	4.3	3.7	3.7	4.5	4.6	5.1	5.0
Finland	0.5	0.4	0.6	0.6	0.5	0.4	0.7	0.8	0.7	0.7	0.8	0.9	1.1	0.9	0.7	0.8	0.9	0.9	0.8	1.1
Portugal	0.1	0.2	0.2	0.2	0.2	0.3	0.3	0.3	0.3	0.4	0.3	0.4	0.5	0.4	0.5	0.5	0.5	0.6	0.7	0.7
Spain	0.7	0.9	0.8	0.9	1.3	1.4	1.4	1.4	1.9	1.8	1.7	2.4	2.0	2.1	2.7	2.4	2.8	3.0	2.5	2.7
Switzerland	1.4	1.3	1.4	1.5	1.6	1.7	1.9	1.9	1.9	2.0	2.1	2.4	2.6	2.5	2.4	2.5	2.6	2.1	2.4	2.2
Austria	0.7	0.7	0.8	0.9	0.9	1.1	1.1	1.1	1.1	1.2	1.4	1.5	1.6	1.6	1.5	1.9	1.5	1.4	1.7	2.0
Greece	neg.	neg.	neg.	neg.	neg.	neg.	neg.	0.1	0.1	0.1	0.1	0.1	0.1	0.1	0.2	0.2	0.2	0.2	0.2	0.2
Turkey	neg.	neg.	neg.	0.1	0.1	0.1	0.1	0.1	0.2	0.1	0.2	0.2	0.2	0.3	0.3	0.3	0.2	0.3	0.3	0.4
Iceland	neg.	neg.	neg.	neg.	neg.	neg.	neg.	neg.	0.1	0.1	0.1	0.1	0.1	0.1	0.1	0.1	0.1	0.1	0.2	0.2
Malta/Gibraltar/Cyprus	-	-	-	-	-	-	-	-	-	-	-	-	-	-	-	-	-	-	-	-
Yugoslavia	0.2	0.2	0.3	0.4	0.4	0.5	0.5	0.6	0.7	0.6	0.7	0.8	0.9	1.1	1.3	1.3	1.4	1.6	1.5	1.7
Canary Islands	-	-	-	-	-	-	-	-	-	-	-	-	-	-	-	-	-	-	-	-
TOTAL W. EUROPE	13.6	14.4	15.1	17.0	17.4	20.3	21.0	21.5	23.5	22.9	25.9	27.7	28.2	29.2	28.6	29.8	30.2	30.1	29.7	33.4
Middle East	neg.	neg.	neg.	neg.	neg.	neg.	neg.	neg.	0.1	0.1	0.1	0.1	0.1	0.1	0.2	0.2	0.3	0.3	0.3	0.3
Africa	0.3	0.3	0.4	0.5	0.5	0.6	0.8	0.9	0.9	1.1	1.2	1.3	1.4	1.7	2.0	2.2	2.3	2.1	2.3	2.5
S. Asia	0.4	0.4	0.5	0.5	0.6	0.7	0.9	1.2	1.4	1.5	1.5	1.6	1.8	2.0	2.3	2.4	2.6	2.7	2.6	2.8
S.E. Asia	0.3	0.3	0.3	0.3	0.4	0.4	0.3	0.3	0.3	0.4	0.4	0.5	0.5	0.5	0.6	0.7	0.7	0.9	1.3	1.5
Japan	4.2	4.6	4.7	5.0	5.7	5.1	5.8	5.6	5.9	6.2	6.2	7.1	6.3	6.3	6.9	7.0	7.4	7.6	5.3	6.3
Australia	0.2	0.3	0.3	0.3	0.3	0.4	0.4	0.5	0.6	0.7	0.6	0.6	0.7	0.6	0.7	0.8	1.1	0.9	0.3	0.3
New Zealand/SWPI	0.4	0.4	0.5	0.5	0.5	0.5	0.6	0.7	0.7	0.8	0.9	1.1	1.1	1.1	1.1	1.2	1.3	1.4	0.4	0.4
TOTAL E.HEMISPHERE (NON-COMMUNIST)	19.4	20.7	21.8	24.1	25.4	28.0	29.8	30.7	33.4	33.7	36.8	40.0	40.1	41.5	42.4	44.3	45.9	46.0	42.2	47.5
TOTAL NON-COMMUNIST WORLD	38.2	41.0	43.0	47.3	49.2	53.3	55.7	58.2	60.8	63.3	68.6	73.4	76.7	78.9	83.9	87.1	91.0	91.2	95.4	103.3
COMMUNIST WORLD																				
U.S.S.R.	2.0	2.5	3.5	4.1	4.2	4.4	5.1	6.3	6.6	6.7	7.1	8.1	7.8	9.1	10.1	10.9	11.0	11.7	12.3	12.9
E. Europe	0.3	0.3	0.4	0.5	0.4	0.5	0.6	0.6	0.5	0.5	0.8	0.8	0.8	0.7	0.8	1.1	1.2	1.7	1.8	1.9
China etc.	0.5	0.7	0.9	1.1	1.4	1.7	2.1	2.2	2.4	2.9	3.5	3.9	3.5	3.6	3.6	3.9	4.4	2.8	3.0	3.2
TOTAL COMMUNIST WORLD	2.8	3.5	4.8	5.7	6.0	6.6	7.8	9.1	9.5	10.1	11.4	12.8	12.1	13.4	14.5	15.9	16.6	16.2	17.1	18.0
TOTAL WORLD	41.0	44.5	47.8	53.0	55.2	59.9	63.5	67.3	70.3	73.4	80.0	86.2	88.8	92.3	98.4	103.0	107.6	107.4	112.5	121.3

*__Netherlands & Denmark__ - In some years there have been small (approx. 0.1 million tonnes oil equivalent or less (low factor)) imports of hydro electricity.

__Note:__ A dash represents zero consumption
Neg. is less than 0.05 million tonnes oil equivalent

Table 8 TOTAL CONSUMPTION—NUCLEAR ELECTRICITY (LOW FACTOR) (Million tonnes oil equivalent. Oil = 10 000 kcal/kg)

	1955	1956	1957	1958	1959	1960	1961	1962	1963	1964	1965	1966	1967	1968	1969	1970	1971	1972	1973	1974
WESTERN HEMISPHERE																				
U.S.A.	-	-	-	-	-	neg.	0.1	0.2	0.3	0.3	0.3	0.5	0.6	1.1	1.3	1.9	3.4	5.0	7.2	9.6
Canada	-	-	-	-	-	-	-	neg.	neg.	neg.	neg.	neg.	neg.	neg.	neg.	0.1	0.3	0.6	1.5	1.7
Mexico	-	-	-	-	-	-	-	-	-	-	-	-	-	-	-	-	-	-		
Caribbean	-	-	-	-	-	-	-	-	-	-	-	-	-	-	-	-	-	-		0.1
S. America	-	-	-	-	-	-	-	-	-	-	-	-	-	-	-	-	-	-		
TOTAL WESTERN HEMISPHERE	-	-	-	-	-	neg.	0.1	0.2	0.3	0.3	0.3	0.5	0.6	1.1	1.3	2.0	3.7	5.6	8.7	11.4
EASTERN HEMISPHERE																				
W. EUROPE																				
W. Germany	-	-	-	-	-	-	neg.	neg.	neg.	neg.	neg.	neg.	0.1	0.1	0.4	0.5	0.5	0.7	0.9	0.9
Netherlands	-	-	-	-	-	-	-	-	-	-	-	-	-	neg.	neg.	neg.	neg.	neg.	0.1	0.2
Belgium/Luxembourg	-	-	-	-	-	-	-	neg.	neg.	neg.	neg.	neg.	neg.	neg.	neg.	neg.	neg.	neg.	neg.	neg.
France	-	-	-	-	-	neg.	neg.	neg.	neg.	neg.	0.1	0.1	0.2	0.3	0.4	0.4	0.7	1.2	1.1	1.1
Italy	-	-	-	-	-	-	-	-	neg.	0.2	0.3	0.3	0.3	0.2	0.1	0.3	0.3	0.3	0.2	0.3
United Kingdom	neg.	neg.	neg.	neg.	0.1	0.2	0.2	0.3	0.5	0.7	1.4	1.9	2.1	2.4	2.5	2.2	2.3	2.2	2.0	2.4
Eire	-	-	-	-	-	-	-	-	-	-	-	-	-	-	-	-	-	-	-	-
Denmark	-	-	-	-	-	-	-	-	-	-	-	-	-	-	-	-	-	-	-	-
Norway	-	-	-	-	-	-	-	-	-	-	-	-	-	-	-	-	-	-	-	-
Sweden	-	-	-	-	-	-	-	-	-	neg.	neg.	neg.	neg.	neg.	neg.	neg.	neg.	0.1	0.2	0.1
Finland	-	-	-	-	-	-	-	-	-	-	-	-	-	-	-	-	-	-	-	-
Portugal	-	-	-	-	-	-	-	-	-	-	-	-	-	-	-	-	-	-	-	-
Spain	-	-	-	-	-	-	-	-	-	-	-	-	-	neg.	0.1	0.1	0.2	0.4	0.5	0.6
Switzerland	-	-	-	-	-	-	-	-	-	-	-	-	-	-	-	0.2	0.2	0.3	0.5	0.5
Austria	-	-	-	-	-	-	-	-	-	-	-	-	-	-	-	-	-	-	-	-
Greece	-	-	-	-	-	-	-	-	-	-	-	-	-	-	-	-	-	-	-	-
Turkey	-	-	-	-	-	-	-	-	-	-	-	-	-	-	-	-	-	-	-	-
Iceland	-	-	-	-	-	-	-	-	-	-	-	-	-	-	-	-	-	-	-	-
Malta/Gibraltar/Cyprus	-	-	-	-	-	-	-	-	-	-	-	-	-	-	-	-	-	-	-	-
Yugoslavia	-	-	-	-	-	-	-	-	-	-	-	-	-	-	-	-	-	-	-	-
Canary Islands	-	-	-	-	-	-	-	-	-	-	-	-	-	-	-	-	-	-	-	-
TOTAL W. EUROPE	neg.	neg.	neg.	neg.	0.1	0.2	0.2	0.3	0.5	0.9	1.8	2.3	2.7	3.0	3.5	3.7	4.2	5.2	5.5	6.1
Middle East	-	-	-	-	-	-	-	-	-	-	-	-	-	-	-	-	-	-	-	-
Africa	-	-	-	-	-	-	-	-	-	-	-	-	-	-	-	-	-	-	-	-
S. Asia	-	-	-	-	-	-	-	-	-	-	-	-	-	-	0.1	0.2	0.1	0.2	0.3	0.3
S.E. Asia	-	-	-	-	-	-	-	-	-	-	-	-	-	-	-	-	-	-	-	-
Japan	-	-	-	-	-	-	-	-	neg.	neg.	neg.	neg.	0.1	0.1	0.1	0.4	0.7	0.8	0.7	1.4
Australia	-	-	-	-	-	-	-	-	-	-	-	-	-	-	-	-	-	-	-	-
New Zealand/SWPI	-	-	-	-	-	-	-	-	-	-	-	-	-	-	-	-	-	-	-	-
TOTAL E.HEMISPHERE (NON-COMMUNIST)	neg.	neg.	neg.	neg.	0.1	0.2	0.2	0.3	0.5	0.9	1.8	2.3	2.8	3.1	3.7	4.3	5.0	6.2	6.5	7.8
TOTAL NON-COMMUNIST WORLD	neg.	neg.	neg.	neg.	0.1	0.2	0.3	0.5	0.8	1.2	2.1	2.8	3.4	4.2	5.0	6.3	8.7	11.8	15.2	19.2
COMMUNIST WORLD																				
U.S.S.R.	-	-	-	-	-	n.a.	n.a.	n.a.	n.a.	n.a.	n.a.	0.1	0.1	0.2	0.2	0.3	0.4	0.7	1.1	1.4
E. Europe	-	-	-	-	-	-	n.a.	n.a.	n.a.	n.a.	n.a.	neg.	neg.	neg.	neg.	neg.	neg.	neg.	neg.	0.1
China et.c	-	-	-	-	-	-	-	-	-	-	-	-	-	-	-	-	-	-	-	-
TOTAL COMMUNIST WORLD	-	-	-	-	-	n.a.	n.a.	n.a.	n.a.	n.a.	n.a.	0.1	0.1	0.2	0.2	0.3	0.4	0.7	1.1	1.5
TOTAL WORLD	neg.	neg.	neg.	neg.	0.1	0.2	0.3	0.5	0.8	1.2	2.1	2.9	3.5	4.4	5.2	6.6	9.1	12.5	16.3	20.7

Note: A dash represents zero consumption
Neg. is less than 0.05 million tonnes oil equivalent
n.a. = not available

Table 9 TOTAL PRIMARY ENERGY CONSUMPTION—HYDRO/NUCLEAR ELECTRICITY (LOW FACTOR)
(Million tonnes oil equivalent. Oil = 10 000 kcal/kg)

	1955	1956	1957	1958	1959	1960	1961	1962	1963	1964
WESTERN HEMISPHERE										
U.S.A.	954.0	992.7	995.7	989.4	1031.0	1075.2	1088.7	1139.7	1190.3	1241.2
Canada	61.7	70.2	69.9	68.7	73.2	77.1	78.3	80.6	87.1	96.2
Mexico						21.3	22.1	22.2	24.7	27.4
Caribbean	} 78.0	} 83.2	} 92.6	} 97.0	} 102.3	39.0	43.6	40.7	44.2	46.7
S. America						45.0	47.3	50.3	52.6	54.6
TOTAL WESTERN HEMISPHERE	1093.7	1146.1	1158.2	1155.1	1206.5	1257.6	1280.0	1333.5	1398.9	1466.1
EASTERN HEMISPHERE										
W. EUROPE										
W. Germany	129.8	137.7	139.7	133.9	134.2	148.0	151.3	162.9	176.9	183.3
Netherlands	} 46.6	} 50.8	} 49.8	} 46.2	} 47.4	24.3	26.1	29.1	32.1	34.1
Belgium/Luxembourg						28.1	29.1	31.7	34.2	34.9
France	70.8	80.1	81.5	79.4	77.5	82.5	86.5	93.4	101.8	106.5
Italy	26.4	29.3	31.1	32.2	34.9	42.4	47.2	54.2	60.2	66.6
United Kingdom	183.3	183.3	178.1	177.5	173.8	188.0	189.1	191.5	200.3	201.0
Eire	← Included in ≠ →					4.0	4.4	4.7	4.8	5.0
Denmark						9.1	9.5	11.2	12.3	12.9
Norway	} 29.5	} 31.0	} 30.1	} 31.0	} 33.6	7.3	7.7	8.1	8.8	9.5
Sweden						18.1	18.3	19.8	21.5	23.3
Finland						5.2	5.7	6.2	6.3	7.8
Portugal						2.6	3.0	3.0	3.2	3.7
Spain						17.2	18.2	19.3	21.3	22.0
Switzerland						7.3	7.6	8.6	10.5	10.5
Austria	} 49.0≠	} 53.6≠	} 58.3≠	} 57.7≠	} 61.7≠	11.1	11.2	12.1	13.3	13.8
Greece						3.3	3.6	4.1	4.8	5.2
Turkey						5.1	5.0	6.0	6.8	7.4
Iceland						0.4	0.3	0.4	0.5	0.5
Malta/Gibraltar/Cyprus						0.7	0.7	0.7	0.8	0.8
Yugoslavia						11.5	12.1	12.4	13.9	15.7
Canary Islands						1.2	1.2	1.5	1.6	2.5
TOTAL W. EUROPE	535.4	565.8	568.6	557.9	563.1	617.4	637.8	680.9	735.9	767.0
Middle East						29.0	32.1	32.5	35.0	39.2
Africa						53.6	56.2	57.0	60.1	63.4
S. Asia	} 111.1	} 122.1	} 132.7	} 137.3	} 146.4	50.8	54.4	58.9	64.7	64.4
S.E. Asia						25.2	26.6	31.5	31.4	34.3
Japan	46.7	53.6	59.6	58.2	67.1	79.4	93.5	102.4	115.7	128.6
Australia	} 26.9	} 27.8	} 28.0	} 29.4	} 31.3	29.6	29.9	30.8	33.0	35.7
New Zealand/SWPI						3.9	4.6	4.6	4.8	5.5
TOTAL E.HEMISPHERE (NON-COMMUNIST)	720.1	769.3	788.9	782.8	807.9	888.9	935.1	998.6	1080.6	1138.1
TOTAL NON-COMMUNIST WORLD	1813.8	1915.4	1947.1	1937.9	2014.4	2146.5	2215.1	2332.1	2479.5	2604.2
COMMUNIST WORLD										
U.S.S.R.						434.4	452.2	481.1	521.4	559.5
E. Europe	} 552.3	} 604.1	} 656.1	} 789.6	} 879.2	} 520.9	} 417.9	} 434.5	249.5	261.9
China etc.									215.3	230.9
TOTAL COMMUNIST WORLD	552.3	604.1	656.1	789.6	879.2	955.3	870.1	915.6	986.2	1052.3
TOTAL WORLD	2366.1	2519.5	2603.2	2727.5	2893.6	3101.8	3085.2	3247.7	3465.7	3656.5

	1965	1966	1967	1968	1969	1970	1971	1972	1973	1974
WESTERN HEMISPHERE										
U.S.A.	1295.9	1370.9	1412.5	1499.5	1579.7	1636.0	1669.4	1709.1	1757.7	1712.9
Canada	103.9	110.7	115.7	125.1	133.2	144.0	149.7	155.9	183.9	188.7
Mexico	28.8	33.4	34.8	36.3	39.3	42.3	44.1	51.8		
Caribbean	49.7	53.7	56.9	60.5	62.0	68.4	72.5	81.3	} 238.3	} 251.4
S. America	57.2	61.0	63.6	69.0	75.3	81.2	88.2	94.8		
TOTAL WESTERN HEMISPHERE	1535.5	1629.7	1683.5	1790.4	1889.5	1971.9	2023.9	2092.9	2179.9	2153.0
EASTERN HEMISPHERE										
W. EUROPE										
W. Germany	186.8	190.3	187.4	204.3	224.0	238.2	241.4	246.6	266.6	252.0
Netherlands	36.6	39.0	41.9	46.9	52.9	60.6	64.2	72.3	76.7	71.8
Belgium/Luxembourg	36.2	35.2	37.3	41.0	45.1	48.6	47.9	50.9	50.7	49.1
France	110.1	111.5	120.0	125.4	138.0	148.5	154.3	164.5	174.0	169.4
Italy	72.0	78.7	86.5	92.6	101.3	112.7	118.9	126.2	129.2	129.8
United Kingdom	206.1	205.9	201.2	208.0	215.7	222.3	224.9	212.8	220.4	208.3
Eire	4.7	5.3	5.7	6.7	6.6	6.8	7.0	6.5	7.3	7.1
Denmark	13.9	15.6	15.7	17.2	19.3	21.1	20.2	21.0	20.2	18.8
Norway	10.4	10.8	11.5	12.9	13.4	14.4	14.6	14.3	14.9	15.5
Sweden	25.2	27.6	27.1	29.9	32.2	35.4	34.6	36.6	38.8	36.3
Finland	8.5	9.7	10.1	11.3	12.9	14.1	14.3	19.1	18.5	17.3
Portugal	3.8	4.1	4.5	4.4	5.0	5.6	6.4	7.3	8.4	8.3
Spain	24.3	26.0	28.7	30.2	34.1	37.7	41.9	45.9	52.0	54.8
Switzerland	11.3	11.9	12.6	13.4	14.2	15.7	16.5	16.7	18.1	16.5
Austria	14.1	14.4	14.4	15.5	16.2	18.7	18.9	20.1	20.6	20.6
Greece	6.0	6.3	7.0	7.6	8.1	9.0	10.4	11.3	13.0	12.4
Turkey	9.0	8.9	9.5	10.8	11.8	12.4	13.6	23.5	26.9	27.7
Iceland	0.6	0.6	0.6	0.6	0.6	0.7	0.6	0.7	0.9	0.8
Malta/Gibraltar/Cyprus	0.7	0.7	0.8	0.9	1.0	1.1	1.2	1.3	1.4	1.4
Yugoslavia	16.6	16.6	15.9	17.3	18.7	20.8	24.0	22.1	24.3	25.9
Canary Islands	2.8	3.7	4.4	3.8	3.9	4.0	3.3	2.8	2.9	2.9
TOTAL W. EUROPE	799.7	822.8	842.8	900.7	975.0	1048.4	1079.1	1122.5	1185.8	1146.7
Middle East	39.8	44.2	45.0	49.1	57.6	67.4	73.5	74.4	87.9	95.9
Africa	67.8	70.7	74.9	78.9	82.6	86.7	92.5	104.3	104.1	107.8
S. Asia	69.6	71.0	73.7	79.0	87.2	85.2	86.0	93.3	99.8	101.8
S.E. Asia	36.3	40.7	53.6	58.7	63.1	70.5	73.5	86.6	100.4	106.1
Japan	141.1	156.3	184.1	207.3	238.9	272.7	287.7	299.9	338.9	332.7
Australia	38.9	39.7	42.7	44.9	47.1	49.9	52.6	53.4	60.0	57.9
New Zealand/SWPI	5.7	6.4	5.8	6.4	6.6	6.8	7.1	7.4	6.8	6.9
TOTAL E.HEMISPHERE (NON-COMMUNIST)	1198.9	1251.8	1322.6	1425.0	1558.1	1687.6	1752.0	1841.8	1983.7	1955.8
TOTAL NON-COMMUNIST WORLD	2734.4	2881.5	3006.1	3215.4	3447.6	3659.5	3775.9	3934.7	4163.6	4108.8
COMMUNIST WORLD										
U.S.S.R.	594.2	632.7	668.0	693.4	726.1	759.0	795.4	850.9	893.3	941.8
E. Europe	264.1	270.1	274.3	288.6	308.2	326.3	339.7	316.8	334.7	348.9
China etc.	241.3	264.3	193.5	250.9	275.1	303.5	331.8	340.5	365.0	387.9
TOTAL COMMUNIST WORLD	1099.6	1167.1	1135.8	1232.9	1309.4	1388.8	1466.9	1508.2	1593.0	1678.6
TOTAL WORLD	3834.0	4048.6	4141.9	4448.3	4757.0	5048.3	5242.8	5442.9	5756.6	5787.4

Table 10 TOTAL CONSUMPTION—HYDRO ELECTRICITY (HIGH FACTOR) (Million tonnes oil equivalent. Oil = 10 000 kcal/kg)

	1955	1956	1957	1958	1959	1960	1961	1962	1963	1964	1965	1966	1967	1968	1969	1970	1971	1972	1973	1974
WESTERN HEMISPHERE																				
U.S.A.	30.6	33.0	35.1	37.8	37.1	39.3	40.8	45.2	44.4	47.4	51.8	52.0	59.1	59.4	66.7	65.9	70.9	72.9	75.6	76.9
Canada	20.1	21.5	21.8	23.7	25.5	27.8	27.3	27.3	26.9	29.7	30.8	34.1	34.9	35.4	39.2	41.2	42.2	37.1	55.5	59.4
Mexico	0.9	1.1	0.9	1.2	1.5	1.4	1.4	1.5	1.5	1.8	2.3	2.6	2.9	3.4	3.6	3.9	3.8	4.9	} 28.5	} 31.3
Caribbean	0.7	0.8	0.7	0.9	0.9	1.2	1.2	1.3	1.6	1.8	1.9	2.0	2.3	2.7	3.2	3.4	3.6	4.1	}	}
S. America	4.0	4.6	5.3	6.1	6.2	6.6	7.0	7.5	7.7	8.3	9.1	10.2	10.7	10.9	11.7	13.8	14.8	16.2	}	}
TOTAL WESTERN HEMISPHERE	56.3	61.0	63.8	69.7	71.2	76.3	77.7	82.8	82.1	89.0	95.9	100.9	109.9	111.8	124.4	128.2	135.3	135.2	159.6	167.6
EASTERN HEMISPHERE																				
W. EUROPE																				
W. Germany	3.2	3.4	3.3	3.5	2.9	3.4	3.4	3.3	3.3	3.2	4.0	4.4	4.3	4.4	3.8	4.6	3.7	6.1	3.0	5.0
Netherlands	•	•	•	•	•	•	•	•	•	•	•	•	•	•	•	•	•	•	•	•
Belgium/Luxembourg	neg.	neg.	neg.	neg.	neg.	neg.	0.1	0.1	0.2	0.2	0.3	0.3	0.3	0.2	0.3	0.3	0.3	0.4	0.1	0.1
France	6.7	6.8	6.5	8.4	8.5	10.6	10.1	9.3	11.3	9.1	12.2	13.5	11.9	13.2	13.9	14.9	12.8	11.0	10.6	12.6
Italy	8.1	8.2	8.4	9.5	10.1	12.7	11.7	10.9	12.7	11.0	12.0	12.3	12.0	12.1	11.8	11.6	11.2	12.2	10.1	11.0
United Kingdom	0.4	0.6	0.7	0.7	0.7	0.8	1.1	1.1	0.9	1.1	1.3	1.2	1.5	1.2	1.2	1.5	1.2	1.2	1.2	1.3
Eire	0.1	0.2	0.2	0.2	0.2	0.2	0.2	0.2	0.2	0.2	0.2	0.2	0.2	0.2	0.1	0.2	0.2	0.1	0.2	0.2
Denmark	•	•	•	•	•	•	•	•	•	•	•	•	•	•	•	•	•	•	•	•
Norway	5.9	6.2	6.7	7.1	7.5	8.1	8.7	9.9	10.3	11.4	12.8	12.6	13.9	15.6	14.9	15.0	16.5	15.1	17.1	21.5
Sweden	5.7	6.3	7.1	7.6	7.6	8.2	9.6	10.3	10.0	11.3	12.2	12.0	12.9	12.8	11.0	10.9	13.7	14.2	15.5	15.2
Finland	1.6	1.4	1.8	1.8	1.5	1.4	2.1	2.5	2.2	2.2	2.4	2.7	3.0	2.7	2.5	2.4	2.7	2.8	2.5	3.1
Portugal	0.4	0.5	0.5	0.6	0.7	0.8	0.9	0.9	1.1	1.2	1.1	1.4	1.5	1.4	1.7	1.5	1.6	2.0	2.1	2.1
Spain	2.3	2.9	2.5	2.9	3.8	4.1	4.2	4.2	5.6	5.5	5.1	7.1	6.0	6.4	8.1	7.4	8.6	9.2	7.5	8.2
Switzerland	4.0	3.9	4.1	4.4	4.7	4.9	5.8	5.6	5.7	6.0	6.3	7.2	7.8	7.7	7.4	7.7	7.8	6.5	7.2	6.8
Austria	2.1	2.3	2.4	2.8	2.8	3.2	3.0	3.2	3.2	3.5	4.2	4.5	4.6	4.7	4.4	5.6	4.4	4.1	5.0	5.9
Greece	0.1	0.1	0.1	0.1	0.1	0.1	0.1	0.2	0.2	0.2	0.2	0.4	0.4	0.3	0.5	0.7	0.7	0.7	0.6	0.6
Turkey	neg.	neg.	0.1	0.2	0.2	0.2	0.3	0.3	0.5	0.4	0.5	0.6	0.6	0.8	0.9	0.8	0.6	0.8	1.0	1.2
Iceland	0.1	0.1	0.1	0.1	0.1	0.1	0.1	0.1	0.2	0.2	0.2	0.2	0.2	0.2	0.2	0.4	0.4	0.3	0.5	0.5
Malta/Gibraltar/Cyprus	-	-	-	-	-	-	-	-	-	-	-	-	-	-	-	-	-	-	-	-
Yugoslavia	0.7	0.7	0.9	1.2	1.3	1.6	1.5	1.8	2.2	2.0	2.4	2.6	2.8	3.0	3.9	3.9	4.1	4.7	4.5	5.0
Canary Islands	-	-	-	-	-	-	-	-	-	-	-	-	-	-	-	-	-	-	-	-
TOTAL W. EUROPE	41.4	43.6	45.4	51.1	52.7	60.4	62.9	63.9	69.8	68.7	77.4	83.2	83.9	86.9	86.4	89.4	90.5	91.4	88.7	100.3
Middle East	neg.	0.1	0.1	0.1	0.1	neg.	0.1	0.1	0.2	0.2	0.2	0.3	0.3	0.4	0.6	0.6	0.9	0.7	0.8	0.9
Africa	0.8	1.1	1.3	1.5	1.6	1.9	2.4	2.8	2.9	3.3	3.6	3.7	4.2	5.0	5.9	6.5	6.9	6.4	6.9	7.5
S. Asia	1.2	1.4	1.5	1.7	2.0	2.2	2.8	3.5	4.0	4.3	4.5	4.7	5.4	6.1	6.8	7.1	7.9	8.1	8.0	8.4
S.E. Asia	0.8	0.9	1.1	1.1	1.2	1.4	0.8	0.8	0.9	1.2	1.3	1.5	1.6	1.7	1.9	2.2	2.2	3.0	3.8	4.3
Japan	12.6	14.0	14.2	15.2	16.9	15.5	17.4	16.8	17.6	18.5	18.7	21.4	18.8	18.8	20.9	21.0	21.9	22.7	15.8	19.0
Australia	0.6	0.8	0.8	0.9	1.1	1.3	1.2	1.6	1.9	2.1	1.9	2.0	2.1	2.0	2.3	2.6	3.3	2.8	1.0	1.0
New Zealand/SWPI	1.3	1.4	1.5	1.5	1.5	1.7	1.8	2.1	2.2	2.4	2.7	2.8	3.0	3.2	3.2	3.4	3.9	4.1	1.2	1.4
TOTAL E.HEMISPHERE (NON-COMMUNIST)	58.7	63.3	65.9	73.1	77.1	84.4	89.4	91.6	99.5	100.7	110.3	119.6	119.3	124.1	128.0	132.8	137.5	139.2	126.2	142.8
TOTAL NON-COMMUNIST WORLD	115.0	124.3	129.7	142.8	148.3	160.7	167.1	174.4	181.6	189.7	206.2	220.5	229.2	235.9	252.4	261.0	272.8	274.4	285.8	310.4
COMMUNIST WORLD																				
U.S.S.R.	6.1	7.6	10.4	12.2	12.5	13.3	15.5	18.9	20.0	20.3	21.4	24.2	23.2	27.3	30.2	32.7	33.1	35.1	36.9	38.7
E. Europe	1.1	1.1	1.2	1.5	1.3	1.7	1.7	1.8	1.7	1.7	2.5	2.5	2.5	2.3	2.4	3.3	3.5	4.9	5.3	5.7
China etc.	1.6	2.1	2.7	3.2	4.1	5.0	6.3	6.5	7.2	8.8	10.5	11.8	10.5	10.6	10.7	11.8	13.2	8.5	9.0	9.5
TOTAL COMMUNIST WORLD	8.8	10.8	14.3	16.9	17.9	20.0	23.5	27.2	28.9	30.8	34.4	38.5	36.2	40.2	43.3	47.8	49.8	48.5	51.2	53.9
TOTAL WORLD	123.8	135.1	144.0	159.7	166.2	180.7	190.6	201.6	210.5	220.5	240.6	259.0	265.4	276.1	295.7	308.8	322.6	322.9	337.0	364.3

• Netherlands & Denmark - In some years there have been small (approx. 0.1 million tonnes oil equivalent or less (low factor) imports of hydro electricity.

Note: A dash represents zero consumption
Neg. is less than 0.05 million tonnes oil equivalent

Table 11 TOTAL CONSUMPTION—NUCLEAR ELECTRICITY (HIGH FACTOR) (Million tonnes oil equivalent. Oil = 10 000 kcal/kg)

	1955	1956	1957	1958	1959	1960	1961	1962	1963	1964	1965	1966	1967	1968	1969	1970	1971	1972	1973	1974
WESTERN HEMISPHERE																				
U.S.A.	-	-	-	-	-	0.1	0.4	0.6	0.8	0.8	0.9	1.5	2.0	3.3	3.7	5.7	10.0	15.2	21.8	28.8
Canada	-	-	-	-	-	-	-	neg.	neg.	neg.	neg.	neg.	neg.	0.2	0.1	0.2	1.1	1.8	4.4	4.9
Mexico	-	-	-	-	-	-	-	-	-	-	-	-	-	-	-	-	-	-	}	}
Caribbean	-	-	-	-	-	-	-	-	-	-	-	-	-	-	-	-	-	-	} -	} 0.3
S. America	-	-	-	-	-	-	-	-	-	-	-	-	-	-	-	-	-	-	}	}
TOTAL WESTERN HEMISPHERE	-	-	-	-	-	0.1	0.4	0.6	0.8	0.8	0.9	1.5	2.0	3.5	3.8	5.9	11.1	17.0	26.2	34.0
EASTERN HEMISPHERE																				
W. EUROPE																				
W. Germany	-	-	-	-	-	-	neg.	neg.	neg.	neg.	neg.	0.1	0.3	0.4	1.3	1.6	1.6	2.3	2.9	2.7
Netherlands	-	-	-	-	-	-	-	-	-	-	-	-	-	neg.	0.1	0.1	0.1	0.1	0.3	0.8
Belgium/Luxembourg	-	-	-	-	-	-	-	neg.	neg.	neg.	neg.	neg.	neg.	neg.	neg.	neg.	neg.	neg.	-	-
France	-	-	-	-	-	neg.	0.1	0.1	0.1	0.1	0.2	0.3	0.6	0.8	1.2	1.4	2.3	3.6	3.0	3.0
Italy	-	-	-	-	-	-	-	-	0.1	0.6	0.9	1.1	0.8	0.6	0.4	0.8	0.8	0.9	0.8	0.9
United Kingdom	neg.	neg.	0.1	0.1	0.3	0.5	0.6	0.9	1.7	2.2	4.2	5.6	6.4	7.1	7.5	6.7	7.0	6.5	5.9	7.3
Eire	-	-	-	-	-	-	-	-	-	-	-	-	-	-	-	-	-	-	-	-
Denmark	-	-	-	-	-	-	-	-	-	-	-	-	-	-	-	-	-	-	-	-
Norway	-	-	-	-	-	-	-	-	-	-	-	-	-	-	-	-	-	-	-	-
Sweden	-	-	-	-	-	-	-	-	-	neg.	neg.	neg.	neg.	neg.	neg.	neg.	neg.	0.3	0.5	0.4
Finland	-	-	-	-	-	-	-	-	-	-	-	-	-	-	-	-	-	-	-	-
Portugal	-	-	-	-	-	-	-	-	-	-	-	-	-	-	-	-	-	-	-	-
Spain	-	-	-	-	-	-	-	-	-	-	-	-	-	neg.	0.2	0.2	0.6	1.3	1.7	1.9
Switzerland	-	-	-	-	-	-	-	-	-	-	-	-	-	-	0.1	0.6	0.7	0.9	1.6	1.5
Austria	-	-	-	-	-	-	-	-	-	-	-	-	-	-	-	-	-	-	-	-
Greece	-	-	-	-	-	-	-	-	-	-	-	-	-	-	-	-	-	-	-	-
Turkey	-	-	-	-	-	-	-	-	-	-	-	-	-	-	-	-	-	-	-	-
Iceland	-	-	-	-	-	-	-	-	-	-	-	-	-	-	-	-	-	-	-	-
Malta/Gibraltar/Cyprus	-	-	-	-	-	-	-	-	-	-	-	-	-	-	-	-	-	-	-	-
Yugoslavia	-	-	-	-	-	-	-	-	-	-	-	-	-	-	-	-	-	-	-	-
Canary Islands	-	-	-	-	-	-	-	-	-	-	-	-	-	-	-	-	-	-	-	-
TOTAL W. EUROPE	neg.	neg.	0.1	0.1	0.3	0.5	0.7	1.0	1.9	2.9	5.3	7.1	8.1	8.9	10.8	11.4	13.1	15.9	16.7	18.5
Middle East	-	-	-	-	-	-	-	-	-	-	-	-	-	-	-	-	-	-	-	-
Africa	-	-	-	-	-	-	-	-	-	-	-	-	-	-	-	-	-	-	-	-
S. Asia	-	-	-	-	-	-	-	-	-	-	-	-	-	-	0.3	0.6	0.4	0.7	1.0	1.0
S.E. Asia	-	-	-	-	-	-	-	-	-	-	-	-	-	-	-	-	-	-	-	-
Japan	-	-	-	-	-	-	-	-	neg.	neg.	neg.	0.1	0.2	0.3	0.3	1.1	2.0	2.5	2.1	4.1
Australia	-	-	-	-	-	-	-	-	-	-	-	-	-	-	-	-	-	-	-	-
New Zealand/SWPI	-	-	-	-	-	-	-	-	-	-	-	-	-	-	-	-	-	-	-	-
TOTAL E. HEMISPHERE (NON-COMMUNIST)	neg.	neg.	0.1	0.1	0.3	0.5	0.7	1.0	1.9	2.9	5.3	7.2	8.3	9.2	11.4	13.1	15.5	19.1	19.8	23.6
TOTAL NON-COMMUNIST WORLD	neg.	neg.	0.1	0.1	0.3	0.6	1.1	1.6	2.7	3.7	6.2	8.7	10.3	12.7	15.2	19.0	26.6	36.1	46.0	57.6
COMMUNIST WORLD																				
U.S.S.R.	-	-	-	-	-	n.a.	n.a.	n.a.	n.a.	n.a.	n.a.	0.4	0.4	0.6	0.7	0.9	0.9	2.1	3.0	4.0
E. Europe	-	-	-	-	-	-	n.a.	n.a.	n.a.	n.a.	n.a.	neg.	0.1	0.1	0.1	0.1	0.1	0.1	0.1	0.3
China etc.	-	-	-	-	-	-	-	-	-	-	-	-	-	-	-	-	-	-	-	-
TOTAL COMMUNIST WORLD	-	-	-	-	-	n.a.	n.a.	n.a.	n.a.	n.a.	n.a.	0.4	0.5	0.7	0.8	1.0	1.0	2.2	3.1	4.3
TOTAL WORLD	neg.	neg.	0.1	0.1	0.3	0.6	1.1	1.6	2.7	3.7	6.2	9.1	10.8	13.4	16.0	20.0	27.6	38.3	49.1	61.9

Note: A dash represents zero consumption
Neg. is less than 0.05 million tonnes oil equivalent
n.a. = not available

Table 12 TOTAL PRIMARY ENERGY CONSUMPTION—HYDRO/NUCLEAR ELECTRICITY (HIGH FACTOR)
(Million tonnes oil equivalent. Oil = 10 000 kcal/kg)

	1955	1956	1957	1958	1959	1960	1961	1962	1963	1964	1965	1966	1967	1968	1969	1970	1971	1972	1973	1974
WESTERN HEMISPHERE																				
U.S.A.	974.4	1014.7	1019.1	1014.6	1055.7	1101.5	1116.1	1170.3	1220.4	1273.3	1331.1	1406.6	1453.3	1541.3	1626.5	1683.8	1723.3	1767.9	1822.7	1783.4
Canada	75.1	84.6	84.5	84.5	90.2	95.7	96.5	98.8	104.9	116.0	124.4	133.5	138.9	148.9	159.5	171.5	178.6	181.8	223.8	231.5
Mexico						22.3	23.1	23.2	25.7	28.6	30.4	35.2	36.8	38.5	41.7	44.9	46.6	55.0		
Caribbean	81.7	87.5	97.2	102.5	108.0	39.8	44.4	41.6	45.3	47.9	51.0	55.1	58.5	62.3	64.1	70.6	74.9	84.0	257.3	272.5
S. America						49.4	52.0	55.3	57.8	60.2	63.3	67.8	70.7	76.2	83.1	90.4	98.1	105.6		
TOTAL WESTERN HEMISPHERE	1131.2	1186.8	1200.8	1201.6	1253.9	1308.7	1332.1	1389.2	1454.1	1526.0	1600.2	1698.2	1758.2	1867.2	1974.9	2061.2	2121.5	2194.3	2303.8	2287.4
EASTERN HEMISPHERE																				
W. EUROPE																				
W. Germany	131.9	139.9	141.9	136.2	136.2	150.2	153.5	165.1	179.1	185.4	189.4	193.3	190.4	207.5	227.4	242.3	244.9	252.3	270.5	257.1
Netherlands	46.6	50.8	49.8	46.2	47.4	24.3	26.1	29.1	32.1	34.1	36.6	39.0	41.9	46.9	53.0	60.7	64.3	72.4	76.9	72.4
Belgium/Luxembourg						28.1	29.2	31.8	34.3	35.0	36.4	35.4	37.5	41.1	45.3	48.8	48.1	51.2	50.8	49.2
France	75.3	84.6	85.8	85.0	83.2	89.5	93.3	99.6	109.4	112.7	118.3	120.7	128.3	134.7	148.1	159.5	164.4	174.2	182.9	179.7
Italy	31.3	34.8	36.7	38.5	41.6	50.9	55.0	61.4	68.8	74.3	80.6	87.7	95.0	101.1	109.5	120.9	126.8	134.9	136.5	137.7
United Kingdom	183.6	183.7	178.7	178.1	174.5	188.8	190.3	192.9	202.1	203.3	209.8	210.4	206.5	213.5	221.5	227.8	230.4	217.9	225.1	214.1
Eire	Included in ≠					4.1	4.5	4.8	4.9	5.1	4.8	5.4	5.8	6.8	6.7	6.9	7.1	6.5	7.4	7.2
Denmark						9.1	9.5	11.2	12.3	12.9	13.9	15.6	15.7	17.2	19.3	21.1	20.2	21.0	20.2	18.7
Norway	35.6	37.9	36.8	38.7	40.7	12.7	13.5	14.7	15.6	17.1	18.9	19.2	20.8	23.2	23.4	24.4	25.6	24.5	26.3	30.0
Sweden						23.6	24.7	26.6	28.1	30.8	33.3	35.6	35.7	38.4	39.5	42.6	43.8	46.4	49.5	46.8
Finland						6.2	7.1	7.9	7.8	9.3	10.1	11.5	12.0	13.1	14.5	15.7	16.1	21.0	20.2	19.3
Portugal						3.1	3.6	3.6	4.0	4.5	4.6	5.1	5.5	5.4	6.2	6.6	7.5	8.7	9.8	9.7
Spain						19.9	21.0	22.1	25.0	25.7	27.7	30.7	32.7	34.5	39.6	42.8	48.1	53.0	58.2	61.6
Switzerland						10.5	11.5	12.3	14.3	14.5	15.5	16.7	17.8	18.6	19.3	21.3	22.2	21.7	24.0	22.1
Austria	58.4≠	63.3≠	69.3≠	69.4≠	75.0≠	13.2	13.1	14.2	15.4	16.1	16.9	17.4	17.4	18.6	19.1	22.4	21.8	22.8	23.9	24.5
Greece						3.4	3.7	4.2	4.9	5.3	6.1	6.6	7.3	7.8	8.4	9.5	10.9	11.8	13.4	12.8
Turkey						5.2	5.2	6.2	7.1	7.7	9.3	9.3	9.9	11.3	12.4	12.9	14.0	24.0	27.6	28.5
Iceland						0.5	0.4	0.5	0.6	0.6	0.7	0.7	0.7	0.7	0.7	1.0	0.9	0.9	1.2	1.1
Malta/Gibraltar/Cyprus						0.7	0.7	0.7	0.8	0.8	0.7	0.7	0.8	0.9	1.0	1.1	1.2	1.3	1.4	1.4
Yugoslavia						12.6	13.1	13.6	15.4	17.1	18.3	18.4	17.8	19.2	21.3	23.4	26.7	25.2	27.3	29.2
Canary Islands						1.2	1.2	1.5	1.6	2.5	2.8	3.7	4.4	3.8	3.9	4.0	3.3	2.8	2.9	2.9
TOTAL W. EUROPE	563.2	595.0	599.0	592.1	598.6	657.8	680.2	724.0	783.6	814.8	854.7	883.1	903.9	964.3	1040.1	1115.7	1148.3	1194.5	1256.0	1226.0
Middle East						29.0	32.2	32.6	35.1	39.3	39.9	44.4	45.2	49.4	58.0	67.8	74.1	74.8	88.4	96.5
Africa	112.9	124.6	135.5	140.4	149.8	54.9	57.8	58.9	62.1	65.6	70.2	73.1	77.7	82.2	86.5	91.0	97.1	108.6	108.7	112.8
S. Asia						52.3	56.3	61.2	67.3	67.2	72.6	74.1	77.3	83.1	91.9	90.3	91.6	99.2	105.9	108.1
S.E. Asia						26.2	27.1	32.0	32.0	35.1	37.2	41.7	54.7	59.9	64.4	72.0	75.0	88.7	102.9	108.9
Japan	55.1	63.0	69.1	68.4	78.3	89.8	105.1	113.6	127.4	140.9	153.6	170.7	196.7	220.0	253.1	287.4	303.5	316.7	350.8	348.1
Australia	28.2	29.3	29.5	31.0	33.1	30.5	30.7	31.9	34.3	37.1	40.2	41.1	44.1	46.3	48.7	51.7	54.8	55.3	60.7	58.6
New Zealand/SWPI						5.1	5.8	6.0	6.3	7.1	7.5	8.1	7.7	8.5	8.7	9.0	9.7	10.1	7.6	7.9
TOTAL E.HEMISPHERE (NON-COMMUNIST)	759.4	811.9	833.1	831.9	859.8	945.6	995.2	1060.2	1148.1	1207.1	1275.9	1336.3	1407.3	1513.7	1651.4	1784.9	1854.1	1947.9	2081.0	2066.9
TOTAL NON-COMMUNIST WORLD	1890.6	1998.7	2033.9	2033.5	2113.7	2254.3	2327.3	2449.4	2602.2	2733.1	2876.1	3034.5	3165.5	3380.9	3626.3	3846.1	3975.6	4142.2	4384.8	4354.3
COMMUNIST WORLD																				
U.S.S.R.						443.3	462.6	493.7	534.8	573.1	608.5	649.1	683.7	712.0	746.7	781.4	818.0	875.7	919.8	970.2
E. Europe	558.3	611.4	665.6	800.8	891.1	525.4	423.2	440.0	250.7	263.1	265.8	271.8	276.1	290.3	309.9	328.6	342.1	320.1	338.3	352.9
China etc.									220.1	236.8	248.3	272.2	200.5	257.9	282.2	311.4	340.6	346.2	371.0	394.2
TOTAL COMMUNIST WORLD	558.3	611.4	665.6	800.8	891.1	968.7	885.8	933.7	1005.6	1073.0	1122.6	1193.1	1160.3	1260.2	1338.8	1421.4	1500.7	1542.0	1629.1	1717.3
TOTAL WORLD	2448.9	2610.1	2699.5	2834.3	3004.8	3223.0	3213.1	3383.1	3607.8	3806.1	3998.7	4227.6	4325.8	4641.1	4965.1	5267.5	5476.3	5684.2	6013.9	6071.6

Table 13 WORLD PRIMARY ENERGY CONSUMPTION (Million tonnes)

	OIL (NET)																			
	1955	1956	1957	1958	1959	1960	1961	1962	1963	1964	1965	1966	1967	1968	1969	1970	1971	1972	1973	1974
U.S.A.	415.6	433.0	434.0	443.0	460.0	473.1	478.1	498.2	513.0	527.2	548.9	575.7	595.8	635.5	667.8	694.6	719.3	775.8	818.0	785.4
W. EUROPE	116.8	133.0	134.0	153.0	173.0	204.7	228.1	263.6	303.3	344.5	388.6	425.6	458.3	504.8	563.6	625.7	655.3	701.0	749.0	699.2
JAPAN	10.2	13.0	16.0	17.0	23.0	30.0	41.0	48.8	62.6	75.2	87.9	100.0	122.9	142.7	169.0	199.1	219.7	234.3	268.3	261.1
REST OF NON-COMMUNIST WORLD	151.4	169.0	183.0	189.0	203.0	212.5	228.9	232.0	248.7	266.7	281.1	304.0	330.7	358.3	388.4	415.5	441.1	473.2	506.4	528.3
TOTAL NON-COMMUNIST WORLD	694.0	748.0	767.0	802.0	859.0	920.3	976.1	1042.6	1127.6	1213.6	1306.5	1405.3	1507.7	1641.3	1788.8	1934.9	2035.4	2184.3	2341.7	2274.0
COMMUNIST WORLD	82.3	94.0	108.0	120.0	132.0	144.7	156.2	174.8	192.8	206.9	222.6	241.7	261.7	284.0	307.6	338.9	361.1	393.7	434.2	468.9
TOTAL WORLD	776.3	842.0	875.0	922.0	991.0	1065.0	1132.3	1217.4	1320.4	1420.5	1529.1	1647.0	1769.4	1925.3	2096.4	2273.8	2396.5	2578.0	2775.9	2742.9

Table 14 WORLD PRIMARY ENERGY CONSUMPTION (Million tonnes oil equivalent. Oil = 10 000 kcal/kg)

	SOLID FUELS																			
U.S.A.	282.6	287.1	273.5	242.6	241.1	251.5	246.8	254.0	268.1	281.9	297.7	313.6	308.3	320.9	327.1	329.5	316.3	316.6	335.0	331.9
W. EUROPE	400.2	412.5	413.1	381.0	363.5	381.2	375.5	380.9	393.3	381.6	363.9	344.1	324.7	323.2	323.9	313.5	291.7	272.5	271.8	261.9
JAPAN	32.2	35.8	38.7	35.9	38.0	43.6	45.8	46.7	45.4	45.3	45.2	47.3	52.8	56.1	60.5	62.6	56.0	52.2	59.8	58.8
REST OF NON-COMMUNIST WORLD	103.4	108.7	109.9	108.8	110.4	115.4	115.4	120.6	126.9	130.2	140.0	141.4	143.5	147.6	152.1	153.2	157.6	179.2	189.7	189.9
TOTAL NON-COMMUNIST WORLD	818.4	844.1	835.2	768.3	753.0	791.7	783.5	802.2	833.7	839.0	846.8	846.4	829.3	847.8	863.6	858.8	821.6	820.5	856.3	842.5
COMMUNIST WORLD	452.6	488.5	518.4	629.2	698.5	751.4	640.1	650.3	685.6	718.0	728.3	759.6	692.8	751.5	788.2	814.9	849.4	870.1	896.8	925.7
TOTAL WORLD	1271.0	1332.6	1353.6	1397.5	1451.5	1543.1	1423.6	1452.5	1519.3	1557.0	1575.1	1606.0	1522.1	1599.3	1651.8	1673.7	1671.0	1690.6	1753.1	1768.2

Table 15 WORLD PRIMARY ENERGY CONSUMPTION (Million tonnes oil equivalent. Oil = 10 000 kcal/kg)

	NATURAL GAS																			
U.S.A.	245.6	261.6	276.5	291.2	317.5	337.5	350.0	372.3	394.1	416.0	431.8	463.8	488.1	522.2	561.2	588.1	606.8	587.4	572.3	560.4
W. EUROPE	4.8	5.9	6.4	6.9	9.1	11.0	13.0	14.6	15.3	17.1	19.5	23.1	28.9	40.5	55.4	75.7	97.7	113.7	129.8	146.1
JAPAN	0.1	0.2	0.2	0.3	0.4	0.7	0.9	1.3	1.8	1.9	1.8	1.9	2.0	2.1	2.4	3.6	3.9	5.0	4.8	5.1
REST OF NON-COMMUNIST WORLD	12.7	14.6	18.8	21.9	26.1	31.8	35.6	40.4	45.4	52.1	57.3	64.8	70.0	78.4	87.3	105.0	110.8	120.8	148.1	158.2
TOTAL NON-COMMUNIST WORLD	263.2	282.3	301.9	320.3	353.1	381.0	399.5	428.6	456.6	487.1	510.4	553.6	589.0	643.2	706.3	772.4	819.2	826.9	855.0	869.8
COMMUNIST WORLD	14.6	18.1	24.9	34.7	42.7	52.6	66.0	81.4	98.3	117.3	137.3	152.9	169.1	183.8	198.9	218.8	239.4	227.5	243.8	264.5
TOTAL WORLD	277.8	300.4	326.8	355.0	395.8	433.6	465.5	510.0	554.9	604.4	647.7	706.5	758.1	827.0	905.2	991.2	1058.6	1054.4	1098.8	1134.3

Table 16 WORLD PRIMARY ENERGY CONSUMPTION GROWTH RATES 1960–1974

	GROWTH RATES													
	5 YEARLY AVERAGES %									10 YEARLY AVERAGES %				
	66/61	67/62	68/63	69/64	70/65	71/66	72/67	73/68	74/69	70/60	71/61	72/62	73/63	74/64
	OIL (NET)													
U.S.A.	3.8	3.6	4.4	4.8	4.8	4.6	5.4	5.2	3.3	3.9	4.2	4.5	4.8	4.1
W. EUROPE	13.3	11.7	10.7	10.3	10.0	9.0	8.9	8.2	4.4	11.8	11.1	10.3	9.5	7.3
JAPAN	19.5	20.3	17.9	17.6	17.8	17.0	13.8	13.5	9.1	20.8	18.3	17.0	15.7	13.3
REST OF NON-COMMUNIST WORLD	5.8	7.3	7.6	7.8	8.1	7.7	7.4	7.2	6.3	6.9	6.8	7.4	7.4	7.1
<u>TOTAL NON-COMMUNIST WORLD</u>	7.6	7.7	7.8	8.1	8.2	7.7	7.7	7.4	4.9	7.7	7.6	7.7	7.6	6.5
COMMUNIST WORLD	8.9	8.2	8.0	8.3	9.0	8.5	8.7	8.9	8.8	8.9	8.7	8.5	8.5	8.5
<u>TOTAL WORLD</u>	7.8	7.7	7.8	8.1	8.3	7.8	7.8	7.6	5.5	7.9	7.8	7.8	7.7	6.8

Table 17 WORLD PRIMARY ENERGY CONSUMPTION GROWTH RATES 1960–1974 (Oil = 10 000 kcal/kg)

	SOLID FUELS													
U.S.A.	4.9	4.0	3.7	3.0	2.1	0.2	0.5	0.9	0.3	2.7	2.5	2.2	2.3	1.6
W. EUROPE	-1.7	-3.1	-3.9	-3.2	-3.0	-3.3	-3.5	-3.4	-4.2	-1.9	-2.5	-3.3	-3.6	-3.7
JAPAN	0.6	2.5	4.3	5.9	6.7	3.5	-0.2	1.3	-0.5	3.7	2.0	1.1	2.8	2.6
REST OF NON-COMMUNIST WORLD	4.1	3.5	3.0	3.1	1.8	2.2	4.5	5.2	4.6	2.9	3.2	4.0	4.1	3.8
<u>TOTAL NON-COMMUNIST WORLD</u>	1.6	0.7	0.3	0.6	0.3	-0.6	-0.2	0.2	-0.5	0.8	0.5	0.2	0.3	-
COMMUNIST WORLD	3.5	1.3	1.8	1.9	2.3	2.3	7.1	3.6	3.3	0.8	2.9	4.1	2.7	2.6
<u>TOTAL WORLD</u>	2.4	0.9	1.0	1.2	1.2	0.8	3.4	1.9	1.4	0.8	1.6	2.1	1.4	1.3

Table 18 WORLD PRIMARY ENERGY CONSUMPTION GROWTH RATES 1960–1974 (Oil = 10 000 kcal/kg)

	NATURAL GAS													
U.S.A.	5.8	5.6	5.8	6.2	6.4	5.5	3.8	1.9	-	5.7	5.7	4.7	3.8	3.0
W. EUROPE	12.0	14.6	21.3	26.3	30.8	33.5	31.7	26.3	21.3	21.1	22.3	22.8	23.8	23.8
JAPAN	14.3	9.5	4.1	4.4	14.6	14.9	20.3	17.3	16.6	17.6	14.6	14.7	10.5	10.3
REST OF NON-COMMUNIST WORLD	12.7	11.5	11.5	10.9	13.0	11.3	11.6	13.6	12.6	12.7	12.0	11.6	12.5	11.8
<u>TOTAL NON-COMMUNIST WORLD</u>	6.7	6.6	7.1	7.7	8.6	8.2	7.0	5.9	4.2	7.3	7.4	6.8	6.5	6.0
COMMUNIST WORLD	18.3	15.8	13.3	11.1	9.8	9.4	6.2	5.8	5.9	15.3	13.7	10.9	9.5	8.5
<u>TOTAL WORLD</u>	8.7	8.2	8.3	8.4	8.9	8.4	6.9	5.8	4.6	8.6	8.6	7.6	7.1	6.5

<u>Note</u>: A dash represents a positive or negative growth rate less than 0.1

Table 19 WORLD PRIMARY ENERGY CONSUMPTION (Million tonnes oil equivalent. Oil = 10 000 kcal/kg)

	HYDRO ELECTRICITY (LOW FACTOR)																			
	1955	1956	1957	1958	1959	1960	1961	1962	1963	1964	1965	1966	1967	1968	1969	1970	1971	1972	1973	1974
U.S.A.	10.2	11.0	11.7	12.6	12.4	13.1	13.7	15.0	14.8	15.8	17.2	17.3	19.7	19.8	22.3	21.9	23.6	24.3	25.2	25.6
W. EUROPE	13.6	14.4	15.1	17.0	17.4	20.3	21.0	21.5	23.5	22.9	25.9	27.7	28.2	29.2	28.6	29.8	30.2	30.1	29.7	33.4
JAPAN	4.2	4.6	4.7	5.0	5.7	5.1	5.8	5.6	5.9	6.2	6.2	7.1	6.3	6.3	6.9	7.0	7.4	7.6	5.3	6.3
REST OF NON-COMMUNIST WORLD	10.2	11.0	11.5	12.7	13.7	14.8	15.2	16.1	16.6	18.4	19.3	21.3	22.5	23.6	26.1	28.4	29.8	29.2	35.2	38.0
TOTAL NON-COMMUNIST WORLD	38.2	41.0	43.0	47.3	49.2	53.3	55.7	58.2	60.8	63.3	68.6	73.4	76.7	78.9	83.9	87.1	91.0	91.2	95.4	103.3
COMMUNIST WORLD	2.8	3.5	4.8	5.7	6.0	6.6	7.8	9.1	9.5	10.1	11.4	12.8	12.1	13.4	14.5	15.9	16.6	16.2	17.1	18.0
TOTAL WORLD	41.0	44.5	47.8	53.0	55.2	59.9	63.5	67.3	70.3	73.4	80.0	86.2	88.8	92.3	98.4	103.0	107.6	107.4	112.5	121.3

Table 20 WORLD PRIMARY ENERGY CONSUMPTION (Million tonnes oil equivalent. Oil = 10 000 kcal/kg)

	NUCLEAR ELECTRICITY (LOW FACTOR)																			
	1955	1956	1957	1958	1959	1960	1961	1962	1963	1964	1965	1966	1967	1968	1969	1970	1971	1972	1973	1974
U.S.A.	-	-	-	-	-	neg.	0.1	0.2	0.3	0.3	0.3	0.5	0.6	1.1	1.3	1.9	3.4	5.0	7.2	9.6
W. EUROPE	neg.	neg.	neg.	neg.	0.1	0.2	0.2	0.3	0.5	0.9	1.8	2.3	2.7	3.0	3.5	3.7	4.2	5.2	5.5	6.1
JAPAN	-	-	-	-	-	-	-	-	neg.	neg.	neg.	neg.	0.1	0.1	0.1	0.4	0.7	0.8	0.7	1.4
REST OF NON-COMMUNIST WORLD	-	-	-	-	-	-	-	-	-	-	-	-	-	-	0.1	0.3	0.4	0.8	1.8	2.1
TOTAL NON-COMMUNIST WORLD	neg.	neg.	neg.	neg.	0.1	0.2	0.3	0.5	0.8	1.2	2.1	2.8	3.4	4.2	5.0	6.3	8.7	11.8	15.2	19.2
COMMUNIST WORLD	-	-	-	-	-	n.a.	n.a.	n.a.	n.a.	n.a.	n.a.	0.1	0.1	0.2	0.2	0.3	0.4	0.7	1.1	1.5
TOTAL WORLD	neg.	neg.	neg.	neg.	0.1	0.2	0.3	0.5	0.8	1.2	2.1	2.9	3.5	4.4	5.2	6.6	9.1	12.5	16.3	20.7

Table 21 WORLD PRIMARY ENERGY CONSUMPTION (Million tonnes oil equivalent. Oil = 10 000 kcal/kg)

	TOTAL PRIMARY ENERGY CONSUMPTION (LOW FACTOR)																			
	1955	1956	1957	1958	1959	1960	1961	1962	1963	1964	1965	1966	1967	1968	1969	1970	1971	1972	1973	1974
U.S.A.	954.0	992.7	995.7	989.4	1031.0	1075.2	1088.7	1139.7	1190.3	1241.2	1295.9	1370.9	1412.5	1499.5	1579.7	1636.0	1669.4	1709.1	1757.7	1712.9
W. EUROPE	535.4	565.8	568.6	557.9	563.1	617.4	637.8	680.9	735.9	767.0	799.7	822.8	842.8	900.7	975.0	1048.4	1079.1	1122.5	1185.8	1146.7
JAPAN	46.7	53.6	59.6	58.2	67.1	79.4	93.5	102.4	115.7	128.6	141.1	156.3	184.1	207.3	238.9	272.7	287.7	299.9	338.9	332.7
REST OF NON-COMMUNIST WORLD	277.7	303.3	323.2	332.4	353.2	374.5	395.1	409.1	437.6	467.4	497.7	531.5	566.7	607.9	654.0	702.4	739.7	803.2	881.2	916.5
TOTAL NON-COMMUNIST WORLD	1813.8	1915.4	1947.1	1937.9	2014.4	2146.5	2215.1	2332.1	2479.5	2604.2	2734.4	2881.5	3006.1	3215.4	3447.6	3659.5	3775.9	3934.7	4163.6	4108.8
COMMUNIST WORLD	552.3	604.1	656.1	789.6	879.2	955.3	870.1	915.6	986.2	1052.3	1099.6	1167.1	1135.8	1232.9	1309.4	1388.8	1466.9	1508.2	1593.0	1678.6
TOTAL WORLD	2366.1	2519.5	2603.2	2727.5	2893.6	3101.8	3085.2	3247.7	3465.7	3656.5	3834.0	4048.6	4141.9	4448.3	4757.0	5048.3	5242.8	5442.9	5756.6	5787.4

A dash represents zero consumption

neg. = less than 0.05 million tonnes oil equivalent

n.a. = not available

Table 22 WORLD PRIMARY ENERGY CONSUMPTION GROWTH RATES 1960–1974 (Oil = 10 000 kcal/kg)

	GROWTH RATES													
	5 YEARLY AVERAGES %									10 YEARLY AVERAGES %				
	66/61	67/62	68/63	69/64	70/65	71/66	72/67	73/68	74/69	70/60	71/61	72/62	73/63	74/64
	HYDRO ELECTRICITY (LOW FACTOR)													
U.S.A.	4.9	5.5	6.0	7.1	5.0	6.4	4.3	4.9	2.9	5.3	5.7	4.9	5.5	5.0
W. EUROPE	5.8	5.6	4.6	4.7	2.9	1.6	1.5	0.3	3.0	4.0	3.7	3.6	2.5	3.9
JAPAN	4.3	2.3	1.3	2.6	2.3	0.4	3.8	-3.4	-1.9	3.1	2.3	3.0	-1.1	0.3
REST OF NON-COMMUNIST WORLD	6.9	6.7	7.1	7.2	7.8	7.0	5.4	8.3	7.9	6.5	6.9	6.1	7.7	7.6
<u>TOTAL NON-COMMUNIST WORLD</u>	5.8	5.6	5.4	5.9	4.8	4.3	3.6	3.9	4.3	5.0	5.0	4.6	4.6	5.1
COMMUNIST WORLD	10.4	5.9	6.9	7.1	6.8	5.3	7.2	4.9	4.4	9.0	7.8	6.6	5.9	5.7
<u>TOTAL WORLD</u>	6.4	5.7	5.6	6.0	5.1	4.5	4.1	4.0	4.3	5.5	5.4	4.9	4.8	5.2

Table 23 WORLD PRIMARY ENERGY CONSUMPTION GROWTH RATES 1960–1974 (Oil = 10 000 kcal/kg)

	NUCLEAR ELECTRICITY (LOW FACTOR)													
U.S.A.	26.6	27.5	31.3	33.0	42.9	47.0	49.9	46.0	51.1	45.3	36.4	38.2	38.4	41.8
W. EUROPE	58.4	49.6	36.7	29.2	16.4	13.7	14.2	12.8	11.3	34.8	34.2	30.7	24.2	19.9
JAPAN	*	*	*	*	185.9	68.8	73.5	50.3	70.2	*	*	*	*	*
REST OF NON-COMMUNIST WORLD	*	50.3	58.2	66.3	95.0	101.1	120.4	88.3	66.7	*	*	82.0	73.4	66.5
<u>TOTAL NON-COMMUNIST WORLD</u>	49.8	43.8	36.0	31.2	24.9	25.5	28.3	28.9	30.4	39.0	37.1	35.8	32.4	30.8
COMMUNIST WORLD	*	*	*	*	*	22.0	29.9	32.5	37.6	*	*	*	n.a.	n.a.
<u>TOTAL WORLD</u>	51.3	45.3	37.5	32.7	26.2	25.3	28.4	29.1	30.8	39.7	37.7	36.6	33.3	31.7

Table 24 WORLD PRIMARY ENERGY CONSUMPTION GROWTH RATES 1960–1974 (Oil = 10 000 kcal/kg)

	TOTAL PRIMARY ENERGY CONSUMPTION (LOW FACTOR)													
U.S.A.	4.7	4.4	4.7	4.9	4.8	4.0	3.9	3.2	1.6	4.3	4.4	4.1	4.0	3.3
W. EUROPE	5.2	4.4	4.1	4.9	5.6	5.6	5.9	5.7	3.3	5.4	5.4	5.1	4.9	4.1
JAPAN	10.8	12.4	12.4	13.2	14.1	13.0	10.3	10.3	6.8	13.1	11.9	11.3	11.3	10.0
REST OF NON-COMMUNIST WORLD	6.1	6.7	6.8	6.9	7.1	6.8	7.2	7.7	7.0	6.5	6.5	7.0	7.3	7.0
<u>TOTAL NON-COMMUNIST WORLD</u>	5.4	5.2	5.3	5.8	6.0	5.6	5.5	5.3	3.6	5.5	5.5	5.4	5.3	4.7
COMMUNIST WORLD	6.0	4.4	4.6	4.5	4.8	4.7	5.8	5.3	5.1	3.8	5.4	5.1	4.9	4.8
<u>TOTAL WORLD</u>	5.6	5.0	5.1	5.4	5.7	5.3	5.6	5.3	4.0	5.0	5.4	5.3	5.2	4.7

n.a. = not available
* = Greater than 300%

Table 25 WORLD PRIMARY ENERGY CONSUMPTION (Million tonnes oil equivalent. Oil = 10 000 kcal/kg)

	HYDRO ELECTRICITY (HIGH FACTOR)																			
	1955	1956	1957	1958	1959	1960	1961	1962	1963	1964	1965	1966	1967	1968	1969	1970	1971	1972	1973	1974
U.S.A.	30.6	33.0	35.1	37.8	37.1	39.3	40.8	45.2	44.4	47.4	51.8	52.0	59.1	59.4	66.7	65.9	70.9	72.9	75.6	76.9
W. EUROPE	41.4	43.6	45.4	51.1	52.7	60.4	62.9	63.9	69.8	68.7	77.4	83.2	83.9	86.9	86.4	89.4	90.5	91.4	88.7	100.3
JAPAN	12.6	14.0	14.2	15.2	16.9	15.5	17.4	16.8	17.6	18.5	18.7	21.4	18.8	18.8	20.9	21.0	21.9	22.7	15.8	19.0
REST OF NON-COMMUNIST WORLD	30.4	33.7	35.0	38.7	41.6	45.5	46.0	48.5	49.8	55.1	58.3	63.9	67.4	70.8	78.4	84.7	89.5	87.4	105.7	114.2
TOTAL NON-COMMUNIST WORLD	115.0	124.3	129.7	142.8	148.3	160.7	167.1	174.4	181.6	189.7	206.2	220.5	229.2	235.9	252.4	261.0	272.8	274.4	285.8	310.4
COMMUNIST WORLD	8.8	10.8	14.3	16.9	17.9	20.0	23.5	27.2	28.9	30.8	34.4	38.5	36.2	40.2	43.3	47.8	49.8	48.5	51.2	53.9
TOTAL WORLD	123.8	135.1	144.0	159.7	166.2	180.7	190.6	201.6	210.5	220.5	240.6	259.0	265.4	276.1	295.7	308.8	322.6	322.9	337.0	364.3

Table 26 WORLD PRIMARY ENERGY CONSUMPTION (Million tonnes oil equivalent. Oil = 10 000 kcal/kg)

	NUCLEAR ELECTRICITY (HIGH FACTOR)																			
U.S.A.	-	-	-	-	-	0.1	0.4	0.6	0.8	0.8	0.9	1.5	2.0	3.3	3.7	5.7	10.0	15.2	21.8	28.8
W. EUROPE	neg.	neg.	0.1	0.1	0.3	0.5	0.7	1.0	1.9	2.9	5.3	7.1	8.1	8.9	10.8	11.4	13.1	15.9	16.7	18.5
JAPAN	-	-	-	-	-	-	-	-	neg.	neg.	neg.	0.1	0.2	0.3	0.3	1.1	2.0	2.5	2.1	4.1
REST OF NON-COMMUNIST WORLD	-	-	-	-	-	-	-	-	-	-	-	-	-	0.2	0.4	0.8	1.5	2.5	5.4	6.2
TOTAL NON-COMMUNIST WORLD	neg.	neg.	0.1	0.1	0.3	0.6	1.1	1.6	2.7	3.7	6.2	8.7	10.3	12.7	15.2	19.0	26.6	36.1	46.0	57.6
COMMUNIST WORLD	-	-	-	-	-	n.a.	n.a.	n.a.	n.a.	n.a.	n.a.	0.4	0.5	0.7	0.8	1.0	1.0	2.2	3.1	4.3
TOTAL WORLD	neg.	neg.	0.1	0.1	0.3	0.6	1.1	1.6	2.7	3.7	6.2	9.1	10.8	13.4	16.0	20.0	27.6	38.3	49.1	61.9

Table 27 WORLD PRIMARY ENERGY CONSUMPTION (Million tonnes oil equivalent. Oil = 10 000 kcal/kg)

	TOTAL PRIMARY ENERGY CONSUMPTION (HIGH FACTOR)																			
U.S.A.	974.4	1014.7	1019.1	1014.6	1055.7	1101.5	1116.1	1170.3	1220.4	1273.3	1331.1	1406.6	1453.3	1541.3	1626.5	1683.8	1723.3	1767.9	1822.7	1783.4
W. EUROPE	563.2	595.0	599.0	592.1	598.6	657.8	680.2	724.0	783.6	814.8	854.7	883.1	903.9	964.3	1040.1	1115.7	1148.3	1194.5	1256.0	1226.0
JAPAN	55.1	63.0	69.1	68.4	78.3	89.8	105.1	113.6	127.4	140.9	153.6	170.7	196.7	220.0	253.1	287.4	303.5	316.7	350.8	348.1
REST OF NON-COMMUNIST WORLD	297.9	326.0	346.7	358.4	381.1	405.2	425.9	441.5	470.8	504.1	536.7	574.1	611.6	655.3	706.6	759.2	800.5	863.1	955.3	996.8
TOTAL NON-COMMUNIST WORLD	1890.6	1998.7	2033.9	2033.5	2113.7	2254.3	2327.3	2449.4	2602.2	2733.1	2876.1	3034.5	3165.5	3380.9	3626.3	3846.1	3975.6	4142.2	4384.8	4354.3
COMMUNIST WORLD	558.3	611.4	665.6	800.8	891.1	968.7	885.8	933.7	1005.6	1073.0	1122.6	1193.1	1160.3	1260.2	1338.8	1421.4	1500.7	1542.0	1629.1	1717.3
TOTAL WORLD	2448.9	2610.1	2699.5	2834.3	3004.8	3223.0	3213.1	3383.1	3607.8	3806.1	3998.7	4227.6	4325.8	4641.1	4965.1	5267.5	5476.3	5684.2	6013.9	6071.6

A dash represents zero consumption

neg. = less than 0.05 million tonnes oil equivalent

n.a. = not available

Table 28 WORLD PRIMARY ENERGY CONSUMPTION GROWTH RATES 1960–1974 (Oil = 10 000 kcal/kg)

	GROWTH RATES													
	5 YEARLY AVERAGES %									10 YEARLY AVERAGES %				
	66/61	67/62	68/63	69/64	70/65	71/66	72/67	73/68	74/69	70/60	71/61	72/62	73/63	74/64
	HYDRO ELECTRICITY (HIGH FACTOR)													
U.S.A.	4.9	5.5	6.0	7.1	5.0	6.4	4.3	4.9	2.9	5.3	5.7	4.9	5.5	5.0
W. EUROPE	5.8	5.6	4.6	4.7	2.9	1.6	1.5	0.3	3.0	4.0	3.7	3.6	2.5	3.9
JAPAN	4.3	2.3	1.3	2.6	2.3	0.4	3.8	-3.4	-1.9	3.1	2.3	3.0	-1.1	0.3
REST OF NON-COMMUNIST WORLD	6.9	6.7	7.1	7.2	7.8	7.0	5.4	8.3	7.9	6.5	6.9	6.1	7.7	7.6
<u>TOTAL NON-COMMUNIST WORLD</u>	5.8	5.6	5.4	5.9	4.8	4.3	3.6	3.9	4.3	5.0	5.0	4.6	4.6	5.1
COMMUNIST WORLD	10.4	5.9	6.9	7.1	6.8	5.3	7.2	4.9	4.4	9.0	7.8	6.6	5.9	5.7
<u>TOTAL WORLD</u>	6.4	5.7	5.6	6.0	5.1	4.5	4.1	4.0	4.3	5.5	5.4	4.9	4.8	5.2

Table 29 WORLD PRIMARY ENERGY CONSUMPTION GROWTH RATES 1960–1974 (Oil = 10 000 kcal/kg)

	NUCLEAR ELECTRICITY (HIGH FACTOR)													
U.S.A.	26.6	27.5	31.3	33.0	42.9	47.0	49.9	46.0	51.1	45.3	36.4	38.2	38.4	41.8
W. EUROPE	58.4	49.6	36.7	29.2	16.4	13.7	14.2	12.8	11.3	34.8	34.2	30.7	24.2	19.9
JAPAN	*	*	*	*	185.9	68.8	73.5	50.3	70.2	*	*	*	*	*
REST OF NON-COMMUNIST WORLD	*	50.3	58.2	66.3	95.0	101.1	120.4	88.3	66.7	*	*	82.0	73.4	66.5
<u>TOTAL NON-COMMUNIST WORLD</u>	49.8	43.8	36.0	31.2	24.9	25.5	28.3	28.9	30.4	39.0	37.1	35.8	32.4	30.8
COMMUNIST WORLD	*	*	*	*	*	22.0	29.9	32.5	37.6	*	*	*	n.a.	n.a.
<u>TOTAL WORLD</u>	51.3	45.3	37.5	32.7	26.2	25.3	28.4	29.1	30.8	39.7	37.7	36.6	33.3	31.7

Table 30 WORLD PRIMARY ENERGY CONSUMPTION GROWTH RATES 1960–1974 (Oil = 10 000 kcal/kg)

	TOTAL PRIMARY ENERGY CONSUMPTION (HIGH FACTOR)													
U.S.A.	4.7	4.4	4.8	5.0	4.8	4.1	4.0	3.4	1.9	4.3	4.4	4.2	4.1	3.4
W. EUROPE	5.4	4.5	4.2	5.0	5.5	5.4	5.7	5.4	3.3	5.4	5.4	5.1	4.8	4.2
JAPAN	10.2	11.6	11.5	12.4	13.4	12.2	10.0	9.8	6.6	12.3	11.2	10.8	10.7	9.5
REST OF NON-COMMUNIST WORLD	6.1	6.7	6.8	7.0	7.2	6.9	7.1	7.8	7.1	6.5	6.5	6.9	7.3	7.1
<u>TOTAL NON-COMMUNIST WORLD</u>	5.4	5.3	5.4	5.8	6.0	5.6	5.5	5.3	3.7	5.5	5.5	5.4	5.4	4.8
COMMUNIST WORLD	6.1	4.4	4.6	4.5	4.8	4.7	5.8	5.3	5.1	3.9	5.5	5.1	4.9	4.8
<u>TOTAL WORLD</u>	5.6	5.0	5.2	5.5	5.7	5.3	5.6	5.3	4.1	5.0	5.5	5.3	5.2	4.8

n.a. = not available

* = Greater than 300%

Table 31 WORLD CONSUMPTION OF MAJOR PETROLEUM PRODUCTS 1973 AND 1974 (Thousand tonnes)

	Gasolines		*Middle Distillates*		*Fuel Oils*		*Others*		*Total*	
	1973	1974	1973	1974	1973	1974	1973	1974	1973	1974
USA	312 700	305 650	199 700	187 830	139 750	129 620	100 650	100 080	752 800	723 180
Canada	26 670	28 080	25 790	27 150	16 550	17 430	7 680	8 080	76 690	80 740
TOTAL WESTERN HEMISPHERE	384 070	380 800	266 930	258 590	203 180	196 400	120 240	120 690	974 420	956 480
Austria	2 500	2 310	2 420	2 280	5 400	4 750	890	1 040	11 210	10 380
Belgium/Luxembourg	4 530	4 090	10 170	9 200	10 290	8 450	1 960	1 760	26 950	23 500
Denmark	2 080	2 040	6 910	5 970	6 800	6 340	760	660	16 550	15 010
Ireland	910	870	1 530	1 460	2 490	2 390	260	250	5 190	4 970
Finland	1 450	1 290	4 790	4 350	5 690	4 550	550	570	12 480	10 760
France	21 530	20 460	46 720	41 850	36 650	36 740	8 050	7 720	112 950	106 770
Germany, West	28 830	23 890	65 690	57 640	29 950	24 960	12 160	15 830	136 630	122 320
Greece	1 150	1 100	3 300	2 720	3 850	3 770	560	430	8 860	8 020
Iceland	130	80	310	310	100	90	10	—	550	480
Italy	16 670	15 820	22 680	22 610	43 140	43 590	6 590	5 980	89 080	88 000
Netherlands	9 160	8 820	9 210	7 260	4 700	3 420	2 350	2 750	25 420	22 250
Norway	1 510	1 440	3 610	3 060	1 640	1 640	820	770	7 580	6 910
Portugal	1 130	980	1 500	1 610	1 780	2 100	850	540	5 260	5 230
Spain	5 410	5 420	8 020	7 430	17 350	20 140	3 290	2 990	34 070	35 980
Sweden	4 430	4 150	9 970	8 380	12 130	11 850	1 240	1 230	27 770	25 610
Switzerland	2 690	2 550	8 460	7 400	2 530	2 150	700	660	14 380	12 760
Turkey	1 420	1 530	3 500	3 610	5 010	4 730	1 330	1 490	11 260	11 360
UK	25 440	24 670	28 180	25 560	39790	37 350	6 080	5 270	99 490	92 850
Yugoslavia	1 720	1 900	3 700	4 070	3 680	4 040	1 230	1 350	10 330	11 360
Gibraltar, Malta, Cyprus	180	180	500	500	520	520	70	70	1 270	1 270
TOTAL WESTERN EUROPE	132 870	123 590	241 170	217 270	233 490	223 570	49 750	51 360	657 280	615 790
TOTAL AFRICA	10 200	10 270	16 380	16 190	8 290	8 060	2 520	2 510	37 390	37 030
India, Burma, Pakistan	4 020	4 240	13 060	12 080	6 390	6 460	3 630	3 230	27 100	26 010
Japan	42 800	42 570	49 490	48 730	125 050	119 110	19 590	18 520	236 930	228 930
Australia	9 400	9 820	6 780	7 570	5 770	5 480	1 800	1 930	23 750	24 800
New Zealand	1 700	1 680	1 120	1 140	1 020	1 110	250	240	4 090	4 170
TOTAL ASIA/AUSTRALASIA	74 970	75 990	105 820	107 310	175 500	171 190	32 420	31 820	388 710	386 310
TOTAL EASTERN HEMISPHERE*	218 040	209 850	363 370	340 770	417 280	402 820	84 690	85 690	1 083 380	1 039 130
TOTAL WORLD INLAND*	602 110	590 650	630 300	599 360	620 460	599 220	204 930	206 380	2 057 800	1 995 610
Bunkers	—	—	15 090	14 170	113 550	113 190	—	—	128 640	127 360
TOTAL WORLD*	602 110	590 650	645 390	613 530	734 010	712 410	204 930	206 380	2 186 440	2 122 970

* Excluding USSR, E. Europe and China

Table 32 QUARTERLY OIL DEMAND IN WESTERN EUROPE—MOTOR GASOLINE (Thousand tonnes)

	U.K.	FRANCE	GERMANY	ITALY	BELGIUM LUX.	NETHER-LANDS	DENMARK	SWEDEN	NORWAY	FINLAND	AUSTRIA	SWITZ-ERLAND	SPAIN	OTHER O.E.C.D. EUROPE *	TOTAL O.E.C.D. EUROPE **
1968															
I	2998	2229	2711	1710	421	561	294	532	152	169	266	365	392	429	13229
II	3282	2756	3325	2004	498	661	355	692	202	221	343	453	477	534	15803
III	3549	3044	3486	2306	529	704	385	741	244	246	414	519	608	627	17402
IV	3183	2653	3268	1942	504	661	348	640	190	203	325	449	473	536	15375
YEAR	13012	10682	12790	7962	1952	2587	1382	2605	788	839	1346	1786	1950	2126	61809
1969															
I	2993	2305	3015	1837	451	581	304	555	167	181	285	385	446	490	13995
II	3488	2958	3601	2158	555	741	382	725	218	239	368	491	550	584	17058
III	3645	3382	3893	2466	577	732	406	783	260	265	443	561	705	695	18813
IV	3318	2701	3577	2145	539	704	358	678	212	226	340	492	547	585	16422
YEAR	13444	11346	14086	8606	2122	2758	1450	2741	857	911	1436	1929	2248	2354	66288
% CHANGE OVER1968	3.3	6.2	10.1	8.1	8.7	6.6	4.9	5.2	8.8	8.6	6.5	8.0	15.3	10.7	7.2
1970															
I	3177	2599	3317	2021	494	608	308	594	184	178	320	414	539	537	15290
II	3654	3139	3918	2340	584	793	398	752	231	258	393	533	630	639	18262
III	3854	3520	4206	2625	625	815	417	810	280	297	485	617	820	749	20120
IV	3553	3021	4051	2285	608	779	375	722	229	263	384	544	633	628	18075
YEAR	14238	12279	15492	9271	2311	2995	1498	2878	924	996	1582	2108	2622	2553	71747
% CHANGE OVER1969	5.9	8.2	10.0	7.7	8.9	8.6	3.3	5.0	7.8	9.3	10.2	9.3	16.6	8.5	8.2
1971															
I	3368	2805	3772	2052	551	712	339	623	194	213	346	491	590	588	16644
II	3818	3423	4376	2432	618	836	400	786	253	285	454	594	709	683	19667
III	4046	3800	4582	2740	632	853	433	810	291	316	538	667	889	820	21417
IV	3731	3295	4476	2403	629	841	384	710	238	262	460	600	710	694	19433
YEAR	14963	13323	17206	9627	2430	3242	1556	2929	976	1076	1798	2352	2898	2785	77161
% CHANGE OVER1970	5.1	8.5	11.1	3.8	5.1	8.2	3.9	1.8	5.6	8.0	13.7	11.6	10.5	9.1	7.5

Table 32—*contd.*

	U.K.	FRANCE	GERMANY	ITALY	BELGIUM LUX.	NETHER-LANDS	DENMARK	SWEDEN	NORWAY	FINLAND	AUSTRIA	SWITZ-ERLAND	SPAIN	OTHER O.E.C.D. EUROPE *	TOTAL O.E.C.D. EUROPE **
1972															
I	3588	3161	4152	2309	616	774	350	676	216	249	402	516	685	665	18359
II	4015	3698	4672	2630	652	876	409	802	263	306	501	630	812	799	21065
III	4293	4110	4678	2944	724	886	448	833	306	330	638	690	1018	953	22851
IV	4003	3553	4628	2622	643	882	394	739	251	286	483	618	810	803	20715
YEAR	15899	14522	18130	10505	2635	3418	1601	3050	1036	1171	2024	2454	3325	3220	82990
% CHANGE OVER1971	6.3	9.0	5.4	9.1	8.4	5.4	2.9	4.1	6.1	8.8	12.6	4.3	14.7	15.6	7.6
1973															
I	3912	3413	4319	2529	602	844	401	694	228	257	467	558	802	790	19816
II	4353	4072	4904	2883	729	961	401	835	279	329	559	643	950	915	22813
III	4557	4327	4718	3168	678	889	456	867	322	351	669	683	1157	1071	23913
IV	4105	3960	4567	2670	721	862	397	817	268	307	523	626	959	911	21693
YEAR	16927	15772	18508	11250	2730	3556	1655	3213	1097	1244	2218	2510	3868	3687	88235
% CHANGE OVER1972	6.5	8.6	2.1	7.1	3.6	4.0	3.4	5.3	5.9	6.2	9.6	2.3	16.3	14.5	6.3
1974															
I	3632	3217	3875	2293	626	675	326	548	205	236	425	515	833	779	18185
II	4174	3996	4755	2654	643	843	383	774	258	298	511	610	955	903	21757
III	4447	4207	4783	2856	698	846	408	871	304	346	604	695	1175	1076	23316
IV	4230	3762	4635	2602	697	838	389	752	255	300	490	591	983	957	21481
YEAR	16483	15182	18048	10405	2664	3202	1506	2945	1022	1180	2030	2411	3946	3715	84739
% CHANGE OVER1973	-2.6	-3.7	-2.5	-7.5	-2.4	-10.0	-9.0	-8.3	-6.8	-5.1	-8.5	-3.9	2.0	.8	-4.0
1975															
I	3607	3395	4381	2405	670	796	354	697	240	284	460	536	935	875	19635
II	4136	4057	5145	2817	773	905	402	897	294	338	526	618	1044	1010	22962
III	4330	4539	5061	3050	745	855	440	909	338	379	630	691	1301	1222	24490
IV	4050	3931	5159	2728	772	919	410	777	280	340	509	599	1030	1048	22552
YEAR	16123	15922	19746	11000	2960	3475	1606	3280	1152	1341	2125	2444	4310	4155	89639
% CHANGE OVER1974	-2.2	4.9	9.4	5.7	11.1	8.5	6.6	11.4	12.7	13.6	4.7	1.4	9.2	11.8	5.8

* Includes Turkey, Greece, Portugal and Eire
** Canary Islands and Iceland not included

Table 33 QUARTERLY OIL DEMAND IN WESTERN EUROPE—GAS OIL (Thousand tonnes)

	U.K.	FRANCE	GERMANY	ITALY	BELGIUM LUX.	NETHER-LANDS	DENMARK	SWEDEN	NORWAY	FINLAND	AUSTRIA	SWITZ-ERLAND	SPAIN	OTHER O.E.C.D. EUROPE *	TOTAL O.E.C.D. EUROPE **
1968															
I	4111	9430	10921	1155	2171	1856	1556	2566	568	1090	246	1136	828	940	38574
II	3014	4737	8598	1327	1144	1051	849	1496	412	616	244	1428	910	935	26761
III	2811	3517	9982	1615	1232	966	757	1301	369	689	295	1536	1004	1008	27082
IV	3884	9285	10952	1918	2153	1861	1419	2312	603	1036	324	1098	920	1130	38895
YEAR	13820	26969	40453	6015	6700	5734	4581	7675	1952	3431	1109	5198	3662	4013	131312
1969															
I	4488	10391	13706	1864	2371	1965	1795	2799	630	1200	276	1243	886	1064	44678
II	3459	5262	11569	1663	1441	1185	1101	1717	449	715	270	1791	1026	1049	32697
III	2960	4775	10117	2061	1351	992	830	1259	421	734	312	1565	1113	1129	29619
IV	4475	9782	12248	3018	2161	1851	1607	2472	691	1122	384	1213	1037	1208	43269
YEAR	15382	30210	47640	8606	7324	5993	5333	8247	2191	3771	1242	5812	4062	4450	150263
% CHANGE OVER1968	11.3	12.0	17.8	43.1	9.3	4.5	16.4	7.5	12.2	9.9	12.0	11.8	10.9	10.9	14.4
1970															
I	5035	12289	15404	3002	2782	2193	1999	3147	724	1277	355	1429	998	1159	51793
II	3886	6267	12451	1988	1706	1424	1224	1863	523	797	317	1970	1131	1154	36701
III	3306	4805	11895	2569	1629	1154	957	1525	491	850	393	1790	1204	1213	33781
IV	4417	10861	13533	3911	2341	1898	1771	2492	753	1142	551	1242	1114	1358	47384
YEAR	16644	34222	53283	11470	8458	6669	5951	9027	2491	4066	1616	6431	4447	4884	169659
% CHANGE OVER1969	8.2	13.3	11.8	33.3	15.5	11.3	11.6	9.5	13.7	7.8	30.1	10.7	9.5	9.8	12.9
1971															
I	5553	14376	16439	4223	3014	2136	1999	2948	821	1250	399	1920	1068	1370	57516
II	3866	5776	13639	2059	1614	1274	1158	1833	564	781	352	1789	1140	1216	37061
III	3262	4422	11494	3051	1595	1239	972	1344	503	825	396	1724	1325	1336	33488
IV	5074	11997	14282	4807	2342	1929	1797	2668	826	1148	531	1424	1217	1502	51544
YEAR	17755	36571	55854	14140	8565	6578	5926	8793	2714	4004	1678	6857	4750	5424	179609
% CHANGE OVER1970	6.7	6.9	4.8	23.3	1.3	-1.4	-.4	-2.6	9.0	-1.5	3.8	6.6	6.8	11.1	5.9

Table 33—*contd.*

	U.K.	FRANCE	GERMANY	ITALY	BELGIUM LUX.	NETHER-LANDS	DENMARK	SWEDEN	NORWAY	FINLAND	AUSTRIA	SWITZ-ERLAND	SPAIN	OTHER O.E.C.D. EUROPE *	TOTAL O.E.C.D. EUROPE **
1972															
I	6450	13358	16238	4735	2943	2291	2083	3070	912	1331	490	1667	1170	1597	58335
II	4675	7348	13838	2721	2100	1593	1227	1820	583	868	381	1911	1327	1413	41805
III	3656	6096	13108	3564	1718	1327	998	1453	537	858	471	1834	1415	1535	38570
IV	5597	12565	15218	5620	2786	2116	1756	2675	813	1197	700	1491	1315	1780	55629
YEAR	20378	39367	58402	16640	9547	7327	6064	9018	2845	4254	2042	6903	5227	6325	194339
% CHANGE OVER 1971	14.8	7.6	4.6	17.7	11.5	11.4	2.3	2.6	4.8	6.2	21.7	.7	10.0	16.6	8.2
1973															
I	6318	15886	17262	5786	3130	2317	2127	2909	846	1284	582	1809	1359	1865	63480
II	4649	8204	15502	3228	2415	1722	1205	2137	625	865	489	1976	1449	1632	46098
III	3780	5677	13892	3515	1539	1269	976	1439	586	904	502	2101	1551	1706	39437
IV	6011	13984	16105	5366	2796	2021	1857	3061	939	1568	801	1896	1605	2052	60062
YEAR	20758	43751	62761	17895	9880	7329	6165	9546	2996	4621	2374	7782	5964	7255	209077
% CHANGE OVER 1972	1.9	11.1	7.5	7.5	3.5	.0	1.7	5.9	5.3	8.6	16.3	12.7	14.1	14.7	7.6
1974															
I	5570	13110	12525	5947	2708	1842	1671	2250	720	1181	575	1400	1356	1776	52631
II	4243	8645	14116	3910	2058	1190	1044	1678	503	765	388	2010	1491	1609	43650
III	3602	6075	14246	3797	1644	1226	977	1524	522	677	584	1845	1633	1708	40060
IV	5684	10482	13889	4746	2509	1706	1653	2479	853	1370	635	1481	1526	1940	50953
YEAR	19099	38312	54776	18400	8919	5964	5345	7931	2598	3993	2182	6736	6006	7033	187294
% CHANGE OVER 1973	-8.0	-12.4	-12.7	2.8	-9.7	-18.6	-13.3	-16.9	-13.3	-13.6	-8.1	-13.4	.7	-3.1	-10.4
1975															
I	5315	10556	14300	5110	2666	1544	1704	2576	785	1275	479	1647	1384	1831	51174
II	4382	8442	13113	3850	2071	1603	1263	2149	641	863	487	1819	1581	1684	43948
III	3445	6217	14303	4080	1580	1348	950	1447	621	911	684	1899	1641	1999	41125
IV	5311	11435	13456	5360	2572	1686	1523	2148	851	1262	670	1376	1690	1985	51325
YEAR	18453	36652	55172	18400	8889	6181	5440	8320	2898	4311	2320	6741	6296	7499	187572
% CHANGE OVER 1974	-3.4	-4.3	.7	.0	-.3	3.6	1.8	4.9	11.5	8.0	6.3	.1	4.8	6.6	.1

* Includes Turkey, Greece, Portugal and Eire
** Canary Islands and Iceland not included

Table 34 QUARTERLY OIL DEMAND IN WESTERN EUROPE—FUEL OIL (Thousand tonnes)

	U.K.	FRANCE	GERMANY	ITALY	BELGIUM LUX.	NETHER-LANDS	DENMARK	SWEDEN	NORWAY	FINLAND	AUSTRIA	SWITZ-ERLAND	SPAIN	OTHER O.E.C.D EUROPE *	TOTAL O.E.C.D EUROPE **
1968															
I	9530	4635	5603	10693	1884	2362	1587	2749	560	838	1117	485	2381	1653	46077
II	6959	3137	4751	5748	1663	1749	1138	1903	465	610	777	467	1911	1468	32746
III	5692	3084	4767	6890	1541	1618	1081	1806	394	553	857	381	2121	1584	32369
IV	8921	5964	6234	10469	2449	2403	1994	3780	599	932	1209	510	2591	1785	49840
YEAR	31102	16820	21355	33800	7537	8132	5800	10238	2018	2933	3960	1843	9004	6490	161032
1969															
I	10078	4995	5778	11210	2308	2638	2080	3219	640	980	1223	459	2527	1744	49879
II	7642	3868	5212	6590	2042	1806	1486	2574	522	689	845	518	2026	1699	37519
III	6397	3577	5090	7440	1860	1613	1418	2095	442	760	942	471	2163	1750	36018
IV	9810	6011	6917	10260	2286	2270	2401	4055	684	1403	1277	563	2669	1992	52598
YEAR	33927	18451	22997	35500	8496	8327	7385	11943	2288	3832	4287	2011	9385	7185	176014
% CHANGE OVER1968	9.1	9.7	7.7	5.0	12.7	2.4	27.3	16.7	13.4	30.7	8.3	9.1	4.2	10.7	9.3
1970															
I	11216	6273	6846	11921	2616	2579	2452	3726	604	1210	1186	497	2705	1951	55782
II	8648	5198	6327	7674	2466	1973	1822	3303	644	905	961	573	2576	1925	44995
III	7557	4691	5772	7847	1914	1581	1736	3053	507	964	955	552	2672	1870	41671
IV	11165	7306	7401	11558	2422	1918	2464	3717	657	1184	1397	621	3417	2341	57568
YEAR	38586	23468	26346	39000	9418	8051	8474	13799	2412	4263	4499	2243	11370	8087	200016
% CHANGE OVER1969	13.7	27.2	14.6	9.9	10.9	-3.3	14.7	15.5	5.4	11.2	4.9	11.5	21.2	12.6	13.6
1971															
I	12068	7568	7335	11750	2392	2086	2414	3350	521	1237	1467	638	3682	2461	58969
II	9148	5652	5625	7750	1944	1319	1744	2644	417	877	1155	534	2754	2228	43791
III	7098	5304	5547	7950	2132	1176	1718	2367	406	940	987	498	2448	2343	40914
IV	11071	8999	7819	11350	3021	1763	2456	3779	485	1396	1499	672	3267	2689	60266
YEAR	39385	27523	26326	38800	9489	6344	8332	12140	1829	4450	5108	2342	12151	9721	203940
% CHANGE OVER1970	2.1	17.3	-.1	-.5	.8	-21.2	-1.7	-12.0	-24.2	4.4	13.5	4.4	6.9	20.2	2.0

Table 34—*contd.*

	U.K.	FRANCE	GERMANY	ITALY	BELGIUM LUX.	NETHER-LANDS	DENMARK	SWEDEN	NORWAY	FINLAND	AUSTRIA	SWITZ-ERLAND	SPAIN	OTHER O.E.C.D EUROPE *	TOTAL O.E.C.D EUROPE **
1972															
I	12540	8459	7592	11150	2862	1764	2492	3373	462	1373	1487	588	3718	2960	60820
II	9828	7084	6455	8050	2441	1308	1850	2691	441	1141	1032	571	2971	2739	48602
III	7497	6509	6335	7650	2123	963	1619	2501	348	1099	959	603	2946	2648	43800
IV	11465	9844	8022	11650	2657	1615	2357	3844	497	1456	1465	652	3638	3114	62276
YEAR	41330	31896	28404	38500	10083	5650	8318	12409	1748	5069	4943	2414	13273	11461	215498
% CHANGE OVER1971	4.9	15.9	7.9	-.8	6.3	-10.9	-.2	2.2	-4.4	13.9	-3.2	3.1	9.2	17.9	5.7
1973															
I	12229	9981	8154	11800	2780	1483	2173	2943	406	1444	1564	587	4278	3348	63170
II	9218	8273	6639	8900	2614	1108	1502	2507	391	1192	1220	642	3727	3023	50956
III	7358	6975	6609	8650	2359	884	1310	2893	392	1340	1287	602	3923	3185	47767
IV	10642	10928	8304	10650	2686	1224	1817	3787	499	1715	1598	694	5297	3522	63363
YEAR	39447	36157	29706	40000	10439	4699	6802	12130	1688	5691	5669	2525	17225	13078	225256
% CHANGE OVER1972	-4.6	13.4	4.6	3.9	3.5	-16.8	-18.2	-2.2	-3.4	12.3	14.7	4.6	29.8	14.1	4.5
1974															
I	11156	9449	6117	11200	2272	954	1548	2432	424	1216	1366	570	4843	3441	56988
II	7947	8272	5671	8600	2098	797	1460	2842	370	969	939	505	4824	2959	48253
III	7085	7577	6005	8800	2078	691	1603	2872	350	887	1208	500	4817	3252	47725
IV	11137	10832	6971	11800	2583	977	1827	3698	498	1478	1381	575	5523	3465	62745
YEAR	37325	36130	24764	40400	9031	3419	6438	11844	1642	4550	4894	2150	20007	13117	215711
% CHANGE OVER1973	-5.4	-.1	-16.6	1.0	-13.5	-27.2	-5.4	-2.4	-2.7	-20.0	-13.7	-14.9	16.2	.3	-4.2
1975															
I	9717	9079	5341	11400	2526	680	1756	2429	362	1281	1170	396	5164	3671	54972
II	7156	6913	5295	8400	1625	577	1404	2678	377	940	913	409	4608	3161	44456
III	5297	5113	5079	6400	1327	544	1226	2475	291	788	1136	416	4312	3184	37588
IV	8280	9063	6721	10300	2337	778	1914	2767	425	1401	1281	416	5120	3758	54561
YEAR	30450	30168	22436	36500	7815	2579	6300	10349	1455	4410	4500	1637	19204	13774	191577
% CHANGE OVER1974	-18.4	-16.5	-9.4	-9.7	-13.5	-24.6	-2.1	-12.6	-11.4	-3.1	-8.1	-23.9	-4.0	5.0	-11.2

* Includes Turkey, Greece, Portugal and Eire
** Canary Islands and Iceland not included

Table 35 QUARTERLY OIL DEMAND IN WESTERN EUROPE—TOTAL INLAND DEMAND (Thousand tonnes)

	U.K.	FRANCE	GERMANY	ITALY	BELGIUM LUX.	NETHER-LANDS	DENMARK	SWEDEN	NORWAY	FINLAND	AUSTRIA	SWITZ-ERLAND	SPAIN	OTHER O.E.C.D. EUROPE *	TOTAL O.E.C.D. EUROPE **
1968															
I	21957	18115	22511	16750	5230	5818	3482	6294	1521	2154	1717	2107	4301	3709	115666
II	17247	12526	20698	11988	4092	4300	2249	4538	1282	1558	1567	2543	3987	3674	92249
III	15657	11930	22652	13699	4162	4147	2607	4295	1197	1673	1836	2630	4477	3938	94900
IV	20915	19326	24795	17493	5983	6052	3985	7179	1654	2299	2051	2186	4711	4372	123001
YEAR	75776	61897	90656	59930	19467	20317	12323	22306	5654	7684	7171	9466	17476	15693	425816
1969															
I	23534	19679	25961	18552	6033	6335	4241	7086	1732	2422	1879	2218	4604	4147	128423
II	19626	14448	24743	13525	4683	4652	2980	5529	1433	1733	1704	3011	4355	4199	106621
III	16562	14314	24050	15224	4498	4141	2641	4650	1353	1935	2020	2807	4794	4457	103446
IV	22634	20768	27360	19148	5918	6019	4478	7718	1913	2871	2243	2408	5046	4771	133295
YEAR	82356	69209	102114	66449	21132	21147	14340	24983	6431	8961	7846	10444	18799	17574	471785
% CHANGE OVER 1968	8.7	11.8	12.6	10.9	8.6	4.1	16.4	12.0	13.7	16.6	9.4	10.3	7.6	12.0	10.8
1970															
I	24727	23393	29489	21273	6623	6728	4784	7966	1820	2767	1975	2500	5117	4541	143703
II	20282	17339	27613	15541	5569	5258	3496	6416	1683	2067	1937	3333	5230	4543	120307
III	18299	15851	27257	17017	4943	5063	3166	5887	1538	2321	2217	3213	5667	4762	117201
IV	24039	23957	30057	22350	6143	6222	4741	7429	1973	2719	2620	2577	6104	5304	146235
YEAR	87347	80540	114416	76181	23278	23271	16187	27698	7014	9874	8749	11623	22118	19150	527446
% CHANGE OVER 1969	6.1	16.4	12.0	14.6	10.2	10.0	12.9	10.9	9.1	10.2	11.5	11.3	17.7	9.0	11.8
1971															
I	25731	27350	31514	23037	6835	6428	4783	7429	1896	2799	2331	3205	6251	5265	154854
II	20510	17682	28583	16571	5090	4829	3464	5770	1523	2151	2215	3167	5535	5059	122149
III	17544	16472	26904	18326	5133	4660	3324	5029	1481	2342	2249	3138	5672	5518	117792
IV	23911	26955	31962	23795	6711	6198	4920	7664	1911	2935	2714	2862	6219	5990	154747
YEAR	87696	88459	118963	81729	23769	22115	16491	25892	6811	10227	9509	12372	23677	21832	549542
% CHANGE OVER 1970	.4	9.8	4.0	7.3	2.1	-5.0	1.9	-6.5	-2.9	3.6	8.7	6.4	7.0	14.0	4.2

Table 35—*contd.*

	U.K.	FRANCE	GERMANY	ITALY	BELGIUM LUX.	NETHER-LANDS	DENMARK	SWEDEN	NORWAY	FINLAND	AUSTRIA	SWITZ-ERLAND	SPAIN	OTHER O.E.C.D. EUROPE *	TOTAL O.E.C.D. EUROPE **
1972															
I	27223	27972	32819	23279	7238	6795	5152	7554	1980	3028	2495	2945	6783	5848	161111
II	22340	21484	30640	18118	6050	5761	3764	5899	1603	2496	2191	3392	6219	5789	135746
III	18692	20167	30066	19090	5404	5156	3401	5383	1483	2566	2394	3406	6547	5986	129741
IV	25490	29500	34019	25493	6962	6939	5031	7839	1943	3120	2918	2946	7014	7010	166224
YEAR	93745	99123	127544	85980	25654	24651	17348	26675	7009	11210	9998	12689	26563	24633	592822
% CHANGE OVER 1971	6.9	12.1	7.2	5.2	7.9	11.5	5.2	3.0	2.9	9.6	5.1	2.6	12.2	12.8	7.9
1973															
I	27227	32683	35139	25501	7315	7026	5005	7052	1879	3098	2794	3095	7760	6960	172534
II	22169	24205	33644	19846	6745	5972	3446	6146	1644	2503	2605	3541	7340	6510	146316
III	19375	20739	31864	20774	5521	5509	3082	5818	1651	2929	2856	3666	8050	7061	138895
IV	26330	32857	35480	24929	7205	6818	4368	8192	2166	3850	3248	3395	9170	7638	175646
YEAR	95101	110484	136127	91050	26786	25325	15901	27208	7340	12380	11503	13697	32320	28169	633391
% CHANGE OVER 1972	1.4	11.5	6.7	5.9	4.4	2.7	-8.3	2.0	4.7	10.4	15.1	7.9	21.7	14.4	6.8
1974															
I	25178	29500	28100	24775	6377	5867	3825	5649	1604	2777	2540	2610	8362	6904	154068
II	20387	24800	30859	20200	5568	5276	3195	5957	1548	2223	2159	3390	8498	6242	140302
III	18656	21600	31601	20782	5160	5242	3299	5918	1491	2201	2789	3295	8805	6891	137730
IV	24915	28500	30982	24443	6760	5564	4148	7498	2025	3322	2828	2808	9145	7609	160547
YEAR	89136	104400	121542	90200	23865	21949	14467	25022	6668	10523	10316	12103	34810	27646	592647
% CHANGE OVER 1973	-6.3	-5.5	-10.7	-.9	-10.9	-13.3	-9.0	-8.0	-9.2	-15.0	-10.3	-11.6	7.7	-1.9	-6.4
1975															
I	22228	26122	28416	23870	6384	4894	4081	6106	1730	2952	2292	2689	8658	7340	147762
II	18608	22627	28388	19130	5097	5119	3335	6318	1639	2353	2236	3071	8316	6704	132941
III	15762	18976	29325	18100	4174	4598	2939	5472	1558	2327	2839	3221	8408	7387	125086
IV	21095	27894	30536	24300	6321	5502	4126	6140	1951	3116	2781	2534	9100	7861	153257
YEAR	77693	95619	116665	85400	21976	20113	14481	24036	6878	10748	10148	11515	34482	29292	559046
% CHANGE OVER 1974	-12.8	-8.4	-4.0	-5.3	-7.9	-8.4	.1	-3.9	3.1	2.1	-1.6	-4.9	-.9	6.0	-5.7

* Includes Turkey, Greece, Portugal and Eire
** Canary Islands and Iceland not included

Table 36 UK INLAND ENERGY CONSUMPTION 1924–1974 (Million tonnes oil equivalent)

Year	*Total Inland Energy Consumption*	*Petroleum**	*Coal*	*Natural Gas*	*Hydro Electricity*	*Nuclear Electricity*
1924	114·5	3·6	110·9	—	neg	—
1925	109·0	3·6	105·4	—	neg	—
1926	77·0	4·5	72·5	—	neg	—
1927	114·4	4·8	109·6	—	neg	—
1928	107·2	4·8	102·4	—	neg	—
1929	113·3	4·8	108·5	—	neg	—
1930	108·2	5·4	102·7	—	0·1	—
1931	100·7	5·3	95·3	—	0·1	—
1932	99·1	5·4	93·6	—	0·1	—
1933	99·4	5·8	93·5	—	0·1	—
1934	106·5	6·1	100·3	—	0·1	—
1935	109·7	6·2	103·3	—	0·2	—
1936	115·6	6·5	108·9	—	0·2	—
1937	120·0	6·8	113·0	—	0·2	—
1938	114·1	7·4	106·4	—	0·3	—
1946	119·7	8·1	111·3	—	0·3	—
1947	121·3	10·7	110·3	—	0·3	—
1948	127·5	11·2	115·9	—	0·4	—
1949	129·5	12·1	117·0	—	0·4	—
1950	134·8	13·3	121·1	—	0·4	—
1951	139·5	14·8	124·2	—	0·5	—
1952	139·5	15·4	123·6	—	0·5	—
1953	141·5	16·7	124·2	neg	0·6	—
1954	147·0	18·5	127·8	neg	0·7	—
1955	149·7	20·6	128·6	neg	0·5	—
1956	153·2	22·5	130·0	neg	0·7	neg
1957	150·1	22·0	127·2	neg	0·8	0·1
1958	150·2	28·3	121·0	neg	0·8	0·1
1959	148·0	33·5	113·2	0·1	0·8	0·4
1960	158·4	39·2	117·6	0·1	0·9	0·6
1961	158·9	42·4	114·6	0·1	1·1	0·7
1962	163·5	46·9	114·3	0·1	1·1	1·1
1963	169·6	51·0	115·9	0·1	1·1	1·5
1964	170·9	55·8	111·9	0·2	1·1	1·9
1965	178·1	61·5	110·3	1·3	1·4	3·6
1966	178·5	66·8	104·4	1·2	1·4	4·7
1967	178·0	71·3	97·9	1·9	1·6	5·3
1968	184·4	75·2	98·3	3·6	1·3	6·0
1969	191·2	81·1	96·3	6·3	1·2	6·3
1970	197·8	87·1	92·3	11·2	1·6	5·6
1971	195·6	88·1	82·9	17·7	1·1	5·8
1972	198·7	94·2	72·3	24·7	1·2	6·3
1973	207·4	95·2	78·5	26·6	1·2	5·9
1974	198·2	89·0	69·3	31·6	1·2	7·1

* The petroleum figures in this table exclude non-energy products

Table 37 UK CONSUMPTION OF PETROLEUM PRODUCTS 1922–1974 (Thousand long tons)

Year	*Motor Spirit*	*Aviation Spirits*	*Aviation Turbine Fuel*	*Light Distillate Feedstocks*	*Middle Distillates*	*Fuel Oils (Inland, excluding Refinery Consumption)*	*Bitumens*	*Lubricants (including Fuel Oil sold as Lubricants)*	*Paraffin Waxes*	*Miscellaneous Products*	*Total Inland Trade*	*Total Consumption (including Refinery Consumption and Bunkers)*
1922	1 040·0				813·7	535·4	67·0	345·9	58·5	19·5	2 880·0	3 480·5
1923	1 207·4				847·2	649·4	146·0	333·4	59·0	19·5	3 261·9	4 162·7
1924	1 563·3				846·8	649·0	194·0	394·6	57·8	19·5	3 725·0	4 727·5
1925	1 764·7				1 054·5	576·7	200·0	404·6	61·0	19·5	4 081·0	5 099·8
1926	2 069·6				1 202·6	1 314·5	238·0	385·4	61·4	19·5	5 291·0	6 353·8
1927	2 304·0	*Included*			1 120·1	734·1	282·0	430·4	61·3	19·5	4 951·4	6 044·1
1928	2 702·0	*with*			1 051·2	743·8	334·7	455·1	62·5	19·5	5 368·8	6 397·5
1929	2 971·2	*Motor Spirit*			1 096·0	843·2	353·2	471·1	57·1	19·5	5 811·3	6 900·5
1930	3 324·0	*up to*			1 024·8	888·0	371·8	472·0	53·2	19·5	6 153·3	7 265·2
1931	3 468·1	1937			1 056·7	911·3	399·5	453·4	54·5	20·2	6 363·7	7 254·2
1932	3 585·7				1 079·4	1 008·6	396·3	412·0	52·2	18·9	6 553·1	7 432·0
1933	3 821·8				1 110·2	1 064·0	439·8	440·7	50·3	19·0	6 945·8	8 245·8
1934	4 066·9				1 154·5	1 095·3	495·0	469·2	48·2	19·4	7 348·5	8 721·5
1935	4 292·9				1 181·1	1 133·0	505·6	502·2	54·5	19·1	7 688·4	9 083·4
1936	4 558·3				1 254·0	1 217·4	555·4	521·7	55·3	20·1	8 182·2	9 597·2
1937	4 734·2				1 368·6	1 220·7	591·3	564·3	58·7	20·9	8 558·7	10 339·7
1938	4 883·3	199·5			1 428·6	1 177·0	607·1	564·0	48·1	60·9	8 968·5	10 575·3
1939	4 580·1	195·5			1 563·6	1 242·4	620·8	580·8	50·3	61·6	8 895·1	10 225·0
1940	3 334·8	404·3			2 187·3	753·0	380·1	511·4	57·4	66·3	7 694·6	9 128·9
1941	3 663·0	710·9			2 396·2	776·0	348·2	641·3	58·9	78·0	8 672·5	9 906·9
1942	3 167·0	1 141·1	0·6		2 417·9	817·3	248·5	622·4	53·6	83·9	8 552·3	10 210·2
1943	2 958·5	2 051·9	2·5		2 448·2	785·7	177·4	586·3	49·5	84·0	9 144·0	10 770·4
1944	3 665·2	4 781·9	7·7		2 710·0	773·2	262·5	692·9	53·3	97·3	13 044·0	15 377·5
1945	3 387·0	2 546·8	25·8		2 852·5	737·7	242·8	546·5	46·8	90·6	10 476·5	12 774·0
1946	4 086·1	321·0	33·7		3 141·8	927·8	376·5	608·1	42·1	92·5	9 629·6	11 638·7
1947	4 593·8	243·6	38·6		3 756·8	2 033·6	423·9	659·3	37·7	104·0	11 891·3	14 445·6
1948	4 265·0	290·7	70·6		3 695·2	2 741·7	492·9	653·0	42·7	148·3	12 400·1	14 932·0
1949	4 671·4	327·6	110·1		3 976·5	2 795·2	545·3	708·3	40·2	240·0	13 414·6	16 225·1
1950	5 195·2	334·2	184·4		4 141·3	3 092·7	621·3	742·8	43·2	367·2	14 722·3	17 916·9
1951	5 454·3	391·0	375·4		4 422·8	3 417·8	745·6	767·2	49·8	492·6	16 116·5	20 339·3
1952	5 440·6	457·9	652·6		4 402·4	3 456·8	721·9	722·0	30·4	527·8	16 412·4	21 527·6
1953	5 740·0	423·6	938·9		4 508·8	3 815·3	749·5	771·5	40·1	699·9	17 687·6	23 193·5
1954	5 922·4	354·2	1 249·8		4 808·7	4 415·7	807·6	845·2	46·3	830·6	19 280·5	25 572·7
1955	6 240·4	377·6	1 342·9		5 290·8	5 384·0	865·4	888·5	46·4	806·1	21 242·1	27 957·5
1956	6 323·7	382·4	1 394·6		5 745·2	6 471·2	940·9	894·5	52·3	927·1	23 131·9	29 944·8
1957	5 745·0	386·8	1 243·0		5 600·0	6 930·7	848·4	826·5	48·8	1 083·2	22 712·4	28 705·6
1958	6 623·5	345·2	1 219·8	40·9	6 757·2	10 581·8	867·7	871·6	51·8	1 216·1	28 575·6	35 321·5
1959	7 123·6	288·0	1 340·4	224·8	7 127·7	13 811·9	990·9	927·6	51·4	1 590·5	33 476·8	40 937·3
1960	7 633·0	263·0	1 500·8	400·1	7 859·9	17 437·8	1 062·4	964·8	53·6	2 227·7	39 403·1	47 521·4
1961	8 143·1	268·0	1 888·6	496·7	8 428·0	18 817·0	1 165·9	979·6	51·7	2 295·7	42 534·3	50 634·0
1962	8 565·3	221·9	1 886·4	714·7	9 446·1	21 329·4	1 237·9	968·1	51·8	2 811·8	47 233·4	55 247·3
1963	9 044·4	239·8	1 998·2	900·4	10 755·9	22 704·0	1 314·4	1 003·5	52·4	3 839·0	51 852·0	60 143·0
1964	10 011·9	193·5	2 109·4	1 347·3	11 396·7	24 804·8	1 491·3	1 081·4	58·8	4 723·6	57 218·7	65 617·1
1965	10 739·0	167·0	2 220·5	2 064·9	12 552·6	27 296·2	1 459·9	1 105·7	55·8	5 122·6	62 784·2	71 731·4
1966	11 322·3	146·7	2 434·7	2 993·1	13 520·1	29 095·1	1 539·6	1 137·5	55·1	5 737·4	67 981·6	77 049·1
1967	12 084·4	127·5	2 783·7	3 971·2	14 339·3	30 608·1	1 761·1	1 099·6	53·0	6 631·9	73 459·8	82 714·0
1968	12 807·4	102·9	3 062·5	5 094·5	15 788·4	30 611·2	1 828·4	1 133·5	54·6	7 322·1	77 805·5	87 733·5
1969	13 231·5	99·3	3 198·7	5 314·5	17 553·7	33 392·6	1 811·7	1 208·1	58·3	7 975·4	83 843·8	94 344·3
1970	14 009·9	73·4	3 352·6	3 486·0	19 501·3	37 975·0	2 036·0	1 156·7	56·1	7 826·2	89 473·2	100 375·7
1971	14 734·3	62·5	3 691·3	1 867·7	20 181·3	38 772·4	2 173·1	1 129·3	60·7	7 478·0	90 150·6	101 332·6
1972	15 647·6	63·5	3 942·9	1 446·0	23 112·6	40 654·3	2 168·6	1 095·0	64·9	8 039·5	96 234·9	107 695·4
1973	16 659·3	62·5	4 205·9	1 624·3	23 751·3	38 823·7	2 419·4	1 166·2	74·2	9 077·4	97 864·2	110 217·3
1974	16 223·1	52·7	3 693·2	980·9	21 656·9	36 757·7	2 205·2	1 028·9	[illegible]	[illegible]	[illegible]	[illegible]

Table 38 WESTERN EUROPE (OECD) PRIMARY ENERGY AND GNP 1955–1975

Year	*GNP** (1963 *dollars*)	*GNP Growth Over Previous Year* (%)	*Primary Energy* (*High Factor*) (*Million tonnes oil equivalent*)	*Primary Energy Growth Rate Over Previous Year* (%)
1955	304	6·25	542	7·5
1956	318	4·5	573	5·7
1957	331	4·0	577	0·7
1958	339	2·25	572	−0·9
1959	355	4·5	578	1·1
1960	378	6·5	637	10·2
1961	398	5·5	659	3·5
1962	416	4·5	702	6·5
1963	436	4·75	761	8·4
1964	462	6·0	793	4·2
1965	482	4·5	833	5·0
1966	502	4·0	862	3·5
1967	518	3·25	883	2·4
1968	544	5·0	943	6·8
1969	576	6·0	1 018	8·0
1970	606	5·25	1 093	7·4
1971	627	3·5	1 125	2·9
1972	653	4·25	1 170	4·0
1973	690	5·5	1 233	5·4
1974	707	2·5	1 200	−2·7
1975	693	−2·0	1 165	−3·0

* From National Accounts OECD

Table 39 US PRODUCTION OF MINERAL FUELS, WATER POWER AND WOOD 1800–1949

Years	Annual Average Bituminous Coal Equivalent (Thousand tons)							Percentage of Total					
	Bituminous Coals	*Anthracite*	*Crude Petroleum**	*Natural Gas*	*Water Power†*	*Fuel Wood*	*Total*	*Bituminous*	*Anthracite*	*Crude Petroleum**	*Natural Gas*	*Water Power†*	*Fuel Wood*
1800–1809	140	neg	—	—	1 144	21 462	22 746	0·6	neg	—	—	5·0	94·4
1810–1819	250	2	—	—	1 320	28 575	30 147	0·0	neg	—	—	4·4	94·9
1820–1829	446	46	—	neg	1 585	38 396	40 473	1·1	0·1	—	—	3·9	94·9
1830–1839	942	686	—	neg	2 007	52 181	55 816	1·7	1·2	—	neg	3·6	93·5
1840–1849	2 124	2 442	—	neg	2 877	70 283	77 726	2·7	3·2	—	neg	3·7	90·4
1850–1859	6 958	7 367	—	neg	4 204	92 381	110 910	6·3	6·6	—	neg	3·8	83·3
1860–1869	12 444	13 258	658	neg	5 638	108 426	140 424	8·9	9·4	0·5	neg	4·0	77·2
1870–1879	31 109	23 038	2 336	neg	6 755	112 556	175 797	17·7	13·1	1·3	neg	3·9	64·0
1880–1889	73 055	37 142	6 043	6 012	7 500	104 300	234 052	31·2	15·9	2·6	2·6	3·2	44·5
1890–1899	138 356	51 819	11 848	9 218	9 014	87 019	307 274	45·0	16·9	3·9	3·0	2·9	28·3
1900–1909	302 479	69 143	27 193	13 924	14 470	67 863	495 072	61·1	14·0	5·5	2·8	2·9	13·7
1910–1919	471 644	87 756	62 862	25 657	24 752	67 625	740 296	63·7	11·9	8·5	3·5	3·3	9·1
1920–1929	510 203	76 775	165 504	47 308	31 324	55 289	886 403	57·6	8·7	18·7	5·3	3·5	6·2
1930–1939	385 272	52 549	233 530	79 258	30 034	51 852	832 495	46·3	6·3	28·1	9·5	3·6	6·2
1940–1949	554 405	54 830	383 884	156 627	50 106	46 444	1 246 296	44·5	4·4	30·8	12·6	4·0	3·7

* Includes natural gas liquids
† Includes hydro electric and direct drive

Table 40 US PRODUCTION OF MINERAL FUELS, WATER POWER AND NUCLEAR POWER 1950–1974

Year	*Bituminous Coal Equivalent (Thousand short tons)*							*Percentage on a Btu Basis*					
	Bituminous Coals & Lignites	*Anthracite Coals*	*Crude Petroleum*	*Natural Gas-Wet*	*Water Power*	*Nuclear Power*	*Total**	*Bituminous Coals & Lignites*	*Anthracite Coals*	*Crude Petroleum*	*Natural Gas-Wet*	*Water Power*	*Nuclear Power*
1950	516 311	42 748	436 984	261 107	60 038	—	1 317 188	39·2	3·2	33·2	19·8	4·6	—
1951	533 665	41 374	497 595	309 389	59 504	—	1 441 527	37·0	2·9	34·5	21·5	4·1	—
1952	466 841	39 351	506 946	332 252	60 343	—	1 405 733	33·2	2·8	36·1	23·6	4·3	—
1953	457 290	30 000	521 793	347 939	58 092	—	1 415 114	32·3	2·1	36·9	24·6	4·1	—
1954	391 706	28 206	512 480	362 137	55 305	—	1 349 834	29·0	2·1	38·0	26·8	4·1	—
1955	464 634	25 382	549 999	389 466	55 229	—	1 484 710	31·3	1·7	37·1	26·2	3·7	—
1956	500 874	28 015	579 427	417 176	58 855	—	1 584 347	31·6	1·8	36·6	26·3	3·7	—
1957	492 704	24 580	579 314	441 642	58 168	—	1 596 408	30·9	1·5	36·3	27·7	3·6	—
1958	410 445	20 534	542 138	455 840	64 618	—	1 493 575	27·5	1·4	36·3	30·5	4·3	—
1959	412 028	20 001	569 993	497 584	62 943	—	1 562 549	26·4	1·3	36·5	31·8	4·0	—
1960	415 512	18 245	570 069	527 586	65 755	208	1 597 375	26·1	1·1	35·8	33·2	3·8	—
1961	402 977	16 909	580 413	547 205	66 880	708	1 615 092	25·0	1·1	35·9	33·9	4·1	—
1962	422 149	16 374	592 442	572 671	73 917	958	1 678 511	25·2	1·0	35·4	34·3	4·0	0·1
1963	458 928	17 710	609 389	608 434	72 542	1 375	1 768 378	26·0	1·0	34·6	34·5	3·8	0·1
1964	486 998	16 641	616 946	648 434	77 542	1 417	1 847 978	26·4	0·9	33·5	35·2	3·9	0·1
1965	512 088	14 870	626 672	694 414	80 999	1 495	1 930 538	26·5	0·8	32·4	36·0	4·2	0·1
1966	533 881	13 004	668 972	750 356	81 502	2 253	2 049 968	26·1	0·6	32·6	36·6	4·0	0·1
1967	552 626	12 361	719 396	798 370	93 283	3 180	2 179 216	25·4	0·6	33·0	36·6	4·3	0·1
1968	545 245	11 612	741 939	859 856	93 735	5 188	2 257 575	24·2	0·5	32·8	38·1	4·2	0·2
1969	560 505	10 683	758 474	917 189	106 345	5 863	2 359 059	23·8	0·4	32·2	38·9	4·5	0·2
1970	602 932	9 928	794 695	970 820	105 707	9 204	2 493 286	24·2	0·4	31·9	38·9	4·2	0·4
1971	552 192	8 883	786 133	996 262	113 706	16 238	2 473 414	22·2	0·4	31·8	40·3	4·6	0·7
1972	595 386	7 106	787 028	995 740	115 356	23 151	2 523 767	23·6	0·3	31·2	39·4	4·6	0·9
1973	591 738	6 830	765 627	993 892	114 591	35 691	2 508 369	23·6	0·3	30·5	39·6	4·6	1·4
1974	597 480		722 238	893 794	116 877	47 616	2 378 005	25·1		30·4	37·6	4·9	2·0

* Conversions made by NCA from total Btu equivalents reported by US Bureau of Mines
Source: Bituminous Coal Annual and Bureau of Mines

Table 41 US ENERGY CONSUMPTION, POPULATION AND ECONOMIC ACTIVITY 1947–1974

Year	*Energy Consumption* (10^{12} *Btu*)	*Population* (*Thousands*)	*GNP* (1963$ × 10^9)	*Total Domestic Oil Demand* (10^9 *bbl*)
1947	32 870	144 083	336	1 990
1948	33 994	146 730	351	2 114
1949	31 604	149 304	351	2 118
1950	34 153	151 868	385	2 375
1951	36 913	153 982	416	2 570
1952	36 576	156 393	428	2 664
1953	37 697	158 956	448	2 775
1954	36 360	161 884	441	2 832
1955	39 956	165 069	475	3 088
1956	42 007	168 088	484	3 213
1957	41 920	171 187	491	3 219
1958	41 493	174 149	485	3 315
1959	43 411	177 135	516	3 450
1960	44 816	179 992	529	3 536
1961	45 573	183 057	539	3 579
1962	47 620	185 890	574	3 736
1963	49 649	188 658	596	3 851
1964	51 515	191 372	628	3 959
1965	53 791	193 815	668	4 125
1966	56 948	195 923	712	4 325
1967	58 873	197 864	732	4 481
1968	62 432	199 870	768	4 788
1969	65 645	201 921	789	5 160
1970	67 465	203 810	782	5 364
1971	68 698	206 212	808	5 553
1972	71 946	208 230	858	5 990
1973	74 741	210 579	909	6 317
1974	73 121	211 900	889	6 081

Table 42 INTERNATIONAL TRADE IN CRUDE OIL AND PRODUCTS (Million tonnes)

<table>
<tr><th>Country</th><th>1955</th><th>1960</th><th>1965</th><th>1970</th><th>1974</th></tr>
<tr><td>EXPORTS</td><td></td><td></td><td></td><td></td><td></td></tr>
<tr><td>USA</td><td>18·3</td><td>9·7</td><td>10·4</td><td>13·0</td><td>12·1</td></tr>
<tr><td>Canada</td><td>6·1</td><td>6·1</td><td>15·7</td><td>35·5</td><td>52·5</td></tr>
<tr><td>Caribbean</td><td>105·7</td><td>143·3</td><td>170·7</td><td>172·3</td><td>169·7</td></tr>
<tr><td>Other Western Hemisphere</td><td>—</td><td>1·0</td><td>3·1</td><td>2·0</td><td>12·7</td></tr>
<tr><td>Western Europe</td><td>7·1</td><td>7·6</td><td>8·4</td><td>31·0</td><td>22·3</td></tr>
<tr><td>Middle East</td><td>147·3</td><td>232·7</td><td>378·5</td><td>631·5</td><td>992·5</td></tr>
<tr><td>N. Africa</td><td>—</td><td rowspan="2">11·2</td><td>87·1</td><td>231·8</td><td>125·3</td></tr>
<tr><td>W. Africa</td><td>—</td><td>17·0</td><td>56·8</td><td>124·6</td></tr>
<tr><td>S.E. Asia</td><td>10·2</td><td>20·3</td><td>19·6</td><td>39·5</td><td>67·1</td></tr>
<tr><td>USSR, E. Europe, China</td><td>5·1</td><td>23·4</td><td>45·5</td><td>47·3</td><td>72·7</td></tr>
<tr><td>Other Eastern Hemisphere</td><td>—</td><td>1·0</td><td>3·0</td><td>2·5</td><td>4·3</td></tr>
<tr><td>Total Exports</td><td>299·8</td><td>456·3</td><td>759·0</td><td>1263·2</td><td>1655·8</td></tr>
<tr><td>IMPORTS</td><td></td><td></td><td></td><td></td><td></td></tr>
<tr><td>USA</td><td>65·0</td><td>94·5</td><td>127·0</td><td>170·0</td><td>315·4</td></tr>
<tr><td>Canada</td><td rowspan="3">51·8</td><td rowspan="3">57·9</td><td>27·9</td><td>36·0</td><td>43·6</td></tr>
<tr><td>Caribbean</td><td>9·9</td><td>19·8</td><td>63·8</td></tr>
<tr><td>Other Western Hemisphere</td><td>24·9</td><td>23·8</td><td>44·9</td></tr>
<tr><td>Western Europe</td><td>116·9</td><td>198·1</td><td>383·8</td><td>643·5</td><td>733·8</td></tr>
<tr><td>Middle East</td><td rowspan="8">66·1</td><td rowspan="4">50·8</td><td>1·5</td><td>—</td><td>6·8</td></tr>
<tr><td>N. & W. Africa</td><td>19·6</td><td>8·0</td><td>8·0</td></tr>
<tr><td>E. &. S. Africa, S. Asia</td><td>26·2</td><td>39·5</td><td>47·2</td></tr>
<tr><td>S.E. Asia</td><td>15·3</td><td>46·3</td><td>67·0</td></tr>
<tr><td>Japan</td><td>29·5</td><td>88·9</td><td>211·3</td><td>268·1</td></tr>
<tr><td>Australasia</td><td>13·2</td><td>20·3</td><td>22·3</td><td>16·4</td></tr>
<tr><td>USSR, E. Europe, China</td><td>0·5</td><td>—</td><td>5·0</td><td>20·9</td></tr>
<tr><td>Destination not known</td><td>11·8</td><td>13·7</td><td>37·7</td><td>19·9</td></tr>
<tr><td>Total Imports</td><td>299·8</td><td>456·3</td><td>759·0</td><td>1263·2</td><td>1655·8</td></tr>
</table>

Table 43 WORLD OIL SUPPLY AND DEMAND 1957–1975

1957 (Million tons)

FROM \ TO	USA	Other Western Hemisphere	Western Europe	Other Eastern	USSR etc.	Total
USA	—	13	11	4	—	28
Caribbean	58	40	30	7	—	135
Other Western Hemisphere	8	2	—	—	—	10
Middle East	11	6	90	46	—	153
Far East	3	—	1	11	—	15
Western Europe	—	—	—	4	1	5
USSR, etc.	—	—	5	2	—	7

1958 (Million tons)

FROM \ TO	USA	Other Western Hemisphere	Western Europe	Other Eastern Hemisphere	USSR etc.	Total
USA	—	9	3	2	—	14
Caribbean	60	39	29	2	—	130
Other Western Hemisphere	7	—	—	—	—	7
Middle East	16	10	114	48	—	188
Far East	3	—	1	12	—	16
Western Europe	—	—	—	9	1	10
USSR, etc.	—	1	7	3	—	11

1959 (Thousand b/d)

FROM \ TO	USA	Other Western Hemisphere	Western Europe	Other Eastern Hemisphere	USSR etc.	Total* Exports
USA	—	118	41	81	—	240
Caribbean	1 215	745	560	77	—	2 617
Other Western Hemisphere	165	31	31	—	—	227
Middle East	355	185	2 528	875	—	4 120
Far East	62	—	20	248	—	330
Other Eastern Hemisphere	—	40	40	126	20	226
USSR, etc.	—	20	280	60	—	360
TOTAL IMPORTS	1 797	1 139	3 500	1 467	20	8 120

* Total exports include 10 million tons, 197 000 b/d, unallocated (largely military) liftings.

TABLE 43—*contd.*

1960 (Thousand b/d)

FROM \ TO	USA	Other Western Hemisphere	Western Europe	Africa	Australasia	Other Eastern Hemisphere	USSR etc.	Total Exports
USA	—	66	44	—	—	98	—	208
Caribbean	1 263	805	555	77	—	—	—	2 700
Other Western Hemisphere	124	—	21	—	—	—	—	145
Western Europe	—	—	—	129	—	21	10	160
Africa	—	—	237	—	—	—	—	237
Far East	82	—	41	—	102	205	—	430
Middle East	347	205	2 724	225	165	1 024	—	4 690
USSR, etc	—	40	337	60	—	20	—	457
TOTAL IMPORTS	1 816	1 116	3 959	491	267	1 368	10	9 027

1961 (Thousand b/d)

FROM \ TO	USA	Canada	Other Western Hemisphere	Western Europe	North and West Africa	Indian Ocean Area	Australasia	Japan	Other Asia Unallocated*	USSR	Total Exports
USA	—	35	35	40	—	—	5	55	5	—	175
Canada	190	—	—	—	—	—	—	—	—	—	190
Mexico	40	—	—	—	—	—	—	—	—	—	40
Caribbean	1 245	285	290	705	90	5	5	5	5	—	2 635
South America	5	—	20	5	—	—	—	—	—	—	30
Western Europe	—	—	10	—	100	10	—	—	40	20	180
North and West Africa	5	—	—	430	—	—	—	—	—	—	435
Middle East	345	135	85	2 855	80	340	180	610	460	—	5 090
East Indies	60	—	—	25	—	15	100	90	130	—	420
USSR, Eastern Europe	—	—	80	405	55	5	—	60	10	—	615
Other Eastern Hemisphere	—	—	—	—	—	10	—	—	50	—	60
TOTAL IMPORTS	1 890	455	520	4 465	325	385	290	820	700	20	9 870

* Military use, increased quantities in stock and in transit, transit losses, etc.

1962 (Thousand b/d)

FROM \ TO	USA	Canada	Other Western Hemisphere	Western Europe	North and West Africa	Indian Ocean Area	Australasia	Japan	Other Eastern Hemisphere Unallocated*	Total Exports
USA	—	25	35	40	—	5	5	50	10	170
Canada	250	—	—	—	—	—	—	—	—	250
Mexico	30	—	—	—	—	—	—	—	—	30
Caribbean	1 415	300	260	870	115	5	15	20	60	3 060
South America	5	5	10	10	—	—	—	—	—	30
Western Europe	5	—	10	—	115	10	—	5	40	185
North and West Africa	25	—	—	740	—	—	—	—	—	765
Middle East	315	130	225	3 060	95	380	195	720	465	5 585
East Indies	70	—	—	20	—	10	110	120	120	450
USSR, Eastern Europe	—	—	100	450	40	15	—	60	5	670
Other Eastern Hemisphere Unallocated*	5	—	—	5	—	10	5	10	55	90
TOTAL IMPORTS†	2 120	460	640	5 190	370	430	335	985	755	11 285

* Increased quantities in stock and in transit, transit losses, minor movements not otherwise shown, military use, etc.
† Due to rounding, totals may not balance against column detail.

Table 43—*contd.*

1963 (Thousand b/d)

FROM \ TO	*USA*	*Canada*	*Other Western Hemisphere*	*Western Europe*	*North Africa*	*West Africa*	*Indian Ocean Area*	*Australasia*	*Japan*	*Other Eastern Hemisphere Unallocated**	*Total Exports*
USA	—	30	40	45	—	5	10	5	65	10	210
Canada	265	—	—	—	—	—	—	—	—	—	265
Caribbean	1 405	320	250	855	—	90	10	10	25	90	3 055
Other Western Hemisphere	65	5	—	10	—	—	—	—	—	10	90
Western Europe	5	—	5	—	45	20	10	—	—	20	105
North Africa	25	—	15	1 000	—	5	—	—	—	—	1 045
West Africa	—	—	—	125	5	—	—	—	—	5	135
Middle East	295	140	210	3 400	95	60	400	235	920	400	6 155
South-East Asia	70	—	—	20	—	—	—	105	130	15	340
USSR, Eastern Europe	—	—	100	510	25	—	20	—	65	20	740
Other Eastern Hemisphere, Unallocated*	—	—	5	10	10	—	5	10	15	10	65
TOTAL IMPORTS†	2 130	495	625	5 975	180	180	455	365	1 220	580	12 205

* Increased quantities in stock and in transit, transit losses, minor movements not otherwise shown, military use, etc.
† Due to rounding, totals may not balance against column detail.

1964 (Thousand b/d)

FROM \ TO	*USA*	*Canada*	*Other Western Hemisphere*	*Western Europe*	*North Africa*	*West Africa*	*Indian Ocean Area*	*Australasia*	*Japan*	*Other Eastern Hemisphere Unallocated**	*Total Exports*
USA	—	30	30	45	—	5	5	5	80	—	200
Canada	300	—	—	—	—	—	—	—	—	—	300
Caribbean	1 450	335	190	880	—	95	—	5	40	95	3 090
Other Western Hemisphere	50	5	5	—	—	—	—	—	—	—	60
Western Europe	10	—	—	—	20	25	5	—	—	130	190
North Africa	55	—	10	1 370	—	10	—	—	—	—	1 445
West Africa	—	—	—	195	10	—	—	—	—	—	205
Middle East	310	130	240	3 740	105	85	470	295	1 115	435	6 925
South-East Asia	85	—	—	35	—	—	5	80	165	—	370
USSR, Eastern Europe	—	—	120	555	20	—	20	—	70	5	790
Other Eastern Hemisphere Unallocated*	—	—	—	10	—	—	5	—	5	30	50
TOTAL IMPORTS	2 260	500	595	6 830	155	220	510	385	1 475	695	13 625

* Includes changes in stock and increased quantities in transit, transit losses, minor movements not otherwise shown, military use, etc.

Table 43—*contd.*

1965 (Thousand b/d)

FROM \ TO	*USA*	*Canada*	*Other Western Hemisphere*	*Western Europe*	*Africa*	*South Asia*	*Japan*	*Australasia*	*Other Eastern Hemisphere*	*Destination* not known*	*Total Exports*
USA	—	25	50	45	10	—	40	5	10	—	185
Canada	325	—	—	—	—	—	—	—	—	—	325
Caribbean	1 565	380	220	835	100	—	30	5	20	95	3 250
Other Western Hemisphere	60	—	—	—	—	—	—	—	—	—	60
Western Europe	5	5	5	—	55	—	5	—	15	75	165
Middle East	350	145	240	4 195	435	215	1 435	305	270	20	7 610
North Africa	55	—	5	1 725	10	—	—	—	—	5	1 800
West Africa	15	—	10	315	—	—	—	—	—	—	340
South East Asia	90	—	—	35	—	5	140	95	—	45	410
USSR, Eastern Europe	—	—	145	570	30	60	90	—	5	—	900
Other Eastern Hemisphere	—	—	—	5	5	—	5	5	20	25	65
TOTAL IMPORTS	2 465	555	675	7 725	645	280	1 745	415	340	265	15 110

* Includes changes in stock and increased quantities in transit, transit losses, minor movements not otherwise shown, military use, etc.

1966 (Thousand b/d)

FROM \ TO	*USA*	*Canada*	*Other Western Hemisphere*	*Western Europe*	*Africa*	*South Asia*	*Japan*	*Australasia*	*Other Eastern Hemisphere*	*Destination* not known*	*Total Exports*
USA	—	50	30	70	10	—	40	—	—	—	200
Canada	385	—	—	—	—	—	—	—	—	—	385
Caribbean	1 625	365	255	795	—	—	45	5	10	75	3 175
Other Western Hemisphere	65	—	—	—	—	—	—	—	—	—	65
Western Europe	25	5	10	—	65	15	10	5	10	65	210
Middle East	320	145	295	4 375	500	210	1 650	340	360	430	8 625
North Africa	75	10	—	2 145	5	—	5	—	5	10	2 255
West Africa	15	25	45	395	—	—	—	—	20	10	510
South East Asia	80	—	—	35	—	5	140	100	—	5	345
USSR, Eastern Europe	—	5	145	685	50	45	100	—	20	—	1 050
Other Eastern Hemisphere	5	—	—	10	5	20	5	5	10	10	70
TOTAL IMPORTS	2 575	605	780	8 510	635	295	1 995	455	435	605	16 890

* Includes changes in stock and increased quantities in transit, transit losses, minor movements not other wise shown, military use, etc.

1967 (Thousand b/d)

FROM \ TO	*USA*	*Canada*	*Other Western Hemisphere*	*Western Europe*	*Africa*	*South Asia*	*Japan*	*Australasia*	*Other Eastern Hemisphere*	*Destination* not known*	*Total Exports*
USA	—	65	5	110	—	—	45	—	—	—	225
Canada	465	—	—	—	—	—	—	—	—	—	465
Caribbean	1 600	430	200	815	15	—	45	—	10	85	3 200
Other Western Hemisphere	100	—	—	40	—	—	—	5	—	—	145
Western Europe	55	15	—	—	50	10	—	5	15	100	250
Middle East	200	100	170	4 605	615	270	2 035	345	455	350	9 145
North Africa	45	—	30	2 310	—	—	—	—	15	60	2 460
West Africa	10	30	20	390	—	—	—	—	—	—	450
South East Asia	60	—	—	40	—	—	180	130	—	—	410
USSR, Eastern Europe	—	—	130	785	45	30	95	—	5	—	1 090
Other Eastern Hemisphere	—	—	—	20	5	—	5	—	15	20	65
TOTAL IMPORTS	2 535	640	555	9 115	730	310	2 405	485	515	615	17905

* Includes changes in stock and increased quantities in transit, transit losses, minor movements not otherwise shown, military use, etc.

Table 43—*contd.*

1968 (Thousand b/d)

FROM \ TO	*USA*	*Canada*	*Other Western Hemisphere*	*Western Europe*	*Africa*	*South East Asia*	*Japan*	*Australasia*	*Other Eastern Hemisphere*	*Destination* not known*	*Total Exports*
USA	—	50	45	40	10	5	35	5	5	—	195
Canada	470	—	—	—	—	—	—	—	—	—	470
Caribbean	1 580	480	260	760	60	—	85	—	20	—	3 245
Other Western Hemisphere	125	—	25	5	—	—	—	—	—	—	155
Western Europe	120	—	—	—	60	5	—	—	10	45	240
Middle East	270	125	175	5 065	465	645	2 650	365	345	340	10 445
North Africa	155	—	30	3 315	10	—	5	5	35	70	3 625
West Africa	15	10	5	220	—	—	—	—	—	—	250
South East Asia	75	—	—	15	—	—	240	105	—	—	435
USSR, Eastern Europe	—	—	115	900	40	—	45	—	—	—	1 100
Other Eastern Hemisphere	—	—	—	10	5	175	5	10	5	30	240
TOTAL IMPORTS	2 810	665	655	10 330	650	830	3 065	490	420	485	20 400

* Includes changes in stock and increased quantities in transit, transit losses, minor movements not otherwise shown, military use, etc.

1969 (Thousand b/d)

FROM \ TO	*USA*	*Canada*	*Other Western Hemisphere*	*Western Europe*	*Africa*	*South East Asia*	*Japan*	*Australasia*	*Other Eastern Hemisphere*	*Destination* not known*	*Total Exports*
USA	—	25	50	60	5	5	10	5	10	—	170
Canada	585	—	—	—	—	—	—	—	—	—	585
Caribbean	1 680	480	335	640	50	—	45	—	—	—	3 230
Other Western Hemisphere	80	—	5	5	—	—	—	—	—	—	90
Western Europe	160	—	25	—	100	—	—	—	10	50	345
Middle East	265	130	150	5 780	375	665	3 055	385	450	270	11 525
North Africa	250	40	25	3 890	15	—	15	—	30	—	4 265
West Africa	80	30	90	480	—	—	—	—	—	—	680
South East Asia	70	—	—	10	—	—	320	140	—	—	540
USSR, Eastern Europe	—	—	105	870	20	—	10	—	10	—	1 015
Other Eastern Hemisphere	—	—	—	5	5	160	—	—	5	20	195
TOTAL IMPORTS	3 170	705	785	11 740	570	830	3 455	530	515	340	22 640

* Includes changes in stock and increased quantities in transit, transit losses, minor movements not otherwise shown, military use, etc.

1970 (Thousand b/d)

FROM \ TO	*USA*	*Canada*	*Other Western Hemisphere*	*Western Europe*	*Africa*	*South East Asia*	*Japan*	*Australasia*	*Other Eastern Hemisphere*	*Destination* not known*	*Total Exports*
USA	—	25	25	65	—	70	70	—	5	—	260
Canada	690	—	—	—	—	—	—	—	—	25	715
Caribbean	2 135	510	225	535	20	20	25	—	—	—	3 470
Other Western Hemisphere	5	—	35	—	—	—	—	—	—	—	40
Western Europe	210	5	30	—	80	—	5	—	5	285	620
Middle East	175	150	240	6 220	420	800	3 485	345	430	450	2 715
North Africa	75	—	55	4 440	5	—	40	5	45	—	4 665
West Africa	50	35	155	890	—	—	10	—	—	—	1 140
South East Asia	70	—	10	10	—	—	605	95	5	—	795
USSR, Eastern Europe	5	—	95	795	30	—	15	—	10	—	950
Other Eastern Hemisphere	5	—	—	—	5	40	—	—	—	—	50
TOTAL IMPORTS	3 420	725	870	12 955	560	930	4 255	445	500	760	25 420

* Includes changes in stock and increased quantities in transit, transit losses, minor movements not otherwise shown, military use, etc.

Table 43—*contd.*

1971 (Thousand b/d)

FROM \ TO	*USA*	*Canada*	*Other Western Hemisphere*	*Western Europe*	*Africa*	*South East Asia*	*Japan*	*Australasia*	*Other Eastern Hemisphere*	*Destination* not known*	*Total Exports*
USA	—	30	70	65	5	5	40	5	5	—	225
Canada	800	—	—	—	—	—	—	—	—	—	800
Caribbean	2 170	410	40	510	5	—	15	—	—	—	3 150
Other Western Hemisphere	30	—	—	—	—	—	—	—	—	—	30
Western Europe	145	—	—	—	45	—	10	—	25	110	335
Middle East	390	300	465	7 685	520	980	3 810	320	415	365	15 250
North Africa	90	—	100	3 305	5	—	20	—	215	135	3 870
West Africa	105	60	140	1 125	—	—	40	—	—	150	1 620
South East Asia	130	—	10	5	5	—	645	60	—	—	855
USSR, Eastern Europe	5	—	135	850	60	—	30	—	10	—	1 090
Other Eastern Hemisphere	10	—	—	10	5	30	5	—	—	—	60
TOTAL IMPORTS	3 875	800	960	13 555	650	1 015	4 615	385	670	760	27 285

* Includes changes in stock and quantities in transit, transit losses, minor movements not otherwise shown, military use, etc.

1972 (Thousand b/d)

FROM \ TO	*USA*	*Canada*	*Other Western Hemisphere*	*Western Europe*	*Africa*	*South East Asia*	*Japan*	*Australasia*	*Other Eastern Hemisphere*	*Destination* not known*	*Total Exports*
USA	—	30	70	65	—	10	40	5	5	—	225
Canada	1 085	—	—	—	—	—	—	—	—	—	1 085
Caribbean	2 270	520	160	545	5	10	10	—	—	100	3 620
Other Western Hemisphere	45	—	25	—	—	—	30	—	—	—	100
Western Europe	160	15	—	—	40	—	—	—	20	90	325
Middle East	475	250	1 115	8 570	460	1 265	3 770	260	605	180	16 950
North Africa	230	30	250	2 730	10	—	5	—	220	—	3 475
West Africa	265	70	290	1 170	—	—	70	—	30	20	1 915
South East Asia	180	—	10	5	10	—	830	55	—	—	1 090
USSR, Eastern Europe	20	—	130	970	75	—	25	—	40	—	1 260
Other Eastern Hemisphere	10	—	—	5	—	30	5	—	—	—	50
TOTAL IMPORTS	4 740	915	2 050	14 060	600	1 315	4 785	320	920	390	30 095

* Includes quantities in transit, transit losses, minor movements not otherwise shown, military use, etc.

1973 (Thousand b/d)

FROM \ TO	*USA*	*Canada*	*Other Western Hemisphere*	*Western Europe*	*Africa*	*South East Asia*	*Japan*	*Australasia*	*Other Eastern Hemisphere*	*Destination* not known*	*Total Exports*
USA	—	30	55	85	5	25	30	—	—	—	230
Canada	1 335	—	—	—	—	—	—	—	—	—	1 335
Caribbean	2 540	500	130	355	5	5	10	—	—	85	3 630
Other Western Hemisphere	80	—	100	—	—	—	—	—	—	20	200
Western Europe	265	—	—	—	60	—	—	—	10	70	405
Middle East	820	320	1 000	10 350	535	1 315	4 385	290	775	180	19 970
North Africa	360	40	175	2 545	5	—	20	—	265	10	3 420
West Africa	515	90	415	1 030	—	—	110	—	5	—	2 165
South East Asia	240	—	—	10	—	—	1 115	40	—	—	1 405
USSR, Eastern Europe	35	—	145	935	75	5	55	—	20	15	1 285
Other Eastern Hemisphere	15	—	—	—	—	25	35	—	5	5	85
TOTAL IMPORTS	6 205	980	2 020	15 310	685	1 375	5 760	330	1 080	385	34 130

* Includes quantities in transit, transit losses, minor movements not otherwise shown, military uses, etc.

Table 43—*contd.*

1974 (Thousand b/d)

FROM \ TO	*USA*	*Canada*	*Other Western Hemisphere*	*Western Europe*	*Africa*	*South East Asia*	*Japan*	*Australasia*	*Other Eastern Hemisphere*	*Destination* not known*	*Total Exports*
USA	—	30	60	75	5	5	35	10	—	—	220
Canada	1 025	—	—	20	—	—	—	—	—	—	1 045
Caribbean	2 320	430	155	320	10	—	10	—	—	—	3 245
Other Western Hemisphere	125	—	120	—	—	—	—	—	—	—	245
Western Europe	240	—	—	—	80	—	15	—	35	95	465
Middle East	1 025	400	1 265	10 195	490	1 320	4 050	285	715	305	20 050
North Africa	230	15	130	1 915	10	—	70	—	255	—	2 625
West Africa	795	15	305	1 300	—	30	100	—	—	—	2 545
South East Asia	315	—	45	5	—	—	990	45	—	—	1 400
USSR, E. Europe & China	30	—	160	1 010	60	—	120	—	20	—	1 400
Other Eastern Hemisphere	20	—	—	—	—	30	40	—	—	—	90
TOTAL IMPORTS	6 125	890	2 240	14 840	655	1 385	5 430	340	1 025	400	33 330

* Includes quantities in transit, transit losses, minor movements not otherwise shown, military use, etc.

1975 (Thousand b/d)

FROM \ TO	*USA*	*Canada*	*Latin America*	*Western Europe*	*Africa*	*South East Asia*	*Japan*	*Australasia*	*Other Eastern Hemisphere*	*Total Exports*
USA	—	20	65	70	5	—	30	10	10	210
Canada	800	—	—	—	—	—	—	—	—	800
Latin America	2 270	270	120	380	40	—	20	—	75	3 175
Western Europe	45	—	—	—	165	—	—	—	35	245
Middle East	1 140	540	1 525	8 815	475	1 450	3 675	255	630	18 505
North Africa	500	—	140	1 530	—	—	70	—	165	2 405
West Africa	850	20	200	835	—	—	70	—	—	1 975
East and South Africa	—	—	—	—	—	35	—	—	—	35
South Asia	—	—	—	5	—	—	—	—	—	5
South East Asia	400	—	80	—	10	—	885	80	—	1 455
Japan	—	—	—	5	—	—	—	—	—	5
Australasia	5	—	—	—	—	10	35	—	—	50
USSR, E. Europe & China	15	—	210	970	15	70	160	—	30	1 470
TOTAL IMPORTS	6 025	850	2 340	12 610	710	1 565	4 945	345	945	30 335

Table 44 WORLD TANKER FLEET—OWNERSHIP (Million long tons dwt)

At 31st Dec.	*Total*	*Liberia*	*UK*	*Japan*	*Norway*
1957	49·6	8·4	8·4	1·3	7·3
1958	55·7	10·5	8·7	1·5	8·3
1959	60·9	11·7	9·7	1·9	9·1
1960	64·0	11·1	9·8	2·2	9·4
1961	67·1	10·8	10·4	2·6	9·8
1962	70·4	10·8	11·2	3·5	10·1
1963	75·0	12·4	11·8	4·1	10·8
1964	81·6	15·3	11·5	5·0	12·2
1965	90·1	19·1	11·8	6·5	13·3
1966	97·1	21·2	12·3	8·3	14·8
1967	105·5	23·6	13·1	9·6	16·5
1968	116·7	26·5	15·3	11·1	16·3
1969	132·1	30·8	18·7	13·3	15·6
1970	151·7	37·4	21·7	15·2	17·0
1971	171·4	43·1	25·1	18·2	18·8
1972	189·7	50·1	25·2	21·8	19·6
1973	215·6	59·2	27·8	26·1	21·2
1974	255·8	74·0	32·2	29·2	23·9

Table 45 WORLD TANKER FLEET—VESSELS PER SIZE CATEGORY

At 31st Dec.	*Tonnage Category (Thousand long tons dwt)*				
	25–50*	50–100	100–150	150–200	200–400
1957	419	8	—	—	—
1958	555	13	—	—	—
1959	694	21	—	—	—
1960	794	32	—	—	—
1961	843	49	2	—	—
1962	910	79	4	—	—
1963	947	145	4	—	—
1964	980	246	6	—	—
1965	953	350	15	—	—
1966	946	449	34	2	—
1967	895	551	59	5	2
1968	887	601	82	17	17
1969	905	630	96	30	61
1970	916	656	110	34	131
1971	939	661	125	37	200
1972	941	668	136	38	270
1973	975	681	152	41	357
1974	1 006	712	193	42	479

* From 1957 to 1961—24 000 to 50 000 long tons dwt

Table 46 MAJOR TANKER ROUTES

Comparative Distances and Steaming Times Between Principal Producing and Consuming Centres Based on a 15–knot Vessel

Consuming Centres	*Producing Centres*									
	Gulf of Mexico (New Orleans)		*Caribbean (Aruba)*		*Middle East*				*Far East (Singapore)*	
					Eastern Mediterranean (Tripoli)		*The Gulf (Mina Al Ahmadi)*			
	Nautical Miles	*Days Sailing*	*Nautical Miles*	*Days Sailing*	*Nautical Miles*	*Days Sailing*	*Nautical Miles*	*Days Sailing*	*Nautical Miles*	*Days Sailing*
UK (Isle of Grain)	4777	13	4 190	12	3 302	9	6 396*	19	8 124*	24
Via Cape							11 216	31	11 678	32
Mediterranean (Naples)	5 549	15	4 718	13	1 201	3	4 307*	13	6 035*	18
Via Cape							11 201	31	11 663	32
W. Africa (Dakar)	4 308	12	3 061	9	3 516	10	6 610*	19	8 338*	24
Via Cape							8 734	24	9 196	27
S. Africa (Cape Town)	7 306	20	5 798	16	5 574*	17	5 149	14	5 611	16
Via Cape					7 108	20				
India (Bombay)	9 442*	27	8 611*	25	3 224*	10	1 517	4	2 435	7
Via Cape	11 873	33	10 366	29	11 690	32				
Australia (Westernport)	9 623⁺	28	8 492⁺	25	7 995*	23	6 921	19	3 822	11
Via Cape					13 147	37				
Japan (Yokohama)	9 090⁺	26	8 319⁺	24	8 084	22	6 717	19	2 892	8
USA (New York)	1 707	5	1 763	5	5 044	14	8 318*	24	9 866*	28
Via Cape							11 918	33	12 380	34
USA (San Francisco)	4 649⁺	14	3 878⁺	12	9 589⁺	28	11 181	31	7 356	20
Argentina (La Plata)	6 197	17	4 695	13	7 281	20	8 782	24	9 244	26

Note: This is for one-way voyages only and steaming times include an allowance of 24 hours for passage through the Suez Canal (marked *) and Panama Canal (marked ⁺) but exclude port time at loading and discharge terminals.

Table 47 WORLD REFINING CAPACITIES—END OF YEAR (Million tonnes)

Country/Area	1952	1955	1960	1965	1970	1974
NORTH AMERICA						
USA	380	425	476	518	640	745
Canada	22	29	46	54	67	98
Mexico	12	15	18	22	27	37
Total	414	469	540	594	734	880
CARIBBEAN						
Venezuela	17	25	44	54	64	74
Trinidad	5	6	13	17	22	23
NWI	31	31	31	37	40	45
Others	4	4	12	23	53	94
Total	57	66	100	131	179	236
SOUTH AMERICA						
Argentina	9	9	16	21	24	35
Brazil	} 4	6	10	18	25	46
Others		4	8	11	15	18
Total	13	19	34	50	64	99
WESTERN EUROPE						
Belgium	3	6	8	15	35	46
France	22	25	40	65	114	164
Germany, West	7	11	40	81	120	149
Netherlands	7	10	22	28	67	97
Italy	8	21	35	78	146	194
UK	24	29	50	71	114	149
Others	7	8	23	48	123	173
Total	78	110	218	386	719	972
MIDDLE EAST						
Aden		5	5	7	8	8
Bahrain		9	9	10	10	12
Iran		12	23	26	30	33
Saudi Arabia	} 42	9	9	15	21	23
Kuwait		2	10	17	23	26
Neutral Zone		2	2	2	3	4
Others		6	11	14	18	26
Total	42	45	69	91	113	132
AFRICA			6	28	36	62
SOUTH ASIA	} 25	} 23	7	18	31	36
SOUTH EAST ASIA			18	32	69	137
AUSTRALASIA		8	11	23	33	37
JAPAN	4	10	35	89	164	254
USSR, E. EUROPE AND CHINA	57	80	174	250	403	563

Table 47—*contd.*

Country/Area	1952	1955	1960	1965	1970	1974
TOTAL EASTERN HEMISPHERE	206	276	538	917	1 568	2 193
TOTAL WESTERN HEMISPHERE	484	554	674	775	977	1215
WORLD (Excl. USSR, etc.)	633	750	1 038	1 442	2 142	2 845
WORLD	690	830	1 212	1 692	2 545	3 408
OPEC	*	*	89*	130	161	201
NON-OPEC	*	*	1 123*	1 562	2 384	3 207

* OPEC formed in September 1960

Table 48 CRUDE OIL PRICES 1860–1975 ($/bbl)

Year	*Price*	*Year*	*Price*	*Year*		*Price*
1860	9·59	1904	0·86	December	1946	1·20
1861	0·49	1905	0·62	March	1947	1·60
1862	1·05	1906	0·73	December	1947	2·20
1863	3·15	1907	0·72	July	1948	1·99
1864	8·06	1908	0·72	April	1949	1·84
1865	6·59	1909	0·70	July	1949	1·71
1866	3·74	1910	0·61		1950	1·71
1867	2·41	1911	0·61		1951	1·71
1868	3·63	1912	0·74		1952	1·71
1869	3·64	1913	0·95	July	1953	1·93
1870	3·86	1914	0·81		1954	1 93
1871	4·34	1915	0·64		1955	1·93
1872	3·64	1916	1·10		1956	1·93
1873	1·83	1917	1·56	June	1957	2·08
1874	1·17	1918	1·98		1958	2·08
1875	1·35	1919	2·01	February	1959	1·90
1876	2·56	1920	3·07	August	1960	1·76
1877	2·42	1921	1·73	September	1960	1·80
1878	1·19	1922	1·61		1961	1·80
1879	0·86	1923	1·34		1962	1·80
1880	0·95	1924	1·43		1963	1·80
1881	0·86	1925	1·68		1964	1·80
1882	0·78	1926	1·88		1965	1·80
1883	1·00	1927	1·30		1966	1·80
1884	0·84	1928	1·17		1967	1·80
1885	0·88	1929	1·27		1968	1·80
1886	0·71	1930	1·19		1969	1·80
1887	0·67	1931	0·65		1970	1·80
1888	0·88	1932	0·87	15 February	1971	2·18
1889	0·94	1933	0·67	1 June	1971	2·29
1890	0·87	1934	1·00	20 January	1972	2·48
1891	0·67	1935	0·97	1 January	1973	2·59
1892	0·56	1936	1·09	1 April	1973	2·74
1893	0·64	1937	1·18	1 June	1973	2·90
1894	0·84	1938	1·13	1 July	1973	2·96
1895	1·36	1939	1·02	1 August	1973	3·07
1896	1·18	1940	1·02	1 October	1973	3·01
1897	0·79	1941	1·14	16 October	1973	5·12
1898	0·91	1942	1·19	1 November	1973	5·18
1899	1·29	1943	1·20	1 December	1973	5·04
1900	1·19	1944	1·21	1 January	1974	11·65
1901	0·96	1945	1·05	November	1974	11·25
1902	0·80	1946	1·05		1975	11·25
1903	0·94					

1860–1899 Pennsylvania Crude Prices
1900–1944 US Average Crude Price
1945–1975 Arabian Light posted at Ras Tanura
From 1957–1972 market prices of crude oils were at a discount on posted prices. In the period 1961–1970 market prices were in the range $1·3–$1·5/bbl.

Table 49 PRODUCT PRICES AT ROTTERDAM, SPOT MARKET FREIGHT RATES AND COSTS OF CRUDE OIL 1967–1974 ($/tonne)

Date	*Premium Motor Spirit*	*Gas Oil*	*Fuel Oil, 3·5% Sulphur*	*Spot Market Freight Rate ME–NWE*	*Arabian light Crude Oil at Ras Tanura*
1967 J	29·5	21·5	11	6	13·28
F	29·5	21	11	5	13·28
M	29·5	20	11	4	13·28
A	30	17·5	11	3	13·28
M	30	17·5	11	4	13·28
J	32·5	22·5	12	12	13·28
J	41·5	31·5	16	19	13·28
A	50	30·5	14	18	13·28
S	43	28	13·5	17	13·28
O	40·5	28·5	13·5	12	13·28
N	39	30	13·5	12	13·28
D	38	30·5	13	11	13·28
1968 J	38	32·5	13	10	13·28
F	36·5	31·5	12·5	8	13·28
M	33	27	12·5	9	13·28
A	32	25	13	9	13·28
M	31	22·5	12·5	11	13·28
J	30	23	12	14	13·28
J	30	22·5	11·5	10	13·28
A	30	24	11	8	13·28
S	29·5	24·5	10·5	8	13·28
O	29·5	24·5	10	9	13·28
N	29	25·5	10	10	13·28
D	27·5	26	10	9	13·28
1969 J	27	25·5	10	7	13·28
F	25·5	22·5	10	7	13·28
M	25	21	10	7	13·28
A	25	20	10	5	13·28
M	25·5	20·5	10·5	5	13·28
J	25	19·5	10	6	13·28
J	24·5	19·5	10	7	13·28
A	24·5	19·5	10	8	13·28
S	25	20	10	8	13·28
O	24	20	9·5	10	13·28
N	24·5	19	9·5	10	13·28
D	25	20	10·5	11	13·28
1970 J	25	22	12	9	13·28
F	24·5	21	12·5	10	13·28
M	24	20	14·5	11	13·28
A	23·5	20·5	15	11	13·28
M	23·5	21·3	15	9	13·28
J	25	21·5	16·5	16	13·28
J	27·5	24	20·5	18	13·28
A	32	29·5	25·5	18	13·28
S	31·5	31·5	22	21	13·28
O	31·5	33	20	23	13·28
N	31	34	21	20	13·28
D	31·5	34	22·5	17	13·28

Table 49—*contd.*

Date	*Premium Motor Spirit*	*Gas Oil*	*Fuel Oil, 3·5% Sulphur*	*Spot Market Freight Rate ME–NWE*	*Arabian Light Crude Oil at Ras Tanura*
1971 J	33	35·5	23·5	20	13·28
F	34	35	23	14	14·69
M	35	34	22	12	16·09
A	35	32	22	9	16·09
M	35	31	20·5	7	16·09
J	35	29	17	6	16·90
J	34	27	14	4	16·90
A	33	26	10·5	5	16·90
S	32	27	10·5	4	16·90
O	32·5	24	11	5	16·90
N	32	23	11	6	16·90
D	32·5	27·5	12	7	16·90
1972 J	32·5	26·5	13	6	17·42
F	32·5	24	14	4	18·30
M	32·5	23·5	14	3	18·30
A	32	25	14	3	18·30
M	35·5	24	14	3	18·30
J	37·5	23·5	14	5	18·30
J	41	24	14	5	18·30
A	43·5	25·5	14·5	6	18·30
S	43	27	14·5	7	18·30
O	43·5	29	14	8	18·30
N	44	32·5	14·5	8	18·30
D	48	36	15	9	18·30
1973 J	53	37·5	16·5	12	19·11
F	55	42	18·5	12	19·11
M	62	42	17·5	12	19·11
A	74	44	16	11	20·22
M	96	54·5	17	16	20·22
J	112	66·5	19	24	21·40
J	102·5	72·5	18·5	18	21·84
A	84	71·5	17	28	22·66
S	85·5	79·5	15·5	31	22·66
O	102	121	21·5	28	30·26
N	150	186·5	48	10	38·23
D	187	186	119·5	9	37·20
1974 J	151	132	103	10	85·98
F	153·5	103·5	76·5	9	85·98
M	185·5	100·5	67·5	9	85·98
A	181	83·5	65	7	85·98
M	174	90·5	67	7	85·98
J	148·5	88·5	64	8	85·98
J	124·5	93	62	5	85·98
A	115	92	61	5	85·98
S	113·5	93	63	5	85·98
O	109	92	65	7	85·98
N	113	91	70	4	83·03
D	115·5	92	70	4	83·03

Table 50 UK PETROL PRICES 1912–1975

Year	*Current Prices (Prices of the day)*			*UK Price Index*	*Constant Prices (1975)*		
	Retail price	*Tax*	*Tax free price*		*Retail price*	*Tax*	*Tax free price*
1912	7·9	1·3	6·6	91	74·7	12·3	62·4
1913	8·8	1·3	7·5	91	83·2	12·3	70·9
1914	8·8	1·3	7·5	91	83·2	12·3	70·9
1915	10·6	2·5	8·1	116	78·6	18·5	60·1
1916	12·9	2·5	10·4	146	76·0	14·7	61·3
1917	16·1	2·5	13·6	193	71·7	11·1	60·6
1918	18·4	2·5	15·9	207	76·4	10·4	66·0
1919	15·7	2·5	13·2	222	60·8	9·7	51·1
1920	19·0	2·5	16·5	270	60·5	8·0	52·5
1921	14·9	0	14·9	167	76·7	0	76·7
1922	10·0	0	10·0	141	61·0	0	61·0
1923	8·3	0	8·3	139	51·4	0	51·4
1924	8·7	0	8·7	150	49·9	0	49·9
1925	8·7	0	8·7	146	51·2	0	51·2
1926	7·9	0	7·9	136	50·0	0	50·0
1927	6·1	0	6·1	131	40·0	0	40·0
1928	6·5	1·2	5·3	129	43·3	8·0	35·3
1929	7·7	1·7	6·0	124	53·4	11·8	41·6
1930	7·5	1·7	5·8	104	62·0	14·1	47·9
1931	6·6	2·5	4·1	89	63·8	24·2	39·6
1932	7·3	3·3	4·0	86	73·0	33·0	40·0
1933	7·5	3·3	4·2	85	75·9	33·4	42·5
1934	7·2	3·3	3·9	103	60·1	27·6	32·5
1935	7·4	3·3	4·1	103	61·8	27·6	34·2
1936	7·4	3·3	4·1	106	60·0	26·8	33·2
1937	7·9	3·3	4·6	110	61·8	25·8	36·0
1938	7·9	3·6	4·3	113	60·1	27·4	32·7
1939	8·0	3·8	4·2	113	60·9	28·9	32·0
1940	9·9	3·8	6·1	152	56·0	21·5	34·5
1941	9·9	3·8	6·1	205	41·5	15·9	25·6
1942	10·7	3·8	6·9	195	47·2	16·8	30·4
1943	10·7	3·8	6·9	177	52·0	18·5	33·5
1944	10·7	3·8	6·9	195	47·2	16·8	30·4
1945	10·3	3·8	6·5	191	46·4	17·1	29·3
1946	10·3	3·8	6·5	191	46·4	17·1	29·3
1947	10·1	3·8	6·3	205	42·4	15·9	26·5
1948	10·5	3·8	6·7	219	41·2	14·9	26·3
1949	10·5	3·8	6·7	227	39·8	14·4	25·4
1950	14·1	6·4	7·7	234	51·8	23·5	28·3
1951	17·2	8·9	8·3	251	58·9	30·5	28·4
1952	20·6	11·9	8·7	269	65·9	38·0	27·9
1953	20·9	12·5	8·4	273	65·8	39·4	26·4
1954	20·3	12·5	7·8	276	63·3	38·9	24·4
1955	20·4	12·5	7·9	287	61·1	37·5	23·6
1956	21·0	12·9	8·1	301	60·0	36·9	23·1

Table 50—*contd.*

Year	*Current Prices (Prices of the day)*			*UK Price Index*	*Constant Prices (1975)*		
	Retail price	*Tax*	*Tax free price*		*Retail price*	*Tax*	*Tax free price*
1957	23·3	13·9	9·4	312	64·2	38·3	25·9
1958	21·0	12·5	8·5	319	56·6	33·7	22·9
1959	21·0	12·5	8·5	319	56·6	33·7	22·9
1960	21·3	12·5	8·8	322	56·9	33·4	23·5
1961	21·8	13·1	8·7	333	56·3	33·8	22·5
1962	22·5	13·8	8·7	347	55·8	34·2	21·6
1963	22·5	13·8	8·7	354	54·7	33·5	21·2
1964	22·9	14·1	8·8	365	54·0	33·2	20·8
1965	25·2	16·3	8·9	379	57·2	37·0	20·2
1966	26·0	17·0	9·0	396	56·5	36·9	19·6
1967	27·1	17·9	9·2	404	57·7	38·1	19·6
1968	29·0	19·5	9·5	425	58·7	39·5	19·2
1969	31·1	22·2	8·9	446	60·0	42·8	17·2
1970	32·7	22·5	10·2	474	59·3	40·8	18·5
1971	34·0	22·5	11·5	513	57·0	37·7	19·3
1972	35·0	22·5	12·5	545	55·2	35·5	19·7
1973	35·5	22·5	13·0	595	51·3	32·5	18·8
1974	51·5	26·5	25·0	691	64·1	33·0	31·1
1975*	71·0	36·5	34·5	860	71·0	36·5	34·5

0 = Duty Cancelled. * = Jan–June.

1. All prices are annual averages in new pence (100 pence = £1) and are rounded to nearest tenth whole penny.
2. 1912–1924 Can Prices in London, Premier Grade.
   1925–1938 Bulk Prices in London, Premier Grade.
   1939–1958 Bulk Prices in London, Pool Grade.
   1953–1975 Bulk Prices in London, Ordinary Grade.
3. The octane quality has increased throughout the period, in recent years it has been about 91 octane.
4. The price index (1961 to 1973) was published in the *Economist*, 13th July 1974, page 62.

Table 51 SEASONAL RATIOS: MOTOR SPIRIT—INLAND CONSUMPTION

Country/Area	*Average Quarterly Weights and Their Standard Deviations*							
	1st Quarter		*2nd Quarter*		*3rd Quarter*		*4th Quarter*	
	Mean	*SD*	*Mean*	*SD*	*Mean*	*SD*	*Mean*	*SD*
OECD European Area[1]	0·87	0·022	1·04	0·010	1·12	0·014	0·96	0·010
EEC								
Belgium	0·89	0·043	1·06	0·016	1·08	0·017	0·99	0·052
France	0·86	0·028	1·06	0·016	1·16	0·037	0·93	0·023
Germany	0·88	0·019	1·06	0·017	1·09	0·020	0·98	0·020
Italy	0·87	0·030	1·00	0·026	1·17	0·071	0·96	0·037
Luxembourg[2]	0·80	0·051	1·07	0·033	1·21	0·036	0·91	0·037
Netherlands	0·89	0·034	1·07	0·020	1·09	0·023	0·98	0·020
EFTA								
Austria	0·80	0·046	1·04	0·031	1·28	0·052	0·89	0·070
Denmark	0·83	0·054	1·07	0·022	1·10	0·034	1·00	0·046
Norway	0·80	0·027	1·02	0·020	1·21	0·010	0·97	0·022
Portugal	0·90	0·050	0·99	0·023	1·13	0·031	0·99	0·027
Sweden	0·90	0·126	1·10	0·022	1·10	0·087	0·92	0·070
Switzerland	0·81	0·058	1·04	0·039	1·17	0·044	0·98	0·063
UK	0·90	0·030	1·05	0·014	1·09	0·020	0·96	0·083
Greece[1]	0·90	0·053	1·02	0·031	1·13	0·043	0·94	0·042
Finland[3]	0·80	0·034	1·07	0·004	1·17	0·011	0·97	0·024
Ireland	0·88	0·025	1·02	0·021	1·15	0·010	0·95	0·020
Spain	0·84	0·030	1·03	0·056	1·21	0·029	0·92	0·032
Turkey	0·86	0·054	0·97	0·038	1·17	0·052	1·00	0·042
Japan[4]	1·04	0·060	1·01	0·039	0·89	0·017	1·04	0·012

Figures based on 1961–1970 except:
[1] 1961–1969
[2] 1964–1970
[3] 1967–1970
[4] 1965–1969

Table 52 SEASONAL RATIOS: MOTOR SPIRIT—TOTAL SUPPLY

Country/Area	*Average Quarterly Weights and Their Standard Deviations*							
	1*st Quarter*		*2nd Quarter*		*3rd Quarter*		*4th Quarter*	
	Mean	*SD*	*Mean*	*SD*	*Mean*	*SD*	*Mean*	*SD*
OECD European Area	0·95	0·023	0·98	0·016	1·06	0·012	1·00	0·016
EEC								
Belgium	0·96	0·046	0·99	0·046	1·03	0·067	1·02	0·050
France	0·99	0·043	0·95	0·031	1·04	0·026	1·03	0·035
Germany	0·93	0·035	1·01	0·025	1·06	0·027	1·00	0·032
Italy	0·97	0·030	0·98	0·026	1·03	0·033	1·00	0·062
Luxembourg[1]	0·81	0·053	1·11	0·059	1·22	0·025	0·86	0·051
Netherlands	1·00	0·082	0·95	0·077	1·04	0·046	1·01	0·049
EFTA								
Austria	0·84	0·069	1·07	0·064	1·23	0·167	0·86	0·121
Denmark	0·91	0·087	1·01	0·054	1·04	0·067	1·04	0·069
Norway	0·86	0·100	0·98	0·042	1·14	0·040	1·02	0·073
Portugal	0·96	0·118	0·95	0·072	1·07	0·101	1·01	0·096
Sweden	0·90	0·079	0·90	0·054	1·14	0·059	1·07	0·080
Switzerland	0·82	0·079	1·02	0·053	1·17	0·066	0·99	0·095
UK	0·96	0·035	0·99	0·031	1·05	0·027	1·00	0·040
Greece	0·94	0·075	1·03	0·076	1·06	0·078	0·97	0·074
Finland[2]	0·94	0·011	0·92	0·040	0·92	0·123	1·24	0·124
Ireland	0·89	0·072	1·04	0·073	1·12	0·052	0·94	0·092
Spain	0·93	0·057	1·07	0·046	1·10	0·069	0·90	0·074
Turkey	0·90	0·118	0·97	0·085	1·11	0·054	1·01	0·051
Japan[3]	1·05	0·044	1·03	0·053	0·86	0·160	1·07	0·122

Figures based on 1961–1970 except:
[1] 1964–1970
[2] 1967–1970
[3] 1965–1969

Table 53 SEASONAL RATIOS: GAS/DIESEL OIL—INLAND CONSUMPTION

Country/Area	*Average Quarterly Weights and Their Standard Deviations*							
	1st Quarter		*2nd Quarter*		*3rd Quarter*		*4th Quarter*	
	Mean	*SD*	*Mean*	*SD*	*Mean*	*SD*	*Mean*	*SD*
OECD European Area[1]	1·21	0·048	0·86	0·046	0·82	0·054	1·12	0·039
EEC								
Belgium	1·40	0·128	0·73	0·106	0·67	0·059	1·21	0·064
France	1·42	0·057	0·71	0·047	0·59	0·049	1·28	0·060
Germany	1·14	0·066	0·91	0·057	0·91	0·069	1·04	0·073
Italy	0·89	0·115	0·94	0·107	1·13	0·099	1·03	0·091
Luxembourg[2]	1·31	0·074	0·81	0·086	0·73	0·089	1·14	0·077
Netherlands	1·31	0·070	0·80	0·056	0·70	0·052	1·18	0·053
EFTA								
Austria	0·82	0·087	0·97	0·044	1·08	0·131	1·13	0·144
Denmark	1·36	0·091	0·83	0·082	0·65	0·079	1·16	0·032
Norway	1·21	0·099	0·88	0·053	0·80	0·095	1·11	0·170
Portugal	0·93	0·070	1·02	0·086	1·05	0·023	1·01	0·031
Sweden	1·34	0·094	0·82	0·084	0·67	0·070	1·16	0·030
Switzerland	0·96	0·097	1·08	0·146	1·14	0·118	0·82	0·077
UK	1·18	0·069	0·92	0·029	0·84	0·043	1·06	0·092
Greece[1]	1·18	0·072	0·86	0·048	0·88	0·055	1·08	0·064
Finland[3]	1·30	0·028	0·76	0·026	0·76	0·033	1·17	0·030
Ireland	1·09	0·037	0·99	0·055	0·89	0·040	1·02	0·091
Spain	0·94	0·041	1·02	0·049	1·06	0·032	0·98	0·040
Turkey	0·78	0·056	1·03	0·060	1·14	0·035	1·05	0·052
Japan[4]	1·03	0·055	1·00	0·022	0·95	0·031	1·01	0·099

Figures based on 1961–1970 except:
[1] 1961–1969
[2] 1964–1970
[3] 1967–1970
[4] 1965–1969

Table 54 SEASONAL RATIOS: GAS/DIESEL OIL—TOTAL SUPPLY

Country/Area	*Average Quarterly Weights and Their Standard Deviations*							
	1*st Quarter*		*2nd Quarter*		*3rd Quarter*		*4th Quarter*	
	Mean	*SD*	*Mean*	*SD*	*Mean*	*SD*	*Mean*	*SD*
OECD European Area	1·06	0·026	0·95	0·016	0·96	0·021	1·02	0·024
EEC								
Belgium	1·13	0·053	0·89	0·095	0·92	0·042	1·05	0·070
France	1·15	0·046	0·86	0·048	0·88	0·035	1·11	0·029
Germany	1·09	0·050	0·95	0·040	0·97	0·055	0·98	0·056
Italy	0·88	0·051	1·04	0·028	1·07	0·046	1·01	0·062
Luxembourg[1]	1·32	0·091	0·82	0·079	0·75	0·071	1·11	0·071
Netherlands	1·10	0·061	0·90	0·024	0·97	0·041	1·02	0·066
EFTA								
Austria	0·81	0·064	1·06	0·094	1·12	0·114	1·01	0·147
Denmark	1·25	0·058	0·88	0·038	0·78	0·062	1·09	0·066
Norway	1·04	0·047	0·94	0·058	0·96	0·057	1·07	0·050
Portugal	0·98	0·102	0·95	0·074	1·04	0·064	1·02	0·063
Sweden	1·10	0·103	0·90	0·068	0·93	0·113	1·07	0·077
Switzerland	0·89	0·139	1·11	0·157	1·16	0·088	0·83	0·095
UK	1·00	0·051	1·00	0·046	0·96	0·052	1·03	0·030
Greece	1·15	0·087	0·91	0·038	0·94	0·058	1·00	0·079
Finland[2]	1·03	0·144	0·97	0·162	0·93	0·139	1·07	0·185
Ireland	0·94	0·145	1·05	0·032	1·03	0·084	0·98	0·121
Spain	0·97	0·067	1·05	0·043	1·02	0·031	0·98	0·048
Turkey	0·79	0·076	1·05	0·033	1·12	0·070	1·04	0·065
Japan[3]	0·94	0·020	1·07	0·037	0·97	0·054	1·03	0·041

Figures based on 1961–1970 except:
[1] 1964–1970
[2] 1967–1970
[3] 1965–1969

Table 55 SEASONAL RATIOS: FUEL OIL—INLAND CONSUMPTION

Country/Area	*Average Quarterly Weights and Their Standard Deviations*							
	1st Quarter		*2nd Quarter*		*3rd Quarter*		*4th Quarter*	
	Mean	*SD*	*Mean*	*SD*	*Mean*	*SD*	*Mean*	*SD*
OECD European Area[1]	1·18	0·030	0·86	0·023	0·82	0·015	1·14	0·026
EEC								
Belgium	1·16	0·077	0·94	0·052	0·80	0·027	1·09	0·069
France	1·16	0·075	0·88	0·053	0·80	0·031	1·17	0·034
Germany	1·08	0·052	0·92	0·030	0·87	0·027	1·14	0·052
Italy	1·24	0·029	0·72	0·053	0·88	0·041	1·15	0·033
Luxembourg[2]	1·08	0·039	0·97	0·088	0·89	0·022	1·06	0·044
Netherlands	1·22	0·048	0·91	0·042	0·75	0·028	1·12	0·033
EFTA								
Austria	1·13	0·058	0·81	0·049	0·86	0·054	1·19	0·043
Denmark	1·27	0·061	0·83	0·044	0·70	0·029	1·10	0·049
Norway	1·15	0·070	0·95	0·094	0·80	0·069	1·09	0·065
Portugal	1·02	0·060	0·90	0·065	1·02	0·078	1·06	0·035
Sweden	1·15	0·057	0·86	0·033	0·72	0·037	1·26	0·047
Switzerland	0·85	0·137	1·04	0·086	1·07	0·180	1·04	0·120
UK	1·24	0·037	0·91	0·027	0·74	0·012	1·11	0·022
Greece[1]	1·03	0·046	0·95	0·042	0·99	0·047	1·02	0·052
Finland[3]	1·19	0·088	0·82	0·041	0·76	0·011	1·23	0·102
Ireland	1·07	0·074	1·00	0·013	0·95	0·082	0·98	0·013
Spain	1·02	0·067	0·95	0·064	0·94	0·053	1·09	0·078
Turkey	1·06	0·079	0·94	0·081	0·94	0·079	1·05	0·027
Japan[4]	1·08	0·058	0·91	0·099	0·94	0·014	1·08	0·018

Figures based on 1961–1970 except:
[1] 1961–1969
[2] 1964–1970
[3] 1967–1970
[4] 1965–1969

Table 56 SEASONAL RATIOS: FUEL OIL—TOTAL SUPPLY

Country/Area	*Average Quarterly Weights and Their Standard Deviations*							
	1st Quarter		*2nd Quarter*		*3rd Quarter*		*4th Quarter*	
	Mean	*SD*	*Mean*	*SD*	*Mean*	*SD*	*Mean*	*SD*
OECD European Area[1]	1·07	0·020	0·95	0·021	0·92	0·021	1·05	0·023
EEC								
Belgium	1·09	0·062	1·01	0·058	0·93	0·051	0·96	0·069
France	1·06	0·072	0·91	0·077	0·96	0·052	1·08	0·064
Germany	1·06	0·039	0·98	0·020	0·93	0·029	1·03	0·045
Italy	1·11	0·034	0·88	0·039	0·92	0·016	1·09	0·025
Luxembourg[2]	1·09	0·039	0·95	0·033	0·90	0·028	1·05	0·052
Netherlands	1·06	0·035	0·94	0·033	0·94	0·059	1·06	0·045
EFTA								
Austria	1·11	0·063	0·89	0·078	0·88	0·051	1·12	0·071
Denmark	1·15	0·057	0·89	0·043	0·84	0·040	1·13	0·044
Norway	1·03	0·070	0·95	0·037	0·99	0·038	1·03	0·100
Portugal	1·00	0·127	0·87	0·106	1·12	0·139	1·00	0·143
Sweden	1·06	0·076	0·91	0·075	0·92	0·069	1·10	0·059
Switzerland	0·88	0·116	1·03	0·114	1·10	0·134	1·00	0·123
UK	1·09	0·036	0·99	0·016	0·88	0·050	1·04	0·038
Greece[1]	0·95	0·086	1·06	0·045	1·02	0·053	0·97	0·097
Finland[3]	0·96	0·070	0·78	0·045	1·05	0·157	1·21	0·136
Ireland	0·99	0·154	1·05	0·104	1·01	0·076	0·94	0·100
Spain	1·00	0·067	1·04	0·048	0·97	0·059	0·99	0·087
Turkey	0·99	0·115	0·97	0·063	0·99	0·118	1·05	0·098
Japan[4]	1·06	0·047	0·96	0·083	0·94	0·037	1·04	0·025

Figures based on 1961–1970 except:
[1] 1961–1969
[2] 1964–1970
[3] 1967–1970
[4] 1965–1969

Table 57 SEASONAL RATIOS: ALL PRODUCTS—INLAND CONSUMPTION

Country/Area	*Average Quarterly Weights and Their Standard Deviations*							
	1st Quarter		*2nd Quarter*		*3rd Quarter*		*4th Quarter*	
	Mean	*SD*	*Mean*	*SD*	*Mean*	*SD*	*Mean*	*SD*
OECD European Area[1]	1·10	0·024	0·91	0·023	0·90	0·026	1·09	0·024
EEC								
Belgium	1·17	0·050	0·89	0·038	0·82	0·031	1·11	0·055
France	1·17	0·038	0·86	0·027	0·82	0·044	1·15	0·043
Germany	1·03	0·036	0·95	0·028	0·96	0·040	1·06	0·045
Italy	1·09	0·027	0·81	0·024	0·98	0·028	1·11	0·036
Luxembourg[2]	1·12	0·078	0·93	0·037	0·88	0·033	1·07	0·033
Netherlands	1·18	0·043	0·89	0·037	0·80	0·022	1·12	0·029
EFTA								
Austria	0·96	0·041	0·93	0·032	1·04	0·036	1·08	0·039
Denmark	1·22	0·060	0·86	0·050	0·76	0·044	1·15	0·030
Norway	1·10	0·052	0·92	0·040	0·88	0·054	1·10	0·081
Portugal	0·97	0·028	0·97	0·015	1·04	0·043	1·23	0·033
Sweden	1·15	0·053	0·89	0·034	0·79	0·045	1·17	0·029
Switzerland	0·88	0·055	1·07	0·082	1·15	0·010	0·90	0·059
UK	1·14	0·025	0·94	0·019	0·85	0·021	1·07	0·018
Greece[1]	1·04	0·038	0·94	0·023	0·98	0·030	1·03	0·028
Finland[3]	1·16	0·041	0·82	0·015	0·85	0·014	1·16	0·040
Ireland	1·01	0·040	1·01	0·074	1·02	0·060	0·96	0·061
Spain	0·98	0·023	0·99	0·045	1·01	0·032	1·02	0·051
Turkey	0·93	0·046	0·97	0·030	1·05	0·030	1·06	0·041
Japan[4]	1·09	0·017	0·93	0·012	0·91	0·038	1·08	0·028

Figures based on 1961–1970 except:
[1] 1961–1969
[2] 1964–1970
[3] 1967–1970
[4] 1965–1969

Table 58 SEASONAL RATIOS: ALL PRODUCTS—TOTAL SUPPLY

Country/Area	*Average Quarterly Weights and Their Standard Deviations*							
	1st Quarter		*2nd Quarter*		*3rd Quarter*		*4th Quarter*	
	Mean	*SD*	*Mean*	*SD*	*Mean*	*SD*	*Mean*	*SD*
OECD European Area[1]	1·03	0·020	0·96	0·016	0·97	0·014	1·03	0·019
EEC								
Belgium	1·06	0·033	0·97	0·044	0·96	0·028	1·01	0·037
France	1·05	0·056	0·92	0·053	0·97	0·023	1·07	0·027
Germany	1·02	0·023	0·98	0·011	0·99	0·029	1·01	0·036
Italy	1·02	0·029	0·94	0·027	0·99	0·014	1·05	0·023
Luxembourg[2]	1·27	0·031	0·92	0·033	0·89	0·024	1·05	0·048
Netherlands	1·06	0·040	0·94	0·021	0·96	0·028	1·04	0·034
EFTA								
Austria	0·94	0·059	1·00	0·018	1·04	0·045	1·01	0·051
Denmark	1·13	0·035	0·91	0·022	0·86	0·085	1·09	0·042
Norway	1·00	0·051	0·93	0·020	1·00	0·077	1·06	0·048
Portugal	1·01	0·078	0·94	0·053	1·04	0·045	1·02	0·049
Sweden	1·04	0·068	0·92	0·062	0·97	0·035	1·08	0·056
Switzerland	0·85	0·068	1·08	0·090	1·17	0·079	0·89	0·070
UK	1·04	0·029	0·99	0·018	0·93	0·039	1·03	0·025
Greece[1]	1·03	0·011	1·02	0·055	1·03	0·075	0·92	0·021
Finland[3]	0·98	0·085	0·90	0·087	1·00	0·130	1·12	0·124
Ireland	0·94	0·110	1·04	0·060	1·07	0·052	0·95	0·084
Spain	0·99	0·054	1·06	0·074	0·99	0·048	0·96	0·071
Turkey	0·92	0·087	1·00	0·056	1·05	0·057	1·04	0·064
Japan[4]	1·05	0·025	0·98	0·014	0·92	0·048	1·05	0·035

Figures based on 1961–1970 except:
[1] 1961–1969
[2] 1964–1970
[3] 1967–1970
[4] 1965–1969

Table 59 EUROPEAN WEATHER—CITY DETAIL (Degree days *ex Platt's Oilgram*)

City	1959/1971 *Average*	*October to April Periods*	
		1973/4	1972/3
Oslo	6 596	6 014	5 592
Helsinki	7 243	6 533	5 969
Stockholm	6 154	5 718	5 096
Copenhagen	5 067	4 703	4 589
London	3 307	3 244	3 444
Amsterdam	3 847	3 457	3 833
Brussels	3 744	3 347	3 719
Paris	3 357	3 026	3 448
Hamburg	4 997	4 295	4 730
Essen	4 504	4 164	4 557
Frankfurt	4 337	4 245	4 942
Munich	5 355	5 150	5 760
Geneva	4 137	3 901	4 385
Vienna	4 207	3 731	4 237
Klagenfurt	5 146	4 801	5 262
Marseilles	2 134	1 945	2 193
Total	74 332	68 274	71 756

Table 60 EUROPEAN WEATHER—PERCENTAGE DEVIATION FROM 1959–1975 DEGREE DAY AVERAGES (*ex Platt's Oilgram*)

Year	*January*	*February*	*March*	*April*	*October*	*November*	*December*	*October–December*	*January–March*	*January–April*	*January–December*	*October–April*
1959	—	—	—	—	−12·4	+1·4	−11·9	−7·5	—	—	−3·0	−3·9
1960	−0·6	+3·6	−8·2	−2·3	−0·2	−11·8	−6·6	−7·1	−1·5	−1·6	−3·8	−11·6
1961	+1·5	−19·7	−20·8	−28·2	−22·4	−12·5	+5·8	−6·0	−11·9	−14·5	−11·2	−1·2
1962	−10·6	+0·3	+21·7	−0·1	−4·8	+11·4	+20·3	+12·2	+2·4	+2·0	+6·0	+17·6
1963	+33·2	+33·9	+8·9	−6·0	+2·6	−19·9	+17·0	+1·7	+26·3	+21·2	+13·5	+3·5
1964	+8·6	+1·1	+12·4	−9·7	+22·0	−5·0	+0·4	+3·0	+7·3	+4·6	+4·0	+2·5
1965	−6·9	+10·4	+1·9	+6·8	+0·6	+20·4	−5·6	+4·4	+1·3	+2·2	+3·0	+3·1
1966	+14·9	−8·1	−1·5	+0·5	−17·9	+10·5	−7·1	−3·4	+2·6	+2·3	+0·1	−4·8
1967	+0·4	−11·5	−15·7	+7·3	−24·9	−1·6	+4·7	−3·5	−8·2	−5·8	−4·8	−2·0
1968	+8·6	+3·6	−9·9	−15·0	−10·7	+4·7	+10·0	+4·0	+1·6	−1·1	+1·0	+5·7
1969	−5·1	+15·9	+14·3	+3·0	−15·7	−1·5	+22·3	+6·6	+7·4	+6·7	+6·7	+10·4
1970	+7·1	+13·1	+12·8	+24·0	+3·5	−9·5	0·0	−2·4	+10·7	+12·8	+6·8	+0·2
1971	−1·8	−6·0	+18·6	−2·9	−1·9	+8·9	−15·2	−4·4	+2·8	+1·9	−0·6	−3·2
1972	+5·2	−6·2	−12·7	+4·9	+16·8	−1·2	−11·7	−2·4	−3·8	−2·4	−2·4	−2·0
1973	−7·2	−4·3	−6·5	+21·0	+22·6	+10·0	+1·4	+8·6	−6·0	−1·7	+2·3	−6·8
1974	−20·8	−19·0	−13·5	−9·6	+42·8	−4·3	−23·8	−3·8	−18·1	−16·7	−11·6	−7·5
1975	−26·5	−7·1	−1·8	+6·3					−12·9	−9·9		
1959/75 Averages	14 270	12 160	10 945	6 985	5 860	9 665	13 335	28 860	37 375	44 360	73 220	73 220

Notes

The above averages are the total degree days for sixteen Western European cities as reported in *Platt's Oilgram*.

Degree days are a measure of the energy requirement for space heating. When the temperature falls below a threshold level, heating is required to maintain an acceptable temperature (about 70 °F) inside buildings. In the above data the threshold or base temperature is 62·5 °F for German and Scandinavian cities and 60 °F for the others. Thus if the average temperature over a 24 hour period in London was 55 °F, 5 degree days would be recorded.

Table 61 US OIL COMPANIES

Rank	*Company*	*Other Brand Names**	1974 *Sales* ($ *million*)	*Employees*
1	Exxon†	Esso, Enco, Humble, Cleveland, Pratts, Stanavo, Essolube	42 061	133 000
2	Texaco†	Caltex, Regent	23 255	76 420
3	Mobil Oil†	Mobilgas, Gargoyle, Mobiloil	18 929	73 100
4	Standard Oil of California†	Chevron, Caltex, Socal	17 191	39 540
5	Gulf Oil		16 458	52 700
6	Standard Oil of Indiana†	Amoco	9 085	47 217
7	Shell Oil		7 633	32 287
8	Continental Oil	Conoco, Jet	7 041	41 174
9	Atlantic Richfield†	Arco	6 740	28 711
10	Occidental Petroleum	VIP	5 719	34 000
11	Tenneco		5 001	81 016
12	Phillips Petroleum		4 981	30 802
13	Union Oil of California		4 419	15 364
14	Sun Oil	Sunoco	3 800	27 707
15	Amerada Hess		3 755	5 779
16	Ashland Oil		3 216	27 000
17	Marathon Oil		2 882	9 465
18	Cities Service		2 806	17 400
19	Getty Oil		2 742	11 364
20	Standard Oil of Ohio†	Sohio	2 166	20 300

* Including companies absorbed.
† Formerly part of Standard Oil Trust or have absorbed companies which were part of the Trust.

Table 62 WESTERN EUROPEAN ENERGY COMPANIES

Rank 1974	*Company*	*Country*	*Employees*
1	Royal Dutch Shell	NL/GB	164 000
2	British Petroleum	GB	68 000
6	Compagnie Française des Pétroles	F	27 800
11	VEBA	D	59 122
12	ENI	I	92 420
15	ICI	GB	201 000
18	Elf Aquitaine	F	22 331
19	Electricity Council	GB	172 483
28	Petrofina	B	22 200
29	EDF, Electricité de France	F	98 040
35	Esso Deutschland	D	4 445
44	NCB, National Coal Board	GB	321 000
51	Esso SAF	F	4 953
57	DSM	NL	29 500
64	Esso Petroleum Company	GB	10 708
79	Esso Italiana	I	3 235
87	Degussa	D	14 478
93	Mobil Oil Deutschland	D	2 322
104	ENPETROL	E	5 570
111	Charbonnages de France	F	86 631
113	Gaz de France	F	27 503
115	Union Explosivos Rio Tinto	E	19 335
118	VIAG	D	32 080
127	CEPSA	E	20 177
138	OMV	A	8 626
143	Esso Nederland	NL	1 659
151	SIBP	B	950
154	Texaco	GB	
166	VEW	D	6 656
177	Chevron Oil Italiana	I	1 039
183	Union Carbide Europe	EUR	22 000
185	Mobil Oil	GB	2 289
188	Liquigas	I	
205	Mobil Oil Française	F	2 405
232	Norsk Hydro	N	9 409
234	Ultramar Group	GB	2 000
246	PETRONOR	E	491
267	Chevron Erdoel Deutschland	D	311
294	SACOR	P	3 411
303	Dansk Esso	DK	1 341
353	Butano	E	3 499

Source: *Vision*, October 1975.

Table 63 CRUDE OIL PROCESSING

Crude oil fraction	*Boiling range*	*Intermediates*	*'Finished' Products*
Gases	Up to 15 °C	Methane Ethane Propane Butane LPG	Bottled gases
Light Distillates	15–175 °C	Primary Flash Distillates (PFD) Straight run Gasolines (SRG) Straight run Benzine (SRB) Light Distillate Feedstocks (LDF) Cat. Reformer Feedstocks	Motor Spirit (Petrol) Aviation Gasolines Jet Fuels (Jet Gasolines)[1]
Middle Distillates	175–350 °C	Naphtha Kerosine Gas Oils	Dervs Kerosines Jet Fuels (Kerosines)[2] Domestic Heating Oils Fuel Oils[3]
Atmospheric Residues	> 350 °C	Fuel Oils	Fuel Oils[4]
Vacuum Distillates	350–550 °C	Spindle Oils Light Lubricating Oils Cat. Cracker Feed Hydro Cracker Feed	Lubricating Oils Paraffin Waxes
Vacuum Residues	Decompose on Heating (crack or pyrolyse)	Bitumens Fuel Oil Blending	Bitumens or Asphalts
Deasphalted Residues	> 550 °C	Bright Stocks Cylinder Stocks	Lubricating Oils Micro-wax
Residue Materials (coal-like)	Decompose on Heating (crack or pyrolyse)	Asphaltenes (such as propane precipitated asphalt, PPA)	Fuel Oils Petroleum Cokes

[1] Jet gasolines may be considered as 50% straight run material in the gasoline boiling range and 50% middle distillates.
[2] Jet kerosines may be considered as 25% material in the gasoline boiling range and 75% middle distillates.
[3] Fuel oils used for domestic heating do not include residual material. Examples are ELFO in Germany and FOD in France.
[4] From many crude oils the atmospheric residue is a complete heavy fuel oil. Light residual fuel oils may be considered a blend of 50% heavy fuel oil and 50% gas oil.

Table 64 MULTIPLES AND SUB-MULTIPLES (SI Designation)

Symbol	*Prefix*	*Multiple or Sub-multiple*
T	tera	10^{12}
G	giga	10^{9}
M	mega	10^{6}
k	kilo	10^{3}
h	hecto	10^{2}
da	deca	10
d	deci	10^{-1}
c	centi	10^{-2}
m	milli	10^{-3}
μ	micro	10^{-6}
n	nano	10^{-9}
p	pico	10^{-12}
f	femto	10^{-15}
a	atto	10^{-18}

Table 65 CONVERSION °API TO SPECIFIC GRAVITY 60/60°F

0 *to* 74·9

°API	TENTHS OF DEGREES									
	0	1	2	3	4	5	6	7	8	9
0	1·0760	1·0752	1·0744	1·0736	1·0728	1·0720	1·0712	1·0703	1·0695	1·0687
1	1·0679	1·0671	1·0663	1·0655	1·0647	1·0639	1·0631	1·0623	1·0615	1·0607
2	1·0599	1·0591	1·0583	1·0575	1·0568	1·0560	1·0552	1·0544	1·0536	1·0528
3	1·0520	1·0513	1·0505	1·0497	1·0489	1·0481	1·0474	1·0466	1·0458	1·0451
4	1·0443	1·0435	1·0427	1·0420	1·0412	1·0404	1·0397	1·0389	1·0382	1·0374
5	1·0366	1·0359	1·0351	1·0344	1·0336	1·0328	1·0321	1·0313	1·0306	1·0298
6	1·0291	1·0283	1·0276	1·0268	1·0261	1·0254	1·0246	1·0239	1·0231	1·0224
7	1·0217	1·0209	1·0202	1·0195	1·0187	1·0180	1·0173	1·0165	1·0158	1·0151
8	1·0143	1·0136	1·0129	1·0122	1·0114	1·0107	1·0100	1·0093	1·0086	1·0078
9	1·0071	1·0064	1·0057	1·0050	1·0043	1·0035	1·0028	1·0021	1·0014	1·0007
10	1·0000	0·9993	0·9986	0·9979	0·9972	0·9965	0·9958	0·9951	0·9944	0·9937
11	0·9930	0·9923	0·9916	0·9909	0·9902	0·9895	0·9888	0·9881	0·9874	0·9868
12	0·9861	0·9854	0·9847	0·9840	0·9833	0·9826	0·9820	0·9813	0·9806	0·9799
13	0·9792	0·9786	0·9779	0·9772	0·9765	0·9759	0·9752	0·9745	0·9738	0·9732
14	0·9725	0·9718	0·9712	0·9705	0·9698	0·9692	0·9685	0·9679	0·9672	0·9665
15	0·9659	0·9652	0·9646	0·9639	0·9632	0·9626	0·9619	0·9613	0·9606	0·9600
16	0·9593	0·9587	0·9580	0·9574	0·9567	0·9561	0·9554	0·9548	0·9541	0·9535
17	0·9529	0·9522	0·9516	0·9509	0·9503	0·9497	0·9490	0·9484	0·9478	0·9471
18	0·9465	0·9459	0·9452	0·9446	0·9440	0·9433	0·9427	0·9421	0·9415	0·9408
19	0·9402	0·9396	0·9390	0·9383	0·9377	0·9371	0·9365	0·9358	0·9352	0·9346
20	0·9340	0·9334	0·9328	0·9321	0·9315	0·9309	0·9303	0·9297	0·9291	0·9285
21	0·9279	0·9273	0·9267	0·9260	0·9254	0·9248	0·9242	0·9236	0·9230	0·9224
22	0·9218	0·9212	0·9206	0·9200	0·9194	0·9188	0·9182	0·9176	0·9170	0·9165
23	0·9159	0·9153	0·9147	0·9141	0·9135	0·9129	0·9123	0·9117	0·9111	0·9106
24	0·9100	0·9094	0·9088	0·9082	0·9076	0·9071	0·9065	0·9059	0·9053	0·9047
25	0·9042	0·9036	0·9030	0·9024	0·9018	0·9013	0·9007	0·9001	0·8996	0·8990
26	0·8984	0·8978	0·8973	0·8967	0·8961	0·8956	0·8950	0·8944	0·8939	0·8933
27	0·8927	0·8922	0·8916	0·8911	0·8905	0·8899	0·8894	0·8888	0·8883	0·8877
28	0·8871	0·8866	0·8860	0·8855	0·8849	0·8844	0·8838	0·8833	0·8827	0·8822
29	0·8816	0·8811	0·8805	0·8800	0·8794	0·8789	0·8783	0·8778	0·8772	0·8767
30	0·8762	0·8756	0·8751	0·8745	0·8740	0·8735	0·8729	0·8724	0·8718	0·8713
31	0·8708	0·8702	0·8697	0·8692	0·8686	0·8681	0·8676	0·8670	0·8665	0·8660
32	0·8654	0·8649	0·8644	0·8639	0·8633	0·8628	0·8623	0·8618	0·8612	0·8607
33	0·8602	0·8597	0·8591	0·8586	0·8581	0·8576	0·8571	0·8565	0·8560	0·8555
34	0·8550	0·8545	0·8540	0·8534	0·8529	0·8524	0·8519	0·8514	0·8509	0·8504
35	0·8498	0·8493	0·8488	0·8483	0·8478	0·8473	0·8468	0·8463	0·8458	0·8453
36	0·8448	0·8443	0·8438	0·8433	0·8428	0·8423	0·8418	0·8413	0·8408	0·8403
37	0·8398	0·8393	0·8388	0·8383	0·8378	0·8373	0·8368	0·8363	0·8358	0·8353
38	0·8348	0·8343	0·8338	0·8333	0·8328	0·8324	0·8319	0·8314	0·8309	0·8304
39	0·8299	0·8294	0·8289	0·8285	0·8280	0·8275	0·8270	0·8265	0·8260	0·8255
40	0·8251	0·8246	0·8241	0·8236	0·8232	0·8227	0·8222	0·8217	0·8212	0·8208
41	0·8203	0·8198	0·8193	0·8189	0·8184	0·8179	0·8174	0·8170	0·8165	0·8160
42	0·8156	0·8151	0·8146	0·8142	0·8137	0·8132	0·8128	0·8123	0·8118	0·8114
43	0·8109	0·8104	0·8100	0·8095	0·8090	0·8086	0·8081	0·8076	0·8072	0·8067
44	0·8063	0·8058	0·8054	0·8049	0·8044	0·8040	0·8035	0·8031	0·8026	0·8022
45	0·8017	0·8012	0·8008	0·8003	0·7999	0·7994	0·7990	0·7985	0·7981	0·7976
46	0·7972	0·7967	0·7963	0·7958	0·7954	0·7949	0·7945	0·7941	0·7936	0·7932
47	0·7927	0·7923	0·7918	0·7914	0·7909	0·7905	0·7901	0·7896	0·7892	0·7887
48	0·7883	0·7879	0·7874	0·7870	0·7865	0·7861	0·7857	0·7852	0·7848	0·7844
49	0·7839	0·7835	0·7831	0·7826	0·7822	0·7818	0·7813	0·7809	0·7805	0·7800
50	0·7796	0·7792	0·7788	0·7783	0·7779	0·7775	0·7770	0·7766	0·7762	0·7758
51	0·7753	0·7749	0·7745	0·7741	0·7736	0·7732	0·7728	0·7724	0·7720	0·7715
52	0·7711	0·7707	0·7703	0·7699	0·7694	0·7690	0·7686	0·7682	0·7678	0·7674
53	0·7669	0·7665	0·7661	0·7657	0·7653	0·7649	0·7645	0·7640	0·7636	0·7632
54	0·7628	0·7624	0·7620	0·7616	0·7612	0·7608	0·7603	0·7599	0·7595	0·7591
55	0·7587	0·7583	0·7579	0·7575	0·7571	0·7567	0·7563	0·7559	0·7555	0·7551
56	0·7547	0·7543	0·7539	0·7535	0·7531	0·7527	0·7523	0·7519	0·7515	0·7511
57	0·7507	0·7503	0·7499	0·7495	0·7491	0·7487	0·7483	0·7479	0·7475	0·7471
58	0·7467	0·7463	0·7459	0·7455	0·7451	0·7447	0·7443	0·7440	0·7436	0·7432
59	0·7428	0·7424	0·7420	0·7416	0·7412	0·7408	0·7405	0·7401	0·7397	0·7393
60	0·7389	0·7385	0·7381	0·7377	0·7374	0·7370	0·7366	0·7362	0·7358	0·7354
61	0·7351	0·7347	0·7343	0·7339	0·7335	0·7332	0·7328	0·7324	0·7320	0·7316
62	0·7313	0·7309	0·7305	0·7301	0·7298	0·7294	0·7290	0·7286	0·7283	0·7279
63	0·7275	0·7271	0·7268	0·7264	0·7260	0·7256	0·7253	0·7249	0·7245	0·7242
64	0·7238	0·7234	0·7230	0·7227	0·7223	0·7219	0·7216	0·7212	0·7208	0·7205
65	0·7201	0·7197	0·7194	0·7190	0·7186	0·7183	0·7179	0·7175	0·7172	0·7168
66	0·7165	0·7161	0·7157	0·7154	0·7150	0·7146	0·7143	0·7139	0·7136	0·7132
67	0·7128	0·7125	0·7121	0·7118	0·7114	0·7111	0·7107	0·7103	0·7100	0·7096
68	0·7093	0·7089	0·7086	0·7082	0·7079	0·7075	0·7071	0·7068	0·7064	0·7061
69	0·7057	0·7054	0·7050	0·7047	0·7043	0·7040	0·7036	0·7033	0·7029	0·7026
70	0·7022	0·7019	0·7015	0·7012	0·7008	0·7005	0·7001	0·6998	0·6995	0·6991
71	0·6988	0·6984	0·6981	0·6977	0·6974	0·6970	0·6967	0·6964	0·6960	0·6957
72	0·6953	0·6950	0·6946	0·6943	0·6940	0·6936	0·6933	0·6929	0·6926	0·6923
73	0·6919	0·6916	0·6913	0·6909	0·6906	0·6902	0·6899	0·6896	0·6892	0·6889
74	0·6886	0·6882	0·6879	0·6876	0·6872	0·6869	0·6866	0·6862	0·6859	0·6856

Table 65—*contd.*

75·0 *to* 149·9

°API	TENTHS OF DEGREES									
	0	1	2	3	4	5	6	7	8	9
75	0·6852	0·6849	0·6846	0·6842	0·6839	0·6836	0·6832	0·6829	0·6826	0·6823
76	0·6819	0·6816	0·6813	0·6809	0·6806	0·6803	0·6800	0·6796	0·6793	0·6790
77	0·6787	0·6783	0·6780	0·6777	0·6774	0·6770	0·6767	0·6764	0·6761	0·6757
78	0·6754	0·6751	0·6748	0·6745	0·6741	0·6738	0·6735	0·6732	0·6728	0·6725
79	0·6722	0·6719	0·6716	0·6713	0·6709	0·6706	0·6703	0·6700	0·6697	0·6693
80	0·6690	0·6687	0·6684	0·6681	0·6678	0·6675	0·6671	0·6668	0·6665	0·6662
81	0·6659	0·6656	0·6653	0·6649	0·6646	0·6643	0·6640	0·6637	0·6634	0·6631
82	0·6628	0·6625	0·6621	0·6618	0·6615	0·6612	0·6609	0·6606	0·6603	0·6600
83	0·6597	0·6594	0·6591	0·6588	0·6584	0·6581	0·6578	0·6575	0·6572	0·6569
84	0·6566	0·6563	0·6560	0·6557	0·6554	0·6551	0·6548	0·6545	0·6542	0·6539
85	0·6536	0·6533	0·6530	0·6527	0·6524	0·6521	0·6518	0·6515	0·6512	0·6509
86	0·6506	0·6503	0·6500	0·6497	0·6494	0·6491	0·6488	0·6485	0·6482	0·6479
87	0·6476	0·6473	0·6470	0·6467	0·6464	0·6461	0·6458	0·6455	0·6452	0·6449
88	0·6446	0·6444	0·6441	0·6438	0·6435	0·6432	0·6429	0·6426	0·6423	0·6420
89	0·6417	0·6414	0·6411	0·6409	0·6406	0·6403	0·6400	0·6397	0·6394	0·6391
90	0·6388	0·6385	0·6382	0·6380	0·6377	0·6374	0·6371	0·6368	0·6365	0·6362
91	0·6360	0·6357	0·6354	0·6351	0·6348	0·6345	0·6342	0·6340	0·6337	0·6334
92	0·6331	0·6328	0·6325	0·6323	0·6320	0·6317	0·6314	0·6311	0·6309	0·6306
93	0·6303	0·6300	0·6297	0·6294	0·6292	0·6289	0·6286	0·6283	0·6281	0·6278
94	0·6275	0·6272	0·6269	0·6267	0·6264	0·6261	0·6258	0·6256	0·6253	0·6250
95	0·6247	0·6244	0·6242	0·6239	0·6236	0·6233	0·6231	0·6228	0·6225	0·6223
96	0·6220	0·6217	0·6214	0·6212	0·6209	0·6206	0·6203	0·6201	0·6198	0·6195
97	0·6193	0·6190	0·6187	0·6184	0·6182	0·6179	0·6176	0·6174	0·6171	0·6168
98	0·6166	0·6163	0·6160	0·6158	0·6155	0·6152	0·6150	0·6147	0·6144	0·6141
99	0·6139	0·6136	0·6134	0·6131	0·6128	0·6126	0·6123	0·6120	0·6118	0·6115
100	0·6112	0·6110	0·6107	0·6104	0·6102	0·6099	0·6097	0·6094	0·6091	0·6089
101	0·6086	0·6083	0·6081	0·6078	0·6076	0·6073	0·6070	0·6068	0·6065	0·6063
102	0·6060	0·6057	0·6055	0·6052	0·6050	0·6047	0·6044	0·6042	0·6039	0·6037
103	0·6034	0·6032	0·6029	0·6026	0·6024	0·6021	0·6019	0·6016	0·6014	0·6011
104	0·6008	0·6006	0·6003	0·6001	0·5998	0·5996	0·5993	0·5991	0·5988	0·5986
105	0·5983	0·5981	0·5978	0·5976	0·5973	0·5970	0·5968	0·5965	0·5963	0·5960
106	0·5958	0·5955	0·5952	0·5950	0·5948	0·5945	0·5943	0·5940	0·5938	0·5935
107	0·5933	0·5930	0·5928	0·5925	0·5923	0·5921	0·5918	0·5916	0·5913	0·5911
108	0·5908	0·5906	0·5903	0·5901	0·5898	0·5896	0·5893	0·5891	0·5888	0·5886
109	0·5884	0·5881	0·5879	0·5876	0·5874	0·5871	0·5869	0·5867	0·5864	0·5862
110	0·5859	0·5857	0·5854	0·5852	0·5850	0·5847	0·5845	0·5842	0·5840	0·5837
111	0·5835	0·5833	0·5830	0·5828	0·5825	0·5823	0·5821	0·5818	0·5816	0·5813
112	0·5811	0·5809	0·5806	0·5804	0·5802	0·5799	0·5797	0·5794	0·5792	0·5790
113	0·5787	0·5785	0·5783	0·5780	0·5778	0·5775	0·5773	0·5771	0·5768	0·5766
114	0·5764	0·5761	0·5759	0·5757	0·5754	0·5752	0·5750	0·5747	0·5745	0·5743
115	0·5740	0·5738	0·5736	0·5733	0·5731	0·5729	0·5726	0·5724	0·5722	0·5719
116	0·5717	0·5715	0·5713	0·5710	0·5708	0·5706	0·5703	0·5701	0·5699	0·5696
117	0·5694	0·5692	0·5690	0·5687	0·5685	0·5683	0·5680	0·5678	0·5676	0·5674
118	0·5671	0·5669	0·5667	0·5665	0·5662	0·5660	0·5658	0·5655	0·5653	0·5651
119	0·5649	0·5646	0·5644	0·5642	0·5640	0·5637	0·5635	0·5633	0·5631	0·5628
120	0·5626	0·5624	0·5622	0·5620	0·5617	0·5615	0·5613	0·5611	0·5608	0·5606
121	0·5604	0·5602	0·5600	0·5597	0·5595	0·5593	0·5591	0·5588	0·5586	0·5584
122	0·5582	0·5580	0·5577	0·5575	0·5573	0·5571	0·5569	0·5566	0·5564	0·5562
123	0·5560	0·5558	0·5556	0·5553	0·5551	0·5549	0·5547	0·5545	0·5542	0·5540
124	0·5538	0·5536	0·5534	0·5532	0·5530	0·5527	0·5525	0·5523	0·5521	0·5519
125	0·5517	0·5514	0·5512	0·5510	0·5508	0·5506	0·5504	0·5502	0·5499	0·5497
126	0·5495	0·5493	0·5491	0·5489	0·5486	0·5484	0·5482	0·5480	0·5478	0·5476
127	0·5474	0·5472	0·5470	0·5468	0·5465	0·5463	0·5461	0·5459	0·5457	0·5455
128	0·5453	0·5451	0·5449	0·5446	0·5444	0·5442	0·5440	0·5438	0·5436	0·5434
129	0·5432	0·5430	0·5428	0·5426	0·5424	0·5421	0·5419	0·5417	0·5415	0·5413
130	0·5411	0·5409	0·5407	0·5405	0·5403	0·5401	0·5399	0·5397	0·5395	0·5393
131	0·5390	0·5388	0·5386	0·5384	0·5382	0·5380	0·5378	0·5376	0·5374	0·5372
132	0·5370	0·5368	0·5366	0·5364	0·5362	0·5360	0·5358	0·5356	0·5354	0·5352
133	0·5350	0·5348	0·5346	0·5344	0·5342	0·5340	0·5338	0·5336	0·5334	0·5332
134	0·5330	0·5328	0·5326	0·5324	0·5322	0·5320	0·5318	0·5316	0·5314	0·5312
135	0·5310	0·5308	0·5306	0·5304	0·5302	0·5300	0·5298	0·5296	0·5294	0·5292
136	0·5290	0·5288	0·5286	0·5284	0·5282	0·5280	0·5278	0·5276	0·5274	0·5272
137	0·5270	0·5268	0·5266	0·5264	0·5262	0·5260	0·5258	0·5256	0·5254	0·5252
138	0·5250	0·5249	0·5247	0·5245	0·5243	0·5241	0·5239	0·5237	0·5235	0·5233
139	0·5231	0·5229	0·5227	0·5225	0·5223	0·5221	0·5219	0·5218	0·5216	0·5214
140	0·5212	0·5210	0·5208	0·5206	0·5204	0·5202	0·5200	0·5198	0·5196	0·5195
141	0·5193	0·5191	0·5189	0·5187	0·5185	0·5183	0·5181	0·5179	0·5177	0·5176
142	0·5174	0·5172	0·5170	0·5168	0·5166	0·5164	0·5162	0·5160	0·5159	0·5157
143	0·5155	0·5153	0·5151	0·5149	0·5147	0·5145	0·5144	0·5142	0·5140	0·5138
144	0·5136	0·5134	0·5132	0·5131	0·5129	0·5127	0·5125	0·5123	0·5121	0·5119
145	0·5118	0·5116	0·5114	0·5112	0·5110	0·5108	0·5106	0·5105	0·5103	0·5101
146	0·5099	0·5097	0·5095	0·5094	0·5092	0·5090	0·5088	0·5086	0·5084	0·5083
147	0·5081	0·5079	0·5077	0·5075	0·5074	0·5072	0·5070	0·5068	0·5066	0·5064
148	0·5063	0·5061	0·5059	0·5057	0·5055	0·5054	0·5052	0·5050	0·5048	0·5046
149	0·5045	0·5043	0·5041	0·5039	0·5037	0·5036	0·5034	0·5032	0·5030	0·5028

Table 66 CONVERSION: SPECIFIC GRAVITY 60/60°F TO °API AND INDUSTRIAL UNITS

Specific gravity [column 1] to: [column 2] degrees API,
[column 3] pounds per UK gallon (imperial gallon) (lb/UKgal),
[column 4] long tons* per 1000 UK gallons,
[column 5] metric tonnes (t) per 1000 UK gallons,
[column 6] barrels per long ton* (brl/ton),
[column 7] cubic metres per long ton* (m^3/ton),
[column 8] long tons* per year (ton/a) equivalent to 1 barrel per day (brl/d),
[column 9] barrels per day equivalent to 1 000 000 long tons* per year.

Note: The following table has been calculated on the basis of the *weight in air* of a UK gallon of water as defined in the Weights and Measures Act of 1963 (*4*) and adjusted to 60°F.

*UK ton of 2240 lb.

0·500 *to* **0·544**

Sp gr 60/60°F [1]	°API [2]	pounds per UK gallon [3]	long tons per 1000 UK gallons [4]	metric tons per 1000 UK gallons [5]	barrels per long ton [6]	cubic metres per long ton [7]	ton/a *per* brl/d [8]	brl/d *per* 10^6 ton/a [9]
0·500	151·50	4·996	2·2302	*2·2659*	12·822	*2·0385*	28·468	35127
0·501	150·94	5·006	2·2346	*2·2705*	12·796	*2·0344*	28·525	35057
0·502	150·37	5·016	2·2391	*2·2750*	12·770	*2·0303*	28·582	34987
0·503	149·81	5·026	2·2436	*2·2796*	12·745	*2·0263*	28·639	34917
0·504	149·25	5·036	2·2480	*2·2841*	12·720	*2·0222*	28·696	34848
0·505	148·70	5·046	2·2525	*2·2886*	12·694	*2·0182*	28·753	34779
0·506	148·14	5·056	2·2570	*2·2932*	12·669	*2·0142*	28·810	34710
0·507	147·59	5·066	2·2614	*2·2977*	12·644	*2·0103*	28·867	34641
0·508	147·04	5·076	2·2659	*2·3023*	12·619	*2·0063*	28·924	34573
0·509	146·50	5·086	2·2704	*2·3068*	12·594	*2·0023*	28·981	34505
0·510	145·95	5·096	2·2749	*2·3114*	12·570	*1·9984*	29·038	34437
0·511	145·41	5·106	2·2793	*2·3159*	12·545	*1·9945*	29·095	34369
0·512	144·87	5·116	2·2838	*2·3204*	12·520	*1·9906*	29·152	34302
0·513	144·33	5·126	2·2883	*2·3250*	12·496	*1·9867*	29·209	34235
0·514	143·79	5·136	2·2927	*2·3295*	12·472	*1·9828*	29·267	34168
0·515	143·26	5·146	2·2972	*2·3341*	12·447	*1·9790*	29·324	34102
0·516	142·72	5·156	2·3017	*2·3386*	12·423	*1·9751*	29·381	34036
0·517	142·19	5·166	2·3061	*2·3431*	12·399	*1·9713*	29·438	33970
0·518	141·67	5·176	2·3106	*2·3477*	12·375	*1·9675*	29·495	33904
0·519	141·14	5·186	2·3151	*2·3522*	12·351	*1·9637*	29·552	33838
0·520	140·62	5·196	2·3196	*2·3568*	12·327	*1·9599*	29·609	33773
0·521	140·09	5·206	2·3240	*2·3613*	12·304	*1·9561*	29·666	33708
0·522	139·57	5·216	2·3285	*2·3659*	12·280	*1·9524*	29·723	33644
0·523	139·05	5·226	2·3330	*2·3704*	12·257	*1·9486*	29·780	33579
0·524	138·54	5·236	2·3374	*2·3749*	12·233	*1·9449*	29·837	33515
0·525	138·02	5·246	2·3419	*2·3795*	12·210	*1·9412*	29·894	33451
0·526	137·51	5·256	2·3464	*2·3840*	12·187	*1·9375*	29·951	33387
0·527	137·00	5·266	2·3508	*2·3886*	12·163	*1·9338*	30·008	33324
0·528	136·49	5·276	2·3553	*2·3931*	12·140	*1·9301*	30·065	33261
0·529	135·99	5·286	2·3598	*2·3976*	12·117	*1·9265*	30·122	33198
0·530	135·48	5·296	2·3643	*2·4022*	12·094	*1·9228*	30·179	33135
0·531	134·98	5·306	2·3687	*2·4067*	12·072	*1·9192*	30·236	33072
0·532	134·48	5·316	2·3732	*2·4113*	12·049	*1·9156*	30·294	33010
0·533	133·98	5·326	2·3777	*2·4158*	12·026	*1·9120*	30·351	32948
0·534	133·48	5·336	2·3821	*2·4204*	12·004	*1·9084*	30·408	32886
0·535	132·99	5·346	2·3866	*2·4249*	11·981	*1·9048*	30·465	32824
0·536	132·49	5·356	2·3911	*2·4294*	11·959	*1·9013*	30·522	32763
0·537	132·00	5·366	2·3955	*2·4340*	11·936	*1·8977*	30·579	32702
0·538	131·51	5·376	2·4000	*2·4385*	11·914	*1·8942*	30·636	32641
0·539	131·02	5·386	2·4045	*2·4431*	11·892	*1·8907*	30·693	32580
0·540	130·54	5·396	2·4090	*2·4476*	11·870	*1·8872*	30·750	32520
0·541	130·05	5·406	2·4134	*2·4521*	11·848	*1·8837*	30·807	32460
0·542	129·57	5·416	2·4179	*2·4567*	11·826	*1·8802*	30·864	32400
0·543	129·09	5·426	2·4224	*2·4612*	11·804	*1·8767*	30·921	32340
0·544	128·61	5·436	2·4268	*2·4658*	11·783	*1·8733*	30·978	32280

Table 66—*contd.*

Sp gr 60/60°F [1]	°API [2]	pounds per UK gallon [3]	long tons per 1000 UK gallons [4]	metric tons per 1000 UK gallons [5]	barrels per long ton [6]	cubic metres per long ton [7]	ton/a *per* brl/d [8]	brl/d *per* 10^6 ton/a [9]
0·545	128·13	5·446	2·4313	*2·4703*	11·761	*1·8698*	31·035	32221
0·546	127·66	5·456	2·4358	*2·4749*	11·739	*1·8664*	31·092	32162
0·547	127·18	5·466	2·4402	*2·4794*	11·718	*1·8630*	31·149	32103
0·548	126·71	5·476	2·4447	*2·4839*	11·696	*1·8596*	31·206	32044
0·549	126·24	5·486	2·4492	*2·4885*	11·675	*1·8562*	31·264	31986
0·550	125·77	5·496	2·4536	*2·4930*	11·654	*1·8528*	31·321	31928
0·551	125·31	5·506	2·4581	*2·4976*	11·633	*1·8494*	31·378	31869
0·552	124·84	5·516	2·4626	*2·5021*	11·611	*1·8461*	31·435	31812
0·553	124·38	5·526	2·4671	*2·5066*	11·590	*1·8427*	31·492	31754
0·554	123·92	5·536	2·4715	*2·5112*	11·569	*1·8394*	31·549	31697
0·555	123·45	5·546	2·4760	*2·5157*	11·549	*1·8361*	31·606	31639
0·556	123·00	5·556	2·4805	*2·5203*	11·528	*1·8328*	31·663	31582
0·557	122·54	5·566	2·4849	*2·5248*	11·507	*1·8295*	31·720	31526
0·558	122·08	5·576	2·4894	*2·5294*	11·486	*1·8262*	31·777	31469
0·559	121·63	5·586	2·4939	*2·5339*	11·466	*1·8229*	31·834	31412
0·560	121·18	5·596	2·4983	*2·5384*	11·445	*1·8196*	31·891	31356
0·561	120·73	5·606	2·5028	*2·5430*	11·425	*1·8164*	31·948	31300
0·562	120·28	5·616	2·5073	*2·5475*	11·404	*1·8131*	32·005	31244
0·563	119·83	5·626	2·5118	*2·5521*	11·384	*1·8099*	32·062	31189
0·564	119·39	5·636	2·5162	*2·5566*	11·364	*1·8067*	32·119	31133
0·565	118·94	5·646	2·5207	*2·5611*	11·344	*1·0835*	32·176	31078
0·566	118·50	5·656	2·5252	*2·5657*	11·324	*1·8003*	32·233	31023
0·567	118·06	5·666	2·5296	*2·5702*	11·304	*1·7971*	32·291	30968
0·568	117·62	5·676	2·5341	*2·5748*	11·284	*1·7940*	32·348	30914
0·569	117·18	5·686	2·5386	*2·5793*	11·264	*1·7908*	32·405	30859
0·570	116·75	5·696	2·5430	*2·5839*	11·244	*1·7877*	32·462	30805
0·571	116·31	5·706	2·5475	*2·5884*	11·224	*1·7845*	32·519	30751
0·572	115·88	5·716	2·5520	*2·5929*	11·205	*1·7814*	32·576	30697
0·573	115·45	5·726	2·5565	*2·5975*	11·185	*1·7783*	32·633	30644
0·574	115·02	5·736	2·5609	*2·6020*	11·166	*1·7752*	32·690	30590
0·575	114·59	5·746	2·5654	*2·6066*	11·146	*1·7721*	32·747	30537
0·576	114·16	5·756	2·5699	*2·6111*	11·127	*1·7690*	32·804	30484
0·577	113·73	5·767	2·5743	*2·6156*	11·107	*1·7659*	32·861	30431
0·578	113·31	5·777	2·5788	*2·6202*	11·088	*1·7629*	32·918	30378
0·579	112·89	5·787	2·5833	*2·6247*	11·069	*1·7598*	32·975	30325
0·580	112·47	5·797	2·5877	*2·6293*	11·050	*1·7568*	33·032	30273
0·581	112·05	5·807	2·5922	*2·6338*	11·031	*1·7537*	33·089	30221
0·582	111·63	5·817	2·5967	*2·6383*	11·012	*1·7507*	33·146	30169
0·583	111·21	5·827	2·6012	*2·6429*	10·993	*1·7477*	33·203	30117
0·584	110·79	5·837	2·6056	*2·6474*	10·974	*1·7447*	33·261	30065
0·585	110·38	5·847	2·6101	*2·6520*	10·955	*1·7417*	33·318	30014
0·586	109·97	5·857	2·6146	*2·6565*	10·937	*1·7388*	33·375	29963
0·587	109·56	5·867	2·6190	*2·6611*	10·918	*1·7358*	33·432	29911
0·588	109·15	5·877	2·6235	*2·6656*	10·899	*1·7328*	33·489	29860
0·589	108·74	5·887	2·6280	*2·6701*	10·881	*1·7299*	33·546	29810
0·590	108·33	5·897	2·6324	*2·6747*	10·862	*1·7269*	33·603	29759
0·591	107·92	5·907	2·6369	*2·6792*	10·844	*1·7240*	33·660	29709
0·592	107·52	5·917	2·6414	*2·6838*	10·825	*1·7211*	33·717	29658
0·593	107·12	5·927	2·6459	*2·6883*	10·807	*1·7182*	33·774	29608
0·594	106·72	5·937	2·6503	*2·6928*	10·789	*1·7153*	33·831	29558
0·595	106·32	5·947	2·6548	*2·6974*	10·771	*1·7124*	33·888	29508
0·596	105·92	5·957	2·6593	*2·7019*	10·753	*1·7095*	33·945	29459
0·597	105·52	5·967	2·6637	*2·7065*	10·735	*1·7067*	34·002	29409
0·598	105·12	5·977	2·6682	*2·7110*	10·717	*1·7038*	34·059	29360
0·599	104·73	5·987	2·6627	*2·7156*	10·699	*1·7009*	34·116	29311
0·600	104·33	5·997	2·6771	*2·7201*	10·681	*1·6981*	34·173	39262
0·601	103·94	6·007	2·6816	*2·7246*	10·663	*1·6953*	34·230	29213
0·602	103·55	6·017	2·6861	*2·7292*	10·645	*1·6925*	34·288	29165
0·603	103·16	6·027	2·6906	*2·7337*	10·628	*1·6896*	34·345	29116
0·604	102·77	6·037	2·6950	*2·7383*	10·610	*1·6868*	34·402	29058

Table 66—*contd.*

0·605 *to* **0·664**

Sp gr 60/60°F [1]	°API [2]	pounds per UK gallon [3]	long tons per 1000 UK gallons [4]	metric tons per 1000 UK gallons [5]	barrels per long ton [6]	cubic metres per long ton [7]	ton/a *per* brl/d [8]	brl/d *per* 10^6 ton/a [9]
0·605	102·38	6·047	2·6995	*2·7428*	10·592	*1·6841*	34·459	29020
0·606	102·00	6·057	2·7040	*2·7473*	10·575	*1·6813*	34·516	28972
0·607	101·61	6·067	2·7084	*2·7519*	10·557	*1·6785*	34·573	28924
0·608	101·23	6·077	2·7129	*2·7564*	10·540	*1·6757*	34·630	28876
0·609	100·85	6·087	2·7174	*2·7610*	10·523	*1·6730*	34·687	28829
0·610	100·47	6·097	2·7218	*2·7655*	10·505	*1·6702*	34·744	28782
0·611	100·09	6·107	2·7263	*2·7701*	10·488	*1·6675*	34·801	28734
0·612	99·71	6·117	2·7308	*2·7746*	10·471	*1·6648*	34·858	28687
0·613	99·33	6·127	2·7353	*2·7791*	10·454	*1·6620*	34·915	28640
0·614	98·96	6·137	2·7397	*2·7837*	10·437	*1·6593*	34·972	28594
0·615	98·58	6·147	2·7442	*2·7882*	10·420	*1·6566*	35·029	28547
0·616	98·21	6·157	2·7487	*2·7928*	10·403	*1·6539*	35·086	28501
0·617	97·84	6·167	2·7531	*2·7973*	10·386	*1·6512*	35·143	28454
0·618	97·46	6·177	2·7576	*2·8018*	10·369	*1·6486*	35·200	28408
0·619	97·09	6·187	2·7621	*2·8064*	10·352	*1·6459*	35·258	28362
0·620	96·73	6·197	2·7665	*2·8109*	10·336	*1·6432*	35·315	28317
0·621	96·36	6·207	2·7710	*2·8155*	10·319	*1·6406*	35·372	28271
0·622	95·99	6·217	2·7755	*2·8200*	10·302	*1·6379*	35·429	28225
0·623	95·63	6·227	2·7799	*2·8246*	10·286	*1·6353*	35·486	28180
0·624	95·26	6·237	2·7844	*2·8291*	10·269	*1·6327*	35·543	28135
0·625	94·90	6·247	2·7889	*2·8336*	10·253	*1·6301*	35·600	28090
0·626	94·54	6·257	2·7934	*2·8382*	10·236	*1·6275*	35·657	28045
0·627	94·18	6·267	2·7978	*2·8427*	10·220	*1·6249*	35·714	28000
0·628	93·82	6·277	2·8023	*2·8473*	10·204	*1·6223*	35·771	27955
0·629	93·46	6·287	2·8068	*2·8518*	10·188	*1·6197*	35·828	27911
0·630	93·10	6·297	2·8112	*2·8563*	10·171	*1·6171*	35·885	27866
0·631	92·75	6·307	2·8157	*2·8609*	10·155	*1·6145*	35·942	27822
0·632	92·39	6·317	2·8202	*2·8654*	10·139	*1·6120*	35·999	27778
0·633	92·04	6·327	2·8246	*2·8700*	10·123	*1·6094*	36·056	27734
0·634	91·69	6·337	2·8291	*2·8745*	10·107	*1·6069*	36·113	27690
0·635	91·33	6·347	2·8336	*2·8791*	10·091	*1·6044*	36·170	27647
0·636	90·98	6·357	2·8381	*2·8836*	10·075	*1·6018*	36·227	27603
0·637	90·64	6·367	2·8425	*2·8881*	10·059	*1·5993*	36·285	27560
0·638	90·29	6·377	2·8470	*2·8927*	10·044	*1·5968*	36·342	27516
0·639	89·94	6·387	2·8515	*2·8972*	10·028	*1·5943*	36·399	27473
0·640	89·59	6·397	2·8559	*2·9018*	10·012	*1·5918*	36·456	27430
0·641	89·25	6·407	2·8604	*2·9063*	9·997	*1·5893*	36·513	27387
0·642	88·91	6·417	2·8649	*2·9108*	9·981	*1·5868*	36·570	27345
0·643	88·56	6·427	2·8693	*2·9154*	9·965	*1·5844*	36·627	27302
0·644	88·22	6·437	2·8738	*2·9199*	9·950	*1·5819*	36·684	27260
0·645	87·88	6·447	2·8783	*2·9245*	9·934	*1·5794*	36·741	27217
0·646	87·54	6·457	2·8828	*2·9290*	9·919	*1·5770*	36·798	27175
0·647	87·20	6·467	2·8872	*2·9336*	9·904	*1·5745*	36·855	27133
0·648	86·86	6·477	2·8917	*2·9381*	9·888	*1·5721*	36·912	27091
0·649	86·53	6·487	2·8962	*2·9426*	9·873	*1·5697*	36·969	27049
0·650	86·19	6·497	2·9006	*2·9472*	9·858	*1·5673*	37·026	27008
0·651	85·86	6·507	2·9051	*2·9517*	9·843	*1·5649*	37·083	26966
0·652	85·52	6·517	2·9096	*2·9563*	9·828	*1·5625*	37·140	26925
0·653	85·19	6·527	2·9140	*2·9608*	9·813	*1·5601*	37·197	26883
0·654	84·86	6·537	2·9185	*2·9653*	9·798	*1·5577*	37·255	26842
0·655	84·53	6·547	2·9230	*2·9699*	9·783	*1·5553*	37·312	26801
0·656	84·20	6·557	2·9275	*2·9744*	9·768	*1·5529*	37·369	26760
0·657	83·87	6·568	2·9319	*2·9790*	9·753	*1·5505*	37·426	26719
0·658	83·55	6·578	2·9364	*2·9835*	9·738	*1·5482*	37·483	26679
0·659	83·22	6·588	2·9409	*2·9881*	9·723	*1·5458*	37·540	26638
0·660	82·89	6·598	2·9453	*2·9926*	9·708	*1·5435*	37·597	26598
0·661	82·57	6·608	2·9498	*2·9971*	9·694	*1·5411*	37·654	26557
0·662	82·25	6·618	2·9543	*3·0017*	9·679	*1·5388*	37·711	26517
0·663	81·92	6·628	2·9587	*3·0062*	9·664	*1·5365*	37·768	26477
0·664	81·60	6·638	2·9632	*3·0108*	9·650	*1·5342*	37·825	26437

Table 66—*contd.*

Sp gr 60/60°F [1]	°API [2]	pounds per UK gallon [3]	long tons per 1000 UK gallons [4]	metric tons per 1000 UK gallons [5]	barrels per long ton [6]	cubic metres per long ton [7]	ton/a *per* brl/d [8]	brl/d *per* 10^6 ton/a [9]
0·665	81·28	6·648	2·9677	*3·0153*	9·635	*1·5319*	37·882	26397
0·666	80·96	6·658	2·9722	*3·0198*	9·621	*1·5296*	37·939	26358
0·667	80·64	6·668	2·9766	*3·0244*	9·606	*1·5273*	37·996	26318
0·668	80·33	6·678	2·9811	*3·0289*	9·592	*1·5250*	38·053	26279
0·669	80·01	6·688	2·9856	*3·0335*	9·577	*1·5227*	38·110	26239
0·670	79·69	6·698	2·9900	*3·0380*	9·563	*1·5204*	38·167	26200
0·671	79·38	6·708	2·9945	*3·0426*	9·549	*1·5181*	38·225	26161
0·672	79·07	6·718	2·9990	*3·0471*	9·535	*1·5159*	38·282	26122
0·673	78·75	6·728	3·0034	*3·0516*	9·520	*1·5136*	38·339	26083
0·674	78·44	6·738	3·0079	*3·0562*	9·506	*1·5114*	38·396	26044
0·675	78·13	6·748	3·0124	*3·0607*	9·492	*1·5091*	38·453	26006
0·676	77·82	6·758	3·0169	*3·0653*	9·478	*1·5069*	38·510	25967
0·677	77·51	6·768	3·0213	*3·0698*	9·464	*1·5047*	38·567	25929
0·678	77·20	6·778	3·0258	*3·0743*	9·450	*1·5024*	38·624	25890
0·679	76·89	6·788	3·0303	*3·0789*	9·436	*1·5002*	38·681	25852
0·680	76·59	6·798	3·0347	*3·0834*	9·422	*1·4980*	38·738	25814
0·681	76·28	6·808	3·0392	*3·0880*	9·408	*1·4958*	38·795	25776
0·682	75·98	6·818	3·0437	*3·0925*	9·395	*1·4936*	38·852	25738
0·683	75·67	6·828	3·0481	*3·0970*	9·381	*1·4914*	38·909	25701
0·684	75·37	6·838	3·0526	*3·1016*	9·367	*1·4892*	38·966	25663
0·685	75·07	6·848	3·0571	*3·1061*	9·353	*1·4871*	39·023	25625
0·686	74·77	6·858	3·0616	*3·1107*	9·340	*1·4849*	39·080	25588
0·687	74·47	6·868	3·0660	*3·1152*	9·326	*1·4827*	39·137	25551
0·688	74·17	6·878	3·0705	*3·1198*	9·313	*1·4806*	39·194	25513
0·689	73·87	6·888	3·0750	*3·1243*	9·299	*1·4784*	39·252	25476
0·690	73·57	6·898	3·0794	*3·1288*	9·286	*1·4763*	39·309	25439
0·691	73·28	6·908	3·0839	*3·1334*	9·272	*1·4741*	39·366	25403
0·692	72·98	6·918	3·0884	*3·1379*	9·259	*1·4720*	39·423	25366
0·693	72·68	6·928	3·0928	*3·1425*	9·245	*1·4699*	39·480	25329
0·694	72·39	6·938	3·0973	*3·1470*	9·232	*1·4678*	39·537	25293
0·695	72·10	6·948	3·1018	*3·1515*	9·219	*1·4656*	39·594	25256
0·696	71·80	6·958	3·1062	*3·1561*	9·205	*1·4635*	39·651	25220
0·697	71·51	6·968	3·1107	*3·1606*	9·192	*1·4614*	39·708	25184
0·698	71·22	6·978	3·1152	*3·1652*	9·179	*1·4593*	39·765	25147
0·699	70·93	6·988	3·1197	*3·1697*	9·166	*1·4572*	39·822	25111
0·700	70·64	6·998	3·1241	*3·1743*	9·153	*1·4552*	39·879	25075
0·701	70·35	7·008	3·1286	*3·1788*	9·140	*1·4531*	39·936	25040
0·702	70·07	7·018	3·1331	*3·1833*	9·127	*1·4510*	39·993	25004
0·703	69·78	7·028	3·1375	*3·1879*	9·114	*1·4489*	40·050	24968
0·704	69·49	7·038	3·1420	*3·1924*	9·101	*1·4469*	40·107	24933
0·705	69·21	7·048	3·1465	*3·1970*	9·088	*1·4448*	40·164	24897
0·706	68·93	7·058	3·1509	*3·2015*	9·075	*1·4428*	40·222	24862
0·707	68·64	7·068	3·1554	*3·2060*	9·062	*1·4407*	40·279	24827
0·708	68·36	7·078	3·1599	*3·2106*	9·049	*1·4387*	40·336	24792
0·709	68·08	7·088	3·1644	*3·2151*	9·036	*1·4367*	40·393	24757
0·710	67·80	7·098	3·1688	*3·2197*	9·024	*1·4346*	40·450	24722
0·711	67·52	7·108	3·1733	*3·2242*	9·011	*1·4326*	40·507	24687
0·712	67·24	7·118	3·1778	*3·2288*	8·998	*1·4306*	40·564	24652
0·713	66·96	7·128	3·1822	*3·2333*	8·986	*1·4286*	40·621	24618
0·714	66·68	7·138	3·1867	*3·2378*	8·973	*1·4266*	40·678	24583
0·715	66·40	7·148	3·1912	*3·2424*	8·960	*1·4246*	40·735	24549
0·716	66·13	7·158	3·1956	*3·2469*	8·948	*1·4226*	40·792	24514
0·717	65·85	7·168	3·2001	*3·2515*	8·935	*1·4206*	40·849	24480
0·718	65·58	7·178	3·2046	*3·2560*	8·923	*1·4186*	40·906	24446
0·719	65·30	7·188	3·2091	*3·2605*	8·910	*1·4166*	40·963	24412
0·720	65·03	7·198	3·2135	*3·2651*	8·898	*1·4147*	41·020	24378
0·721	64·76	7·208	3·2180	*3·2696*	8·886	*1·4127*	41·077	24344
0·722	64·48	7·218	3·2225	*3·2742*	8·873	*1·4107*	41·134	24310
0·723	64·21	7·228	3·2269	*3·2787*	8·861	*1·4088*	41·191	24277
0·724	63·94	7·238	3·2314	*3·2833*	8·849	*1·4068*	41·249	24243

Table 66—*contd.*

0·725 *to* **0·784**

Sp gr 60/60°F [1]	°API [2]	pounds per UK gallon [3]	long tons per 1000 UK gallons [4]	metric tons per 1000 UK gallons [5]	barrels per long ton [6]	cubic metres per long ton [7]	ton/a *per* brl/d [8]	brl/d *per* 10^6 ton/a [9]
0·725	63·67	7·248	3·2359	*3·2878*	8·837	*1·4049*	41·306	24210
0·726	63·40	7·258	3·2403	*3·2923*	8·824	*1·4030*	41·363	24176
0·727	63·14	7·268	3·2448	*3·2969*	8·812	*1·4010*	41·420	24143
0·728	62·87	7·278	3·2493	*3·3014*	8·800	*1·3991*	41·477	24110
0·729	62·60	7·288	3·2538	*3·3060*	8·788	*1·3972*	41·534	24076
0·730	62·34	7·298	3·2582	*3·3105*	8·776	*1·3953*	41·591	24043
0·731	62·07	7·308	3·2627	*3·3150*	8·764	*1·3934*	41·648	24011
0·732	61·81	7·318	3·2672	*3·3196*	8·752	*1·3914*	41·705	23978
0·733	61·54	7·328	3·2716	*3·3241*	8·740	*1·3895*	41·762	23945
0·734	61·28	7·338	3·2761	*3·3287*	8·728	*1·3876*	41·819	23912
0·735	61·02	7·348	3·2806	*3·3332*	8·716	*1·3858*	41·876	23880
0·736	60·76	7·358	3·2850	*3·3378*	8·704	*1·3839*	41·933	23847
0·737	60·49	7·369	3·2895	*3·3423*	8·693	*1·3820*	41·990	23815
0·738	60·23	7·379	3·2940	*3·3468*	8·681	*1·3801*	42·047	23782
0·739	59·98	7·389	3·2985	*3·3514*	8·669	*1·3782*	42·104	23750
0·740	59·72	7·399	3·3029	*3·3559*	8·657	*1·3764*	42·161	23718
0·741	59·46	7·409	3·3074	*3·3605*	8·646	*1·3745*	42·219	23686
0·742	59·20	7·419	3·3119	*3·3650*	8·634	*1·3727*	42·276	23654
0·743	58·94	7·429	3·3163	*3·3695*	8·622	*1·3708*	42·333	23622
0·744	58·69	7·439	3·3208	*3·3741*	8·611	*1·3690*	42·390	23590
0·745	58·43	7·449	3·3253	*3·3786*	8·599	*1·3671*	42·447	23559
0·746	58·18	7·459	3·3297	*3·3832*	8·588	*1·3653*	42·504	23527
0·747	57·92	7·469	3·3342	*3·3877*	8·576	*1·3635*	42·561	23495
0·748	57·67	7·479	3·3387	*3·3923*	8·565	*1·3616*	42·618	23464
0·749	57·42	7·489	3·3432	*3·3968*	8·553	*1·3598*	42·675	23433
0·750	57·17	7·499	3·3476	*3·4013*	8·542	*1·3580*	42·732	23401
0·751	56·92	7·509	3·3521	*3·4059*	8·530	*1·3562*	42·789	23370
0·752	56·67	7·519	3·3566	*3·4104*	8·519	*1·3544*	42·846	23339
0·753	56·42	7·529	3·3610	*3·4150*	8·508	*1·3526*	42·903	23308
0·754	56·17	7·539	3·3655	*3·4195*	8·496	*1·3508*	42·960	23277
0·755	55·92	7·549	3·3700	*3·4240*	8·485	*1·3490*	43·017	23246
0·756	55·67	7·559	3·3744	*3·4286*	8·474	*1·3472*	43·074	23215
0·757	55·42	7·569	3·3789	*3·4331*	8·463	*1·3454*	43·131	23185
0·758	55·18	7·579	3·3834	*3·4377*	8·451	*1·3437*	43·188	23154
0·759	54·93	7·589	3·3879	*3·4422*	8·440	*1·3419*	43·246	23123
0·760	54·68	7·599	3·3923	*3·4468*	8·429	*1·3401*	43·303	23093
0·761	54·44	7·609	3·3968	*3·4513*	8·418	*1·3383*	43·360	23063
0·762	54·20	7·619	3·4013	*3·4558*	8·407	*1·3366*	43·417	23032
0·763	53·95	7·629	3·4057	*3·4604*	8·396	*1·3348*	43·474	23002
0·764	53·71	7·639	3·4102	*3·4649*	8·385	*1·3331*	43·531	22972
0·765	53·47	7·649	3·4147	*3·4695*	8·374	*1·3313*	43·588	22942
0·766	53·23	7·659	3·4191	*3·4740*	8·363	*1·3296*	43·645	22912
0·767	52·99	7·669	3·4236	*3·4785*	8·352	*1·3279*	43·702	22882
0·768	52·74	7·679	3·4281	*3·4831*	8·341	*1·3261*	43·759	22852
0·769	52·51	7·689	3·4325	*3·4876*	8·330	*1·3244*	43·816	22822
0·770	52·27	7·699	3·4370	*3·4922*	8·319	*1·3227*	43·873	22793
0·771	52·03	7·709	3·4415	*3·4967*	8·309	*1·3210*	43·930	22763
0·772	51·79	7·719	3·4460	*3·5013*	8·298	*1·3192*	43·987	22734
0·773	51·55	7·729	3·4504	*3·5058*	8·287	*1·3175*	44·044	22704
0·774	51·32	7·739	3·4549	*3·5103*	8·276	*1·3158*	44·101	22675
0·775	51·08	7·749	3·4594	*3·5149*	8·266	*1·3141*	44·158	22645
0·776	50·85	7·759	3·4638	*3·5194*	8·255	*1·3124*	44·216	22616
0·777	50·61	7·769	3·4683	*3·5240*	8·244	*1·3107*	44·273	22587
0·778	50·38	7·779	3·4728	*3·5285*	8·234	*1·3091*	44·330	22558
0·779	50·14	7·789	3·4772	*3·5330*	8·223	*1·3074*	44·387	22529
0·780	49·91	7·799	3·4817	*3·5376*	8·213	*1·3057*	44·444	22500
0·781	49·68	7·809	3·4862	*3·5421*	8·202	*1·3040*	44·501	22471
0·782	49·45	7·819	3·4907	*3·5467*	8·192	*1·3024*	44·558	22442
0·783	49·22	7·829	3·4951	*3·5512*	8·181	*1·3007*	44·615	22414
0·784	48·98	7·839	3·4996	*3·5557*	8·171	*1·2990*	44·672	22385

Table 66—*contd.*

Sp gr 60/60°F [1]	°API [2]	pounds per UK gallon [3]	long tons per 1000 UK gallons [4]	metric tons per 1000 UK gallons [5]	barrels per long ton [6]	cubic metres per long ton [7]	ton/a *per* brl/d [8]	brl/d *per* 10^6 ton/a [9]
0·785	48·75	7·849	3·5041	*3·5603*	8·160	*1·2974*	44·729	22357
0·786	48·53	7·859	3·5085	*3·5648*	8·150	*1·2957*	44·786	22328
0·787	48·30	7·869	3·5130	*3·5694*	8·140	*1·2941*	44·843	22300
0·788	48·07	7·879	3·5175	*3·5739*	8·129	*1·2924*	44·900	22271
0·789	47·84	7·889	3·5219	*3·5785*	8·119	*1·2908*	44·957	22243
0·790	47·61	7·899	3·5264	*3·5830*	8·109	*1·2892*	45·014	22215
0·791	47·39	7·909	3·5309	*3·5875*	8·098	*1·2875*	45·071	22187
0·792	47·16	7·919	3·5354	*3·5921*	8·088	*1·2859*	45·128	22159
0·793	46·94	7·929	3·5398	*3·5966*	8·078	*1·2843*	45·185	22131
0·794	46·71	7·939	3·5443	*3·6012*	8·068	*1·2826*	45·243	22103
0·795	46·49	7·949	3·5488	*3·6057*	8·058	*1·2810*	45·300	22075
0·796	46·26	7·959	3·5532	*3·6102*	8·047	*1·2794*	45·357	22047
0·797	46·04	7·969	3·5577	*3·6148*	8·037	*1·2778*	45·414	22020
0·798	45·82	7·979	3·5622	*3·6193*	8·027	*1·2762*	45·471	21992
0·799	45·60	7·989	3·5666	*3·6239*	8·017	*1·2746*	45·528	21964
0·800	45·38	7·999	3·5711	*3·6284*	8·007	*1·2730*	45·585	21937
0·801	45·15	8·009	3·5756	*3·6330*	7·997	*1·2714*	45·642	21909
0·802	44·93	8·019	3·5801	*3·6375*	7·987	*1·2698*	45·699	21882
0·803	44·71	8·029	3·5845	*3·6420*	7·977	*1·2683*	45·756	21855
0·804	44·50	8·039	3·5890	*3·6466*	7·967	*1·2667*	45·813	21828
0·805	44·28	8·049	3·5935	*3·6511*	7·957	*1·2651*	45·870	21800
0·806	44·06	8·059	3·5979	*3·6557*	7·947	*1·2635*	45·927	21773
0·807	43·84	8·069	3·6024	*3·6602*	7·938	*1·2620*	45·984	21746
0·808	43·62	8·079	3·6069	*3·6647*	7·928	*1·2604*	46·041	21719
0·809	43·41	8·089	3·6113	*3·6693*	7·918	*1·2588*	46·098	21692
0·810	43·19	8·099	3·6158	*3·6738*	7·908	*1·2573*	46·155	21666
0·811	42·98	8·109	3·6203	*3·6784*	7·898	*1·2557*	46·213	21639
0·812	42·76	8·119	3·6248	*3·6829*	7·889	*1·2542*	46·270	21612
0·813	42·55	8·129	3·6292	*3·6875*	7·879	*1·2526*	46·327	21586
0·814	42·33	8·139	3·6337	*3·6920*	7·869	*1·2511*	46·384	21559
0·815	42·12	8·149	3·6382	*3·6965*	7·860	*1·2496*	46·441	21533
0·816	41·91	8·159	3·6426	*3·7011*	7·850	*1·2480*	46·498	21506
0·817	41·69	8·170	3·6471	*3·7056*	7·840	*1·2465*	46·555	21480
0·818	41·48	8·180	3·6516	*3·7102*	7·831	*1·2450*	46·612	21453
0·819	41·27	8·190	3·6560	*3·7147*	7·821	*1·2434*	46·669	21427
0·820	41·06	8·200	3·6605	*3·7192*	7·812	*1·2419*	46·726	21401
0·821	40·85	8·210	3·6650	*3·7238*	7·802	*1·2404*	46·783	21375
0·822	40·64	8·220	3·6695	*3·7283*	7·792	*1·2389*	46·840	21349
0·823	40·43	8·230	3·6739	*3·7329*	7·783	*1·2374*	46·897	21323
0·824	40·22	8·240	3·6784	*3·7374*	7·774	*1·2359*	46·954	21297
0·825	40·02	8·250	3·6829	*3·7420*	7·764	*1·2344*	47·011	21271
0·826	39·81	8·260	3·6873	*3·7465*	7·755	*1·2329*	47·068	21245
0·827	39·60	8·270	3·6918	*3·7510*	7·745	*1·2314*	47·125	21220
0·828	39·39	8·280	3·6963	*3·7556*	7·736	*1·2299*	47·183	21194
0·829	39·19	8·290	3·7007	*3·7601*	7·727	*1·2284*	47·240	21168
0·830	38·98	8·300	3·7052	*3·7647*	7·717	*1·2269*	47·297	21143
0·831	38·78	8·310	3·7097	*3·7692*	7·708	*1·2255*	47·354	21117
0·832	38·57	8·320	3·7142	*3·7737*	7·699	*1·2240*	47·411	21092
0·833	38·37	8·330	3·7186	*3·7783*	7·689	*1·2225*	47·468	21067
0·834	38·16	8·340	3·7231	*3·7828*	7·680	*1·2211*	47·525	21041
0·835	37·96	8·350	3·7276	*3·7874*	7·671	*1·2196*	47·582	21016
0·836	37·76	8·360	3·7320	*3·7919*	7·662	*1·2181*	47·639	20991
0·837	37·56	8·370	3·7365	*3·7965*	7·653	*1·2167*	47·696	20966
0·838	37·35	8·380	3·7410	*3·8010*	7·644	*1·2152*	47·753	20941
0·839	37·15	8·390	3·7454	*3·8055*	7·634	*1·2138*	47·810	20916
0·840	36·95	8·400	3·7499	*3·8101*	7·625	*1·2123*	47·867	20891
0·841	36·75	8·410	3·7544	*3·8146*	7·616	*1·2109*	47·924	20866
0·842	36·55	8·420	3·7588	*3·8192*	7·607	*1·2094*	47·981	20841
0·843	36·35	8·430	3·7633	*3·8237*	7·598	*1·2080*	48·038	20816
0·844	36·15	8·440	3·7678	*3·8282*	7·589	*1·2066*	48·095	20792

Table 66—*contd.*

0·845 *to* 0·904

Sp gr 60/60°F [1]	°API [2]	pounds per UK gallon [3]	long tons per 1000 UK gallons [4]	metric tons per 1000 UK gallons [5]	barrels per long ton [6]	cubic metres per long ton [7]	ton/a *per* brl/d [8]	brl/d *per* 10^6 ton/a [9]
0·845	35·96	8·450	3·7723	*3·8328*	7·580	*1·2051*	48·152	20767
0·846	35·76	8·460	3·7767	*3·8373*	7·571	*1·2037*	48·210	20743
0·847	35·56	8·470	3·7812	*3·8419*	7·562	*1·2023*	48·267	20718
0·848	35·36	8·480	3·7857	*3·8464*	7·553	*1·2009*	48·324	20694
0·849	35·17	8·490	3·7901	*3·8510*	7·544	*1·1994*	48·381	20669
0·850	34·97	8·500	3·7946	*3·8555*	7·535	*1·1980*	48·438	20645
0·851	34·77	8·510	3·7991	*3·8600*	7·527	*1·1966*	48·495	20621
0·852	34·58	8·520	3·8035	*3·8646*	7·518	*1·1952*	48·552	20596
0·853	34·39	8·530	3·8080	*3·8691*	7·509	*1·1938*	48·609	20572
0·854	34·19	8·540	3·8125	*3·8737*	7·500	*1·1924*	48·666	20548
0·855	34·00	8·550	3·8170	*3·8782*	7·491	*1·1910*	48·723	20524
0·856	33·80	8·560	3·8214	*3·8827*	7·483	*1·1896*	48·780	20500
0·857	33·61	8·570	3·8259	*3·8873*	7·474	*1·1882*	48·837	20476
0·858	33·42	8·580	3·8304	*3·8918*	7·465	*1·1869*	48·894	20452
0·859	33·23	8·590	3·8348	*3·8964*	7·456	*1·1855*	48·951	20428
0·860	33·03	8·600	3·8393	*3·9009*	7·448	*1·1841*	49·008	20404
0·861	32·84	8·610	3·8438	*3·9055*	7·439	*1·1827*	49·065	20381
0·862	32·65	8·620	3·8482	*3·9100*	7·430	*1·1813*	49·122	20357
0·863	32·46	8·630	3·8527	*3·9145*	7·422	*1·1800*	49·180	20333
0·864	32·27	8·640	3·8572	*3·9191*	7·413	*1·1786*	49·237	20310
0·865	32·08	8·650	3·8617	*3·9236*	7·405	*1·1772*	49·294	20286
0·866	31·89	8·660	3·8661	*3·9282*	7·396	*1·1759*	49·351	20263
0·867	31·71	8·670	3·8706	*3·9327*	7·388	*1·1745*	49·408	20240
0·868	31·52	8·680	3·8751	*3·9372*	7·379	*1·1732*	49·465	20216
0·869	31·33	8·690	3·8795	*3·9418*	7·371	*1·1718*	49·522	20193
0·870	31·14	8·700	3·8840	*3·9463*	7·362	*1·1705*	49·579	20170
0·871	30·96	8·710	3·8885	*3·9509*	7·354	*1·1691*	49·636	20146
0·872	30·77	8·720	3·8929	*3·9554*	7·345	*1·1678*	49·693	20123
0·873	30·58	8·730	3·8974	*3·9600*	7·337	*1·1664*	49·750	20100
0·874	30·40	8·740	3·9019	*3·9645*	7·328	*1·1651*	49·807	20077
0·875	30·21	8·750	3·9064	*3·9690*	7·320	*1·1638*	49·864	20054
0·876	30·03	8·760	3·9108	*3·9736*	7·312	*1·1624*	49·921	20031
0·877	29·85	8·770	3·9153	*3·9781*	7·303	*1·1611*	49·978	20008
0·878	29·66	8·780	3·9198	*3·9827*	7·295	*1·1598*	50·035	19986
0·879	29·48	8·790	3·9242	*3·9872*	7·287	*1·1585*	50·092	19963
0·880	29·30	8·800	3·9287	*3·9917*	7·278	*1·1571*	50·149	19940
0·881	29·11	8·810	3·9332	*3·9963*	7·270	*1·1558*	50·207	19917
0·882	28·93	8·820	3·9376	*4·0008*	7·262	*1·1545*	50·264	19895
0·883	28·75	8·830	3·9421	*4·0054*	7·254	*1·1532*	50·321	19872
0·884	28·57	8·840	3·9466	*4·0099*	7·245	*1·1519*	50·378	19850
0·885	28·39	8·850	3·9511	*4·0144*	7·237	*1·1506*	50·435	19827
0·886	28·21	8·860	3·9555	*4·0190*	7·229	*1·1493*	50·492	19805
0·887	28·03	8·870	3·9600	*4·0235*	7·221	*1·1480*	50·549	19783
0·888	27·85	8·880	3·9645	*4·0281*	7·213	*1·1467*	50·606	19760
0·889	27·67	8·890	3·9689	*4·0326*	7·205	*1·1454*	50·663	19738
0·890	27·49	8·900	3·9734	*4·0372*	7·196	*1·1441*	50·720	19716
0·891	27·31	8·910	3·9779	*4·0417*	7·188	*1·1428*	50·777	19694
0·892	27·13	8·920	3·9823	*4·0462*	7·180	*1·1416*	50·834	19672
0·893	26·95	8·930	3·9868	*4·0508*	7·172	*1·1403*	50·891	19650
0·894	26·78	8·940	3·9913	*4·0553*	7·164	*1·1390*	50·948	19628
0·895	26·60	8·950	3·9958	*4·0599*	7·156	*1·1377*	51·005	19606
0·896	26·42	8·960	4·0002	*4·0644*	7·148	*1·1365*	51·062	19584
0·897	26·25	8·971	4·0047	*4·0689*	7·140	*1·1352*	51·119	19562
0·898	26·07	8·981	4·0092	*4·0735*	7·132	*1·1339*	51·177	19540
0·899	35·90	8·991	4·0136	*4·0780*	7·124	*1·1327*	51·234	19518
0·900	25·72	9·001	4·0181	*4·0826*	7·116	*1·1314*	51·291	19497
0·901	25·55	9·011	4·0226	*4·0871*	7·108	*1·1301*	51·348	19475
0·902	25·37	9·021	4·0270	*4·0917*	7·101	*1·1289*	51·405	19453
0·903	25·20	9·031	4·0315	*4·0962*	7·093	*1·1276*	51·462	19432
0·904	25·03	9·041	4·0360	*4·1007*	7·085	*1·1264*	51·519	19410

Table 66—*contd.*

Sp gr 60/60°F [1]	°API [2]	pounds per UK gallon [3]	long tons per 1000 UK gallons [4]	metric tons per 1000 UK gallons [5]	barrels per long ton [6]	cubic metres per long ton [7]	ton/a *per* brl/d [8]	brl/d *per* 10^6 ton/a [9]
0·905	24·85	9·051	4·0405	*4·1053*	7·077	*1·1251*	51·576	19389
0·906	24·68	9·061	4·0449	*4·1098*	7·069	*1·1239*	51·633	19367
0·907	24·51	9·071	4·0494	*4·1144*	7·061	*1·1227*	51·690	19346
0·908	24·34	9·081	4·0539	*4·1189*	7·054	*1·1214*	51·747	19325
0·909	24·17	9·091	4·0583	*4·1234*	7·046	*1·1202*	51·804	19303
0·910	23·99	9·101	4·0628	*4·1280*	7·038	*1·1190*	51·861	19282
0·911	23·82	9·111	4·0673	*4·1325*	7·030	*1·1177*	51·918	19261
0·912	23·65	9·121	4·0717	*4·1371*	7·023	*1·1165*	51·975	19240
0·913	23·48	9·131	4·0762	*4·1416*	7·015	*1·1153*	52·032	19219
0·914	23·31	9·141	4·0807	*4·1462*	7·007	*1·1140*	52·089	19198
0·915	23·14	9·151	4·0851	*4·1507*	7·000	*1·1128*	52·146	19177
0·916	22·98	9·161	4·0896	*4·1552*	6·992	*1·1116*	52·204	19156
0·917	22·81	9·171	4·0941	*4·1598*	6·984	*1·1104*	52·261	19135
0·918	22·64	9·181	4·0986	*4·1643*	6·977	*1·1092*	52·318	19114
0·919	22·47	9·191	4·1030	*4·1689*	6·969	*1·1080*	52·375	19093
0·920	22·30	9·201	4·1075	*4·1734*	6·961	*1·1068*	52·432	19072
0·921	22·14	9·211	4·1120	*4·1779*	6·954	*1·1056*	52·489	19051
0·922	21·97	9·221	4·1164	*4·1825*	6·946	*1·1044*	52·546	19031
0·923	21·80	9·231	4·1209	*4·1870*	6·939	*1·1032*	52·603	19010
0·924	21·64	9·241	4·1254	*4·1916*	6·931	*1·1020*	52·660	18990
0·925	21·47	9·251	4·1298	*4·1961*	6·924	*1·1008*	52·717	18969
0·926	21·31	9·261	4·1343	*4·2007*	6·916	*1·0996*	52·774	18948
0·927	21·14	9·271	4·1388	*4·2052*	6·909	*1·0984*	52·831	18928
0·928	20·98	9·281	4·1433	*4·2097*	6·901	*1·0972*	52·888	18908
0·929	20·81	9·291	4·1477	*4·2143*	6·894	*1·0960*	52·945	18887
0·930	20·65	9·301	4·1522	*4·2188*	6·887	*1·0949*	53·002	18867
0·931	20·49	9·311	4·1567	*4·2234*	6·879	*1·0937*	53·059	18847
0·932	20·32	9·321	4·1611	*4·2279*	6·872	*1·0925*	53·116	18826
0·933	20·16	9·331	4·1656	*4·2324*	6·864	*1·0913*	53·174	18806
0·934	20·00	9·341	4·1701	*4·2370*	6·857	*1·0902*	53·231	18786
0·935	19·84	9·351	4·1745	*4·2415*	6·850	*1·0890*	53·288	18766
0·936	19·68	9·361	4·1790	*4·2461*	6·842	*1·0878*	53·345	18746
0·937	19·51	9·371	4·1835	*4·2506*	6·835	*1·0867*	53·402	18726
0·938	19·35	9·381	4·1880	*4·2552*	6·828	*1·0855*	53·459	18706
0·939	19·19	9·391	4·1924	*4·2597*	6·820	*1·0844*	53·516	18686
0·940	19·03	9·401	4·1969	*4·2642*	6·813	*1·0832*	53·573	18666
0·941	18·87	9·411	4·2014	*4·2688*	6·806	*1·0820*	53·630	18646
0·942	18·71	9·421	4·2058	*4·2733*	6·799	*1·0809*	53·687	18626
0·943	18·55	9·431	4·2103	*4·2779*	6·791	*1·0798*	53·744	18606
0·944	18·39	9·441	4·2148	*4·2824*	6·784	*1·0786*	53·801	18587
0·945	18·24	9·451	4·2192	*4·2869*	6·777	*1·0775*	53·858	18567
0·946	18·08	9·461	4·2237	*4·2915*	6·770	*1·0763*	53·915	18547
0·947	17·92	9·471	4·2282	*4·2960*	6·763	*1·0752*	53·972	18528
0·948	17·76	9·481	4·2327	*4·3006*	6·756	*1·0740*	54·029	18508
0·949	17·60	9·491	4·2371	*4·3051*	6·748	*1·0729*	54·086	18489
0·950	17·45	9·501	4·2416	*4·3097*	6·741	*1·0718*	54·143	18469
0·951	17·29	9·511	4·2461	*4·3142*	6·734	*1·0707*	54·201	18450
0·952	17·13	9·521	4·2505	*4·3187*	6·727	*1·0695*	54·258	18430
0·953	16·98	9·531	4·2550	*4·3233*	6·720	*1·0684*	54·315	18411
0·954	16·82	9·541	4·2595	*4·3278*	6·713	*1·0673*	54·372	18392
0·955	16·67	9·551	4·2639	*4·3324*	6·706	*1·0662*	54·429	18372
0·956	16·51	9·561	4·2684	*4·3369*	6·699	*1·0651*	54·486	18353
0·957	16·36	9·571	4·2729	*4·3414*	6·692	*1·0639*	54·543	18334
0·958	16·20	9·581	4·2774	*4·3460*	6·685	*1·0628*	54·600	18315
0·959	16·05	9·591	4·2818	*4·3505*	6·678	*1·0617*	54·657	18296
0·960	15·90	9·601	4·2863	*4·3551*	6·671	*1·0606*	54·714	18277
0·961	15·74	9·611	4·2908	*4·3596*	6·664	*1·0595*	54·771	18258
0·962	15·59	9·621	4·2952	*4·3642*	6·657	*1·0584*	54·828	18239
0·963	15·44	9·631	4·2997	*4·3687*	6·650	*1·0573*	54·885	18220
0·964	15·28	9·641	4·3042	*4·3732*	6·643	*1·0562*	54·942	18201

Table 66—*contd.*

0·965 *to* **1·024**

Sp gr 60/60°F [1]	°API [2]	pounds per UK gallon [3]	long tons per 1000 UK gallons [4]	metric tons per 1000 UK gallons [5]	barrels per long ton [6]	cubic metres per long ton [7]	ton/a *per* brl/d [8]	brl/d *per* 10^6 ton/a [9]
0·965	15·13	9·651	4·3086	*4·3778*	6·636	*1·0551*	54·999	18182
0·966	14·98	9·661	4·3131	*4·3823*	6·630	*1·0540*	55·056	18163
0·967	14·83	9·671	4·3176	*4·3869*	6·623	*1·0529*	55·113	18144
0·968	14·68	9·681	4·3221	*4·3914*	6·616	*1·0518*	55·171	18125
0·969	14·53	9·691	4·3265	*4·3959*	6·609	*1·0507*	55·228	18107
0·970	14·38	9·701	4·3310	*4·4005*	6·602	*1·0497*	55·285	18088
0·971	14·23	9·711	4·3355	*4·4050*	6·595	*1·0486*	55·342	18069
0·972	14·08	9·721	4·3399	*4·4096*	6·589	*1·0475*	55·399	18051
0·973	13·93	9·731	4·3444	*4·4141*	6·582	*1·0464*	55·456	18032
0·974	13·78	9·741	4·3489	*4·4187*	6·575	*1·0453*	55·513	18014
0·975	13·63	9·751	4·3533	*4·4232*	6·568	*1·0443*	55·570	17995
0·976	13·48	9·761	4·3578	*4·4277*	6·562	*1·0432*	55·627	17977
0·977	13·33	9·772	4·3623	*4·4323*	6·555	*1·0421*	55·684	17958
0·978	13·18	9·782	4·3668	*4·4368*	6·548	*1·0411*	55·741	17940
0·979	13·04	9·792	4·3712	*4·4414*	6·541	*1·0400*	55·798	17922
0·980	12·89	9·802	4·3757	*4·4459*	6·535	*1·0389*	55·855	17903
0·981	12·74	9·812	4·3802	*4·4504*	6·528	*1·0379*	55·912	17885
0·982	12·59	9·822	4·3846	*4·4550*	6·521	*1·0368*	55·969	17867
0·983	12·45	9·832	4·3891	*4·4595*	6·515	*1·0358*	56·026	17849
0·984	12·30	9·842	4·3936	*4·4641*	6·508	*1·0347*	56·083	17830
0·985	12·15	9·852	4·3980	*4·4686*	6·502	*1·0337*	56·140	17812
0·986	12·01	9·862	4·4025	*4·4731*	6·495	*1·0326*	56·198	17794
0·987	11·86	9·872	4·4070	*4·4777*	6·488	*1·0316*	56·255	17776
0·988	11·72	9·882	4·4114	*4·4822*	6·482	*1·0305*	56·312	17758
0·989	11·57	9·892	4·4159	*4·4868*	6·475	*1·0295*	56·369	17740
0·990	11·43	9·902	4·4204	*4·4913*	6·469	*1·0284*	56·426	17722
0·991	11·29	9·912	4·4249	*4·4959*	6·462	*1·0274*	56·483	17704
0·992	11·14	9·922	4·4293	*4·5004*	6·456	*1·0264*	56·540	17686
0·993	11·00	9·932	4·4338	*4·5049*	6·449	*1·0253*	56·597	17669
0·994	10·85	9·942	4·4383	*4·5095*	6·443	*1·0243*	56·654	17651
0·995	10·71	9·952	4·4427	*4·5140*	6·436	*1·0233*	56·711	17633
0·996	10·57	9·962	4·4472	*4·5186*	6·430	*1·0222*	56·768	17615
0·997	10·43	9·972	4·4517	*4·5231*	6·423	*1·0212*	56·825	17598
0·998	10·28	9·982	4·4561	*4·5276*	6·417	*1·0202*	56·882	17580
0·999	10·14	9·992	4·4606	*4·5322*	6·410	*1·0192*	56·939	17562
1·000	10·00	10·002	4·4651	*4·5367*	6·404	*1·0181*	56·996	17545
1·001	9·86	10·012	4·4696	*4·5413*	6·398	*1·0171*	57·053	17527
1·002	9·72	10·022	4·4740	*4·5458*	6·391	*1·0161*	57·110	17510
1·003	9·58	10·032	4·4785	*4·5504*	6·385	*1·0151*	57·168	17492
1·004	9·44	10·042	4·4830	*4·5549*	6·378	*1·0141*	57·225	17475
1·005	9·30	10·052	4·4874	*4·5594*	6·372	*1·0131*	57·282	17457
1·006	9·16	10·062	4·4919	*4·5640*	6·366	*1·0121*	57·339	17440
1·007	9·02	10·072	4·4964	*4·5685*	6·359	*1·0111*	57·396	17423
1·008	8·88	10·082	4·5008	*4·5731*	6·353	*1·0101*	57·453	17405
1·009	8·74	10·092	4·5053	*4·5776*	6·347	*1·0090*	57·510	17388
1·010	8·60	10·102	4·5098	*4·5821*	6·340	*1·0080*	57·567	17371
1·011	8·46	10·112	4·5143	*4·5867*	6·334	*1·0070*	57·624	17354
1·012	8·32	10·122	4·5187	*4·5912*	6·328	*1·0061*	57·681	17337
1·013	8·18	10·132	4·5232	*4·5958*	6·322	*1·0051*	57·738	17319
1·014	8·05	10·142	4·5277	*4·6003*	6·315	*1·0041*	57·795	17302
1·015	7·91	10·152	4·5321	*4·6049*	6·309	*1·0031*	57·852	17285
1·016	7·77	10·162	4·5366	*4·6094*	6·303	*1·0021*	57·909	17268
1·017	7·63	10·172	4·5411	*4·6139*	6·297	*1·0011*	57·966	17251
1·018	7·50	10·182	4·5455	*4·6185*	6·291	*1·0001*	58·023	17234
1·019	7·36	10·192	4·5500	*4·6230*	6·284	*0·9991*	58·080	17217
1·020	7·23	10·202	4·5545	*4·6276*	6·278	*0·9982*	58·138	17200
1·021	7·09	10·212	4·5590	*4·6321*	6·272	*0·9972*	58·195	17184
1·022	6·95	10·222	4·5634	*4·6366*	6·266	*0·9962*	58·252	17167
1·023	6·82	10·232	4·5679	*4·6412*	6·260	*0·9952*	58·309	17150
1·024	6·68	10·242	4·5724	*4·6457*	6·254	*0·9943*	58·366	17133

Table 66—*contd.*

Sp gr 60/60°F [1]	°API [2]	pounds per UK gallon [3]	long tons per 1000 UK gallons [4]	metric tons per 1000 UK gallons [5]	barrels per long ton [6]	cubic metres per long ton [7]	ton/a *per* brl/d [8]	brl/d *per* 10^6 ton/a [9]
1·025	6·55	10·252	4·5768	*4·6503*	6·248	*0·9933*	58·423	17116
1·026	6·41	10·262	4·5813	*4·6548*	6·241	*0·9923*	58·480	17100
1·027	6·28	10·272	4·5858	*4·6594*	6·235	*0·9913*	58·537	17083
1·028	6·15	10·282	4·5902	*4·6639*	6·229	*0·9904*	58·594	17066
1·029	6·01	10·292	4·5947	*4·6684*	6·223	*0·9894*	58·651	17050
1·030	5·88	10·302	4·5992	*4·6730*	6·217	*0·9885*	58·708	17033
1·031	5·75	10·312	4·6037	*4·6775*	6·211	*0·9875*	58·765	17017
1·032	5·61	10·322	4·6081	*4·6821*	6·205	*0·9865*	58·822	17000
1·033	5·48	10·332	4·6126	*4·6866*	6·199	*0·9856*	58·879	16984
1·034	5·35	10·342	4·6171	*4·6911*	6·193	*0·9846*	58·936	16967
1·035	5·22	10·352	4·6215	*4·6957*	6·187	*0·9837*	58·993	16951
1·036	5·08	10·362	4·6260	*4·7002*	6·181	*0·9827*	59·050	16934
1·037	4·95	10·372	4·6305	*4·7048*	6·175	*0·9818*	59·107	16918
1·038	4·82	10·382	4·6349	*4·7093*	6·169	*0·9808*	59·165	16902
1·039	4·69	10·392	4·6394	*4·7139*	6·163	*0·9799*	59·222	16886
1·040	4·56	10·402	4·6439	*4·7184*	6·157	*0·9789*	59·279	16869
1·041	4·43	10·412	4·6484	*4·7229*	6·151	*0·9780*	59·336	16853
1·042	4·30	10·422	4·6528	*4·7275*	6·146	*0·9771*	59·393	16837
1·043	4·17	10·432	4·6573	*4·7320*	6·140	*0·9761*	59·450	16821
1·044	4·04	10·442	4·6618	*4·7366*	6·134	*0·9752*	59·507	16805
1·045	3·91	10·452	4·6662	*4·7411*	6·128	*0·9743*	59·564	16788
1·046	3·78	10·462	4·6707	*4·7456*	6·122	*0·9733*	59·621	16772
1·047	3·65	10·472	4·6752	*4·7502*	6·116	*0·9724*	59·678	16756
1·048	3·52	10·482	4·6796	*4·7547*	6·110	*0·9715*	59·735	16740
1·049	3·39	10·492	4·6841	*4·7593*	6·105	*0·9705*	59·792	16724
1·050	3·26	10·502	4·6886	*4·7638*	6·099	*0·9696*	59·849	16708
1·051	3·13	10·512	4·6931	*4·7684*	6·093	*0·9687*	59·906	16693
1·052	3·01	10·522	4·6975	*4·7729*	6·087	*0·9678*	59·963	16677
1·053	2·88	10·532	4·7020	*4·7774*	6·081	*0·9668*	60·020	16661
1·054	2·75	10·542	4·7065	*4·7820*	6·076	*0·9659*	60·077	16645
1·055	2·62	10·552	4·7109	*4·7865*	6·070	*0·9650*	60·135	16629
1·060	1·99	10·603	4·7333	*4·8092*	6·041	*0·9605*	60·420	16551
1·065	1·36	10·653	4·7556	*4·8319*	6·013	*0·9559*	60·705	16473
1·070	0·74	10·703	4·7780	*4·8546*	5·985	*0·9515*	60·990	16396
1·075	0·13	10·753	4·8003	*4·8774*	5·957	*0·9470*	61·276	16319
1·080	−0·48	10·803	4·8227	*4·9001*	5·929	*0·9426*	61·561	16244
1·085	−1·09	10·853	4·8450	*4·9228*	5·902	*0·9383*	61·846	16169
1·090	−1·68	10·903	4·8674	*4·9455*	5·875	*0·9340*	62·132	16095
1·095	−2·28	10·953	4·8897	*4·9682*	5·848	*0·9297*	62·417	16021
1·100	−2·86	11·003	4·9121	*4·9909*	5·821	*0·9255*	62·702	15948
1·105	−3·45	11·053	4·9344	*5·0136*	5·795	*0·9213*	62·987	15876
1·110	−4·02	11·103	4·9568	*5·0363*	5·769	*0·9171*	63·273	15804

Table 67 WEIGHT/VOLUME: MULTIPLICATION FACTORS TO FOUR SIGNIFICANT FIGURES

S.G. 60°F./60°F.	A.P.I. G.	L. TONS TO U.S. BBL.	CU. FT. TO L. TONS	CU. METRES TO L. TONS	IMP. GALS. TO L. TONS	U.S. GALS. TO L. TONS	U.S. BBL. TO L. TONS	I.G./DAY TO L. TON/YEAR	BBL./DAY TO L. TON/YEAR
0.650	86.19	9.859	0.01807	0.6380	0.002900	0.002415	0.1014	1.059	37.02
0.655	84.53	9.783	0.01821	0.6429	0.002923	0.002434	0.1022	1.067	37.31
0.660	82.89	9.709	0.01834	0.6478	0.002945	0.002452	0.1030	1.075	37.60
0.665	81.28	9.636	0.01848	0.6528	0.002967	0.002471	0.1038	1.083	37.88
0.670	79.69	9.564	0.01862	0.6577	0.002990	0.002490	0.1046	1.091	38.16
0.675	78.13	9.493	0.01876	0.6626	0.003012	0.002508	0.1053	1.099	38.45
0.680	76.59	9.423	0.01890	0.6675	0.003035	0.002527	0.1061	1.108	38.73
0.685	75.07	9.354	0.01904	0.6724	0.003057	0.002545	0.1069	1.116	39.02
0.690	73.57	9.286	0.01918	0.6773	0.003079	0.002564	0.1077	1.124	39.31
0.695	72.10	9.219	0.01932	0.6822	0.003102	0.002583	0.1085	1.132	39.59
0.700	70.64	9.153	0.01946	0.6872	0.003124	0.002601	0.1093	1.140	39.88
0.705	69.21	9.088	0.01960	0.6921	0.003146	0.002620	0.1100	1.148	40.16
0.710	67.80	9.024	0.01974	0.6970	0.003169	0.002638	0.1108	1.157	40.45
0.715	66.40	8.961	0.01988	0.7019	0.003191	0.002657	0.1116	1.165	40.73
0.720	65.03	8.899	0.02002	0.7068	0.003213	0.002676	0.1124	1.173	41.02
0.725	63.67	8.837	0.02016	0.7118	0.003236	0.002694	0.1132	1.181	41.30
0.730	62.34	8.776	0.02029	0.7167	0.003258	0.002713	0.1139	1.189	41.59
0.735	61.02	8.717	0.02043	0.7216	0.003280	0.002732	0.1147	1.197	41.87
0.740	59.72	8.658	0.02057	0.7265	0.003303	0.002750	0.1155	1.205	42.16
0.745	58.43	8.599	0.02071	0.7314	0.003325	0.002769	0.1163	1.214	42.45
0.750	57.17	8.542	0.02085	0.7364	0.003347	0.002787	0.1171	1.222	42.73
0.755	55.92	8.485	0.02099	0.7413	0.003370	0.002806	0.1179	1.230	43.02
0.760	54.68	8.429	0.02113	0.7462	0.003392	0.002825	0.1186	1.238	43.30
0.765	53.47	8.374	0.02127	0.7511	0.003415	0.002843	0.1194	1.246	43.58
0.770	52.27	8.320	0.02141	0.7560	0.003437	0.002862	0.1202	1.254	43.87
0.775	51.08	8.266	0.02155	0.7609	0.003459	0.002880	0.1210	1.263	44.16
0.780	49.91	8.213	0.02169	0.7658	0.003482	0.002899	0.1218	1.271	44.44
0.785	48.75	8.161	0.02183	0.7707	0.003504	0.002918	0.1225	1.279	44.73
0.790	47.61	8.109	0.02197	0.7757	0.003526	0.002936	0.1233	1.287	45.01
0.795	46.49	8.058	0.02210	0.7806	0.003549	0.002955	0.1241	1.295	45.30
0.800	45.38	8.007	0.02224	0.7855	0.003571	0.002974	0.1249	1.303	45.58
0.805	44.28	7.957	0.02238	0.7904	0.003593	0.002992	0.1257	1.312	45.87
0.810	43.19	7.908	0.02252	0.7953	0.003616	0.003011	0.1265	1.320	46.15
0.815	42.12	7.860	0.02266	0.8003	0.003638	0.003029	0.1272	1.328	46.44
0.820	41.06	7.812	0.02280	0.8052	0.003660	0.003048	0.1280	1.336	46.72
0.825	40.02	7.764	0.02294	0.8101	0.003683	0.003067	0.1288	1.344	47.01
0.830	38.98	7.717	0.02308	0.8150	0.003705	0.003085	0.1296	1.352	47.30
0.835	37.96	7.671	0.02322	0.8199	0.003727	0.003104	0.1304	1.361	47.58
0.840	36.95	7.625	0.02336	0.8248	0.003750	0.003122	0.1311	1.369	47.87
0.845	35.96	7.580	0.02350	0.8298	0.003772	0.003141	0.1319	1.377	48.15

Table 67—*contd.*

S.G. 60°F./60°F.	A.P.I. G	L. TONS TO U.S. BBL.	CU. FT. TO L. TONS	CU. METRES TO L. TONS	IMP. GALS. TO L. TONS	U.S. GALS. TO L. TONS	U.S. BBL. TO L. TONS	I.G./DAY TO L. TON/YEAR	BBL./DAY TO L. TON/YEAR
0.850	34.97	7.536	0.02364	0.8347	0.003795	0.003160	0.1327	1.385	48.44
0.855	34.00	7.491	0.02378	0.8396	0.003817	0.003178	0.1335	1.393	48.72
*0.860	33.03	7.448	0.02391	0.8445	0.003839	0.003197	0.1343	1.401	49.01
0.865	32.08	7.405	0.02405	0.8494	0.003862	0.003215	0.1351	1.409	49.29
0.870	31.14	7.362	0.02419	0.8543	0.003884	0.003234	0.1358	1.418	49.58
0.875	30.21	7.320	0.02433	0.8592	0.003906	0.003253	0.1366	1.426	49.86
0.880	29.30	7.278	0.02447	0.8642	0.003929	0.003271	0.1374	1.434	50.15
0.885	28.39	7.237	0.02461	0.8691	0.003951	0.003290	0.1382	1.442	50.44
0.890	27.49	7.196	0.02475	0.8740	0.003973	0.003309	0.1390	1.450	50.72
0.895	26.60	7.156	0.02489	0.8789	0.003996	0.003327	0.1397	1.458	51.01
0.900	25.72	7.116	0.02503	0.8839	0.004018	0.003346	0.1405	1.467	51.29
0.905	24.85	7.077	0.02517	0.8888	0.004040	0.003364	0.1413	1.475	51.57
0.910	23.99	7.038	0.02531	0.8937	0.004063	0.003383	0.1421	1.483	51.86
0.915	23.14	7.000	0.02545	0.8986	0.004085	0.003402	0.1429	1.491	52.15
0.920	22.30	6.961	0.02559	0.9035	0.004107	0.003420	0.1437	1.499	52.43
0.925	21.47	6.924	0.02572	0.9084	0.004130	0.003439	0.1444	1.507	52.72
0.930	20.65	6.887	0.02586	0.9133	0.004152	0.003457	0.1452	1.516	53.00
0.935	19.84	6.850	0.02600	0.9183	0.004175	0.003476	0.1460	1.524	53.29
0.940	19.03	6.813	0.02614	0.9232	0.004197	0.003495	0.1468	1.532	53.57
0.945	18.24	6.777	0.02628	0.9281	0.004219	0.003513	0.1476	1.540	53.86
0.950	17.45	6.741	0.02642	0.9330	0.004242	0.003532	0.1483	1.548	54.14
0.955	16.67	6.706	0.02656	0.9379	0.004264	0.003551	0.1491	1.556	54.43
0.960	15.90	6.671	0.02670	0.9428	0.004286	0.003569	0.1499	1.564	54.71
0.965	15.13	6.636	0.02684	0.9478	0.004309	0.003588	0.1507	1.573	55.00
0.970	14.38	6.602	0.02698	0.9527	0.004331	0.003606	0.1515	1.581	55.28
0.975	13.63	6.568	0.02712	0.9576	0.004353	0.003625	0.1523	1.589	55.57
0.980	12.89	6.535	0.02726	0.9625	0.004376	0.003644	0.1530	1.597	55.86
0.985	12.15	6.502	0.02740	0.9675	0.004398	0.003662	0.1538	1.605	56.14
0.990	11.43	6.469	0.02753	0.9724	0.004420	0.003681	0.1546	1.613	56.43
0.995	10.71	6.436	0.02767	0.9773	0.004443	0.003699	0.1554	1.622	56.71
1.000	10.00	6.404	0.02781	0.9822	0.004465	0.003718	0.1562	1.630	57.00
1.005	9.30	6.372	0.02795	0.9871	0.004487	0.003737	0.1569	1.638	57.28
1.010	8.60	6.340	0.02809	0.9920	0.004510	0.003755	0.1577	1.646	57.57
1.015	7.91	6.309	0.02823	0.9969	0.004532	0.003774	0.1585	1.654	57.85
1.020	7.23	6.278	0.02837	1.0018	0.004555	0.003792	0.1593	1.662	58.14
1.025	6.55	6.248	0.02851	1.0068	0.004577	0.003811	0.1601	1.671	58.42
1.030	5.88	6.217	0.02865	1.0117	0.004599	0.003830	0.1609	1.679	58.71
1.035	5.21	6.187	0.02879	1.0166	0.004622	0.003848	0.1616	1.687	58.99
1.040	4.56	6.157	0.02893	1.0215	0.004644	0.003867	0.1624	1.695	59.28
1.045	3.91	6.128	0.02907	1.0264	0.004666	0.003886	0.1632	1.703	59.56
1.050	3.26	6.099	0.02921	1.0314	0.004689	0.003904	0.1640	1.711	59.85

* 33° API   7·33 bbl./m.ton

Table 68 DIRECT VISCOSITY COMPARISON (AT SAME TEMPERATURES)

CENTI-STOKES	REDWOOD No.1	SAYBOLT U.S.	ENGLER°	CENTI-STOKES	REDWOOD No.1	SAYBOLT U.S.	ENGLER°	CENTI-STOKES	REDWOOD No.1	SAYBOLT U.S.	ENGLER°
2.0	30.76	32.64	1.119	26.0	108.5	123.4	3.575	170.0	692.4	786.6	22.38
2.1	31.01	33.00	1.129	27.0	112.4	127.8	3.695	175.0	712.6	809.7	23.03
2.2	31.26	33.36	1.139	28.0	116.3	132.3	3.820	180.0	733.1	832.9	23.70
2.3	31.51	33.72	1.149	29.0	120.2	136.7	3.945	185.0	753.5	856.1	24.35
2.4	31.76	34.08	1.159	30.0	124.2	141.1	4.070	190.0	774.0	879.3	25.00
2.5	32.01	34.44	1.169	31.0	128.1	145.5	4.195	195.0	794.3	902.5	25.67
2.6	32.26	34.76	1.178	32.0	132.1	149.9	4.320	200.0	814.6	925.6	26.32
2.7	32.51	35.08	1.188	33.0	136.1	154.4	4.445	210.0	855.2	971.8	27.65
2.8	32.76	35.41	1.198	34.0	140.0	158.9	4.570	220.0	896.3	1018	28.95
2.9	33.01	35.73	1.207	35.0	144.0	163.4	4.695	230.0	936.9	1065	30.28
3.0	33.26	36.05	1.217	36.0	147.9	167.9	4.825	240.0	978.0	1111	31.60
3.1	33.51	36.37	1.226	37.0	151.9	172.4	4.955	250.0	1018	1157	32.90
3.2	33.76	36.69	1.235	38.0	155.9	176.9	5.080	260.0	1059	1203	34.25
3.3	34.01	37.01	1.244	39.0	160.0	181.4	5.205	270.0	1099	1249	35.55
3.4	34.27	37.33	1.253	40.0	164.0	185.9	5.335	280.0	1140	1296	36.85
3.5	34.52	37.65	1.264	41.0	168.0	190.5	5.465	290.0	1181	1342	38.18
3.6	34.77	37.95	1.274	42.0	172.0	195.0	5.590	300.0	1222	1388	39.50
3.7	35.03	38.25	1.283	43.0	176.1	199.5	5.720	310.0	1263	1434	40.80
3.8	35.28	38.55	1.291	44.0	180.1	204.1	5.845	320.0	1303	1480	42.12
3.9	35.53	38.85	1.300	45.0	184.2	208.7	5.975	330.0	1344	1527	43.45
4.0	35.78	39.15	1.308	46.0	188.2	213.3	6.105	340.0	1385	1574	44.75
4.5	37.03	40.76	1.354	47.0	192.2	217.9	6.235	350.0	1425	1620	46.10
5.0	38.31	42.36	1.400	48.0	196.3	222.5	6.365	360.0	1465	1666	47.40
5.5	39.65	43.96	1.441	49.0	200.3	227.1	6.495	370.0	1505	1712	48.70
6.0	40.91	45.57	1.481	50.0	204.3	231.7	6.620	380.0	1546	1759	50.00
6.5	42.26	47.17	1.521	55.0	224.6	254.8	7.258	390.0	1587	1805	51.35
7.0	43.57	48.77	1.563	60.0	244.8	277.8	7.896	400.0	1628	1851	52.65
7.5	44.89	50.42	1.605	65.0	265.2	300.8	8.554	450.0	1832	2082	59.25
8.0	46.26	52.07	1.653	70.0	285.5	323.8	9.212	500.0	2036	2314	65.80
8.5	47.66	53.77	1.700	75.0	305.7	347.0	0.870	550.0	2239	2545	72.40
9.0	49.04	55.48	1.746	80.0	326.0	370.2	10.53	600.0	2443	2777	79.00
9.5	50.47	57.18	1.791	85.0	346.3	393.3	11.19	650.0	2646	3008	85.60
10.0	51.92	58.88	1.837	90.0	366.6	416.5	11.85	700.0	2850	3239	92.20
11.0	54.94	62.39	1.928	95.0	386.8	439.5	12.51	750.0	3054	3471	98.80
12.0	58.05	66.00	2.020	100.0	407.3	462.6	13.16	800.0	3258	3702	105.3
13.0	61.24	69.70	2.120	105.0	427.7	485.8	13.82	850.0	3462	3934	111.9
14.0	64.50	73.50	2.219	110.0	447.9	509.0	14.47	900.0	3666	4165	118.5
15.0	67.89	77.31	2.323	115.0	468.2	532.1	15.14	950.0	3871	4396	125.0
16.0	71.34	81.21	2.434	120.0	488.6	555.3	15.80	1000.0	4074	4628	131.6
17.0	74.80	85.22	2.540	125.0	509.0	578.5	16.45				
18.0	78.36	89.32	2.644	130.0	529.3	601.6	17.11				
19.0	82.00	93.43	2.755	135.0	549.6	624.7	17.76				
20.0	85.66	97.64	2.870	140.0	570.1	647.9	18.43				
21.0	89.42	101.8	2.984	145.0	590.4	671.1	19.08				
22.0	93.16	106.1	3.100	150.0	610.7	694.2	19.75				
23.0	96.96	110.4	3.215	155.0	631.0	717.2	20.40				
24.0	100.8	114.7	3.335	160.0	651.4	740.4	21.05				
25.0	104.6	119.0	3.455	165.0	671.8	763.4	21.72				

For comparison and conversion of Kinematic Centistokes, Redwood No.1 Seconds, Saybolt U.S., and Engler Degrees, all values are calculated at 122°F. (50°C.)

The chart may be used with practical accuracy for conversion purposes between 70°F. (21.1°C.) and 212°F. (100°C.)

For viscosity values above 1,000 Centistokes use multipliers as follows:-

Centistokes	=	0.246	x Redwood 1 Seconds
		7.60	x Engler Degrees
		0.216	x Saybolt U. Seconds
Engler Degrees	=	0.0323	x Redwood 1 Seconds
		0.0285	x Saybolt U. Seconds
Redwood 1 Seconds	=	0.881	x Saybolt U. Seconds

Table 69 EXPORT CRUDES IN WORLD TRADE

Area Country	Crude name	Loading terminal	Crude gravity °API	Sulphur content % wt	% wt. residue >343°C
MIDDLE EAST					
ABU DHABI	Abu Dhabi Land - Murban	Jebel Dhanna	39.4	0.74	40
	Abu Dhabi Marine - Umm Shaif	Das Island	37.0	1.38	43
	Abu Dhabi Marine - Zakum	Das Island	40.1	0.98	39
DUBAI	Dubai Export	Fateh	32.5	1.70	49
EGYPT	El Alamein	Mersa el Hamra	33.0	1.05	48
	El Morgan	Ras Shukheir	31.7	1.66	54
IRAN	Cyrus	Cyrus	18.8	3.46	72
	Darius	Kharg Island	33.8	2.54	50
	Fereidcon		31.4	2.50	56
	Iranian Bahrgan Sar	Bahrgan	32.8	1.39	51
	Iranian Light	Kharg Island	33.6	1.36	49
	Iranian Heavy	Kharg Island	31.0	1.60	53
	Iranian Nowruz	Bahrgan	18.3	4.08	72
	Iranian Rostam	Lavan Island	38.0	1.50	38
	Iranian Sassan	Lavan Island	33.4	1.95	48
IRAQ	Iraq (Basra)	Khor al Amaya	34.0	1.95	51
	Iraq Mediterranean - Ain Zalah/Butmah	Tripoli	31.1	2.60	52
	Iraq Mediterranean - Jambur/Bai Hassan	Banias	34.0	2.44	49
	Iraq Mediterranean - Kirkuk	Banias/Tripoli	36.0	1.88	44
	North Rumaila	Fao	33.7	1.99	51
KUWAIT	Kuwait Export	Mina al Ahmadi	31.3	2.50	55
NEUTRAL ZONE	Hout	Ras al Khafji	35.1	1.39	49
	Khafji	Ras al Khafji	28.6	2.85	59
	Wafra - Burgan	Mina Abdullah/Mina Saud	23.8	3.15	67
	Wafra - Eocene	Mina Abdullah/Mina Saud	16.5	4.47	76
	Wafra - Ratawi	Mina Abdullah/Mina Saud	25.6	3.86	62
OMAN	Oman Export	Mina al Fahal	34.6	0.97	48
QATAR	Qatar Export	Umm Said	41.4	1.10	37
	Qatar Marine	Halul Island	37.0	1.50	42
SAUDI ARABIA	Arabian Light	Sidon/Ras Tanura	34.2	1.65	50
	Arabian Medium	Ras Tanura	30.3	2.65	55
	Arabian Heavy (Safaniya)	Ras Tanura	28.0	2.80	59
	Berri (Offshore)	Ras Tanura	38.5	1.13	43
FAR EAST					
AUSTRALIA	Barrow Island	Barrow Island	37.3	0.06	21
	Gippsland Blend - '74/'77	Westernport Bay	44.3	0.09	30
	Moonie	Brisbane	45.7	0.04	28
BRUNEI	Seria Light	Lutong	38.8	0.05	24
CHINA	Taching	Lu Ta (Dairen)	33.0	0.09	71
INDONESIA - JAVA	Ardjuna	Ardjuna	37.7	0.12	35
INDONESIA - SUMATRA	Cinta	Cinta	33.9	0.08	68
	Duri	Dumai	22.8	0.18	74
	Minas (Sumatran Light)	Dumai	36.5	0.10	60
	Rantau	Pangkalan Susu	47.9	0.05	8

Table 69—*contd.*

Area / Country	Crude name	Loading terminal	Crude Gravity °API	Sulphur content % wt	% wt. residue >343°C
MEDITERRANEAN					
ALGERIA	Arzew Blend	Arzew	43.6	0.10	33
	Hassi Messaoud	Bougie	45.1	0.12	29
	Zarzaitine Blend	La Skhirra	42.5	0.08	38
LIBYA	Amna (Libyan High Pour)	Ras Lanuf	36.1	0.15	56
	Brega	Marsa al Brega	39.2	0.24	42
	Bu-Attifel	Zueitina	44.7	0.008	50
	Es Sider	Es Sider	36.4	0.50	46
	Libyan Export (Sarir)	Marsa al Hariga	37.6	0.14	53
	Libyan Light (Sirtica Blend)	Ras Lanuf	37.8	0.38	43
	Zueitina	Zueitina	41.8	0.21	38
SYRIA	Karachok/Hamzeh	Tartus	20.2	4.85	64
	Souedia	Tartus	24.9	3.50	59
TUNISIA	El Borma	La Skhirra	42.0	0.07	40
EUROPE/RUSSIA					
NORTH SEA	Argyll	Argyll (SBM)	36.2/38.6	0.20	48
	Auk	Auk (SBM)	37.5	0.40	45
	Ekofisk	Ekofisk	35.0	0.20	48
	Forties	Forth	36.6	0.28	45
	Montrose	Forth	41.5	0.23	39
	Ninian	Sullom Voe	35.9	0.41	49
	Piper	Flotta Island	35.5	0.92	47
RUSSIA	Mukhanovo	Novorossijsk	36.1	1.17	47
	Romashkino	Ventspils	32.7	1.55	50
	Tujmaza	Novorossijsk	34.0	1.50	48
WEST AFRICA					
ANGOLA	Cabinda - High Pour	Malongo	36.7	0.05	58
	Cabinda - Low Pour	Malongo	25.6	0.47	69
CONGO	Emeraude Marine	Riviere Rouge	23.7	0.77	67
GABON	Gamba	Gamba	31.8	0.11	66
	Mandji Blend	Cap Lopez	31.4	0.68	56
NIGERIA	Brass River	Brass River	43.1	0.09	28
	Escravos	Escravos	36.4	0.12	36
	Export Blend - Forcados	Forcados	30.5	0.18	41
	Light Export - Bonny	Bonny	37.6	0.13	35
	Medium Export - Bonny	Bonny	26.1	0.23	49
	Qua Iboe	Qua Iboe	37.4	0.09	32
NORTH AMERICA					
UNITED STATES - ALASKA	Drift River Blend	Drift River	33.8	0.09	44
	Middle Ground Shoal	Nikiski	37.8	0.07	38
	Swanson River	Nikiski	37.9	0.08	38
	Prudhoe Bay	Valdez	27.5	0.96	57
SOUTH AMERICA					
BRAZIL	Bahia	Candeias	30.5	0.22	65
	Carmopolis	Aracaju	23.4	0.41	72
COLOMBIA	Orito	Tumaco	34.9	0.52	44
	Rio Zulia	Santa Marta	39.8	0.30	46
ECUADOR	Ecuador Export	Balao	28.6	0.94	56

Table 70 TANKER RATE SCHEDULES

Location	*Dates*	*Scale*
London	1939–1952	MOT
	1st November 1952–30th June 1954	Scale 1
	1st July 1954–14th December 1958	Scale 2
	15th December 1958–14th May 1962	Scale 3
	15th May 1962–14th September 1969	Intascale
	1st September 1969	Worldscale introduced
New York	1939–1956	USMC
	Early 1956	ATRS introduced
	1st February 1961	Enlarged ATRS

MOT; the UK Ministry of Transport
USMC; the US Maritime Commission
ATRS; American Tanker Rate Schedule

Table 71 TANKER RATES FOR KEY ROUTES

Route	Intascale Flat $ September 1969	Worldscale 100 Rate $				
		September 1969	1970	1972	1974	1975
Caribbean - USNH (Aruba - New York)	1.86	1.85	1.85	2.05	2.09	2.77
Caribbean - UKC (Curacao - Rotterdam)	4.11	3.68	3.68	4.21	4.45	6.16
Mediterranean - USNH (Milazzo - New York)	4.13	3.69	3.69	4.24	4.24	5.90
Mediterranean - UKC (La Skinrha - London)	2.74	2.41	2.38	2.85	2.86	3.74
PG - Japan (Ras Tanura - Yokohama)	5.86	5.47	5.47	5.91	6.01	8.57
PG - UKC (Ras Tanura (Mina al Ahmadi av. - Rotterdam	9.70	8.87	8.87	10.08	10.28	14.57
PG - USG (Ras Tanura (Mina al Ahmadi av. - Beaumont	-	-	-	10.66	10.90	15.58
USG - NY *	-	2.85	2.85	2.85	2.85	
Rotterdam - USNH (Rotterdam - New York)	-	-	-	-	3.81	5.21

ATRS scale rate

The nine routes quoted above are monitored in Platt's Oilgram Price Service.

For distances see BP World-Wide Marine Distance Tables
BP Tanker Company 1958

Table 72 WORLDSCALE COMPARISONS (RV RATE MINA AL AHMADI–ROTTERDAM)

W100 = $	1/1/68 9.82*(I)	15/9/69 8.87*(W)	1/11/69 8.93*	1/1/71 9.17*	1/1/72 9.90	8/6/72 10.00	15/7/72 10.08	1/1/73 9.93	1/1/74 10.37	1/1/75 14.70	
	%	%	%	%	%	%	%	%	%	%	
15/9/69 (8.87)	90.3	-	-	-	-	-	-	-	-	-	Reprint
1/11/69 (8.93)	90.9	100.6	-	-	-	-	-	-	-	-	P. Costs
1/1/71 (9.17)	93.4	103.4	102.7	-	-	-	-	-	-	-	Reprint
1/1/72 (9.90)	100.8	111.6	110.9	108.0	-	-	-	-	-	-	Reprint
8/6/72 (10.00)	101.8	112.7	112.0	109.1	101.0	-	-	-	-	-	P. Costs
15/7/72 (10.08)	102.6	113.6	112.9	109.9	101.8	100.8	-	-	-	-	P. Costs
1/1/73 (9.93)	101.1	112.0	111.2	108.3	100.3	99.3	98.5	-	-	-	Reprint
1/1/74 (10.37)	105.6	116.9	116.1	113.1	104.7	103.7	102.9	104.4	-	-	Reprint
1/1/75 (14.70)	149.7	165.7	164.6	160.3	148.5	147.0	145.8	148.0	141.8	-	Reprint

* @ $2.40 = £1

"Reprint" indicates complete revision of the scale
"P. Costs" indicates increase in certain port costs only

Table 73 VOLUME AND WEIGHT OF OIL IN PIPELINES PER LINEAR MILE

These figures are based on A.P.I. standard outside diameters and selected pipewall thicknesses and may be used for APPROXIMATE CALCULATIONS ONLY. FOR ACCURATE figures the actual inside diameter of the pipeline must be known when

$$\text{Volume in gallons per mile} = 179.4 \times (\text{inches I.D.})^2$$

NOMINAL SIZE (Inches)	OUTSIDE DIAM. (Inches)	WALL THICKNESS (Inches)	IMPERIAL GALLONS PER MILE	BARRELS PER MILE	LONG TONS (M.E. CRUDE) PER MILE
3	3.5	0.216	1,690	48	6
4	4.5	0.237	2,910	83	11
6	6.625	0.280	6,600	189	25
8	8.625	0.277	11,700	334	45
10	10.75	0.279	18,600	533	71
12	12.75	0.330	26,200	750	100
14	14	0.375	31,500	901	120
16	16	0.375	41,700	1,190	159
18	18	0.375	53,400	1,530	204
20	20	0.375	66,500	1,900	254
22	22	0.375	81,000	2,320	309
24	24	0.375	97,000	2,770	370
26	26	0.375	114,000	3,270	437
28	28	0.375	133,000	3,810	508
30	30	0.4375	152,000	4,350	581
32	32	0.4375	174,000	4,970	663
34	34	0.4375	197,000	5,630	751
36	36	0.4375	221,000	6,330	845

Table 74 PRACTICAL OIL PIPELINE CAPACITIES—MIDDLE EAST CRUDE

Diameter (inches)	*Capacity—Useful Range*	
	Million tons/year	*Thousands b/d*
6	0·3–0·6	6–12
8	0·7–1·3	15–27
10	1·4–2·5	29–52
12	2·3–4·1	40–85
16	4·4–8·0	91–165
20	8·0–14·5	165–310
*22/20†	9·0–15·0	185–310
22†	9·5–15·5	200–320
24	12·0–18·0	250–375
30	18·0–28·0	375–575
*32/30†	19·5–30·5	400–630
32	21·0–33·0	435–680
34	25·5–36·5	525–750
*36/34†	27·5–40·5	600–840
36	30·0–45·0	620–950

* Composite, equal length of each diameter
† Interpolated
Note: Accurate pipeline capacities depend on length of line, difference in head between the ends of the line, the Specific Gravity and Viscosity of the fluid, etc. The figures above *must be used as a general guide only.*

Table 75 CAPACITY OF SELECTED STORAGE TANKS

BY NOMINAL SIZE

NOMINAL SIZE (m.g.)	DIAM. (ft.)	HEIGHT (ft.)	TOTAL # CAPACITY (water tons)	USEFUL CAPACITY # * MOTOR SPIRIT (tons)	GAS OIL (tons)	FUEL OIL (tons)
.105	35	17½	470	290	330	370
.125	36	20	560	340	400	440
.235	40	30	1,050	650	750	830
.500	58½	30	2,230	1,375	1,590	1,760
.700	65¾	33	3,125	1,930	2,230	2,470
1.0	82½	30	4,460	2,750	3,180	3,525
1.250	78	42	5,580	3,440	3,980	4,410
2.010	112	32¾	8,970	5,530	6,400	7,090
2.715	112	44¼	12,100	7,470	8,640	9,570
3.415	120	48½	15,200	9,395	10,900	12,000
4.695	144	46¼	21,000	12,900	14,900	16,600
5.860	144	57¾	26,200	16,100	18,700	20,700

BY DIAMETER

DIAM. (ft.)	HEIGHT (ft.)	NOMINAL SIZE (m.g.)	TOTAL # CAPACITY (water tons)	USEFUL CAPACITY # * MOTOR SPIRIT (tons)	GAS OIL (tons)	FUEL OIL (tons)
35	17½	.105	470	290	330	370
"	20	.120	540	330	380	420
40	15	.120	540	330	380	420
"	30	.235	1,050	650	750	830
"	35	.275	1,230	760	880	970
45	30	.295	1,320	810	940	1,040
50	40	.490	2,190	1,350	1,560	1,730
60	46½	.815	3,640	2,240	2,590	2,870
73	40	1.045	4,665	2,875	3,330	3,680
82½	30	1.000	4,460	2,750	3,180	3,525
"	39½	1.315	5,870	3,620	4,190	4,640
92	34	1.410	6,295	3,880	4,490	4,970
102	46½	2.365	10,600	6,510	7,530	8,340
112	32¾	2.010	8,970	5,530	6,400	7,090
"	44¼	2.715	12,100	7,470	8,640	9,570
120	42	2.960	13,200	8,140	9,420	10,400
"	48½	3.415	15,200	9,395	10,900	12,000
144	30	3.045	13,600	8,380	9,695	10,700
"	42	4.260	19,000	11,700	13,600	15,000
"	46¼	4.695	21,000	12,900	14,900	16,600
"	57¾	5.860	26,200	16,100	18,700	20,700

Capacities are based on nominal sizes

* Allows 15% ullage/bottoms

Table 76 APPROXIMATE CALORIFIC VALUES (EFFICIENCY AND UTILISATION FACTORS NOT TAKEN INTO ACCOUNT)

Fuels	*Calorific Values*			
Petroleum Fuels	*Btus/lb*		*Btus/gallon*	
	gross	*net*	*gross*	*net*
Liquefied Petroleum Gas	21 250	19 620	94 650	91 630
Motor Spirit	20 400	18 900	148 000	140 000
Kerosine	20 000	18 750	159 000	149 000
Vaporising Oil	20 000	18 750	156 000	146 000
Gas Oil	19 600	18 400	170 000	160 000
Light Fuel Oil	18 750	18 100	167 500	162 000
Heavy Fuel Oil	18 500	17 800	175 000	169 000
Other Liquid Fuels				
Methyl Alcohol	9 605	8 515	76 000	67 500
Ethyl Alcohol	12 820	11 690	101 500	92 500
Motor Benzole	18 050	17 320	158 000	152 000
Coal Tar Fuels	16 560–16 850	15 840–16 200	190–194 000	182–186 000
Solid Fuels	*Btus/lb*			
	gross	*net*		
Coal (British)				
Anthracite	14 500–15 000	14 250–14 750		
Welsh Steam	14 600	14 200		
Bituminous	10 000–13 500	9 530–13 030		
Charcoal	11 000–13 500	10 900–13 400		
Coke	11 500–12 500	11 400–12 400		
Lignite 45–50% H_2O	5 600–6 350	5 100–5 824		
15% H_2O	9 000	8 472		
Peat 50% H_2O	4 630	4 065		
25% H_2O	6 200–6 900	5 636–6 336		
Wood 40% H_2O	5 100–5 500	4 520–4 920		
Gas Fuels	*Btus per ft³ at 30″ Hg and 60°F*			
	gross	*net*		
Methane	1 015	895		
Ethane	1 790	1 580		
Propane	2 570	2 300		
Butane	3 395	2 940		
Oil Gas (depending on process used)	1 050	940		
Natural Gas	1 153	1 000		
Towns Gas	425–550	375–490		
Carburetted Water Gas (depending on amount of oil used)	300–500	265–458		
Water Gas (depending on coke used)	280–310	255–285		
Producer Gas (depending on coal used)	120–180	110–170		

Very approximate equivalents

1 lb oil = 1·5 lb best coal
= 1·6–1·9 lb ordinary coal or coke
= 1·3 lb anthracite
= 35–40 feet³ manufactured gas
= 18 feet³ natural gas
= 5·5 units of electricity

1 bbl oil = 6000 feet³ natural gas
1 therm = 100 000 Btus = roughly 100 feet³ natural gas
5 lb oil
7½ lb coal

Table 77 MAIN OIL ECONOMIC/ECONOMIC JOURNALS

Rank	*Publication*
1	Petroleum Economist
2	Platt's News
3	Oil and Gas Journal
4	Petroleum Intelligence Weekly
5	Financial Times
6	The Times
7	The Guardian
8	Business Week
9	The Economist
10	Oil Daily
11	Wall Street Journal
12	Far Eastern Economic Review
13	New Scientist
14	Lloyds List
15	Petroleum Times
16	Sunday Times
17	Middle East Economic Survey
18	Europe Oil Prices (London)
19	Coal Week
20	Daily Telegraph
21	Energy Policy
22	Investors Chronicle
23	Commerce Today
24	Fortune
25	Trade and Industry
26	Forbes
27	Energy International
28	Euromoney
29	Institute of Fuel Journal
30	Challenge
31	Technology Review
32	Sunday Telegraph
33	Coal and Energy Quarterly
34	Oil Week
35	Platt's Prices

The survey was carried out in the first nine months of 1975 and covered about 3000 abstracts.
A total of 316 journals were extracted but a much larger number were covered by using abstracting services in related fields. Seventy percent of the abstracts taken came from journals in the above list.